CW00594500

An Introduction to Thermogeology:
Ground Source Heating and Cooling

For Jenny 'the Bean'

An Introduction to Thermogeology: Ground Source Heating and Cooling

David Banks

Holymoor Consultancy, Chesterfield, Derbyshire
and
Senior Research Associate in Thermogeology
University of Newcastle-upon-Tyne

Blackwell
Publishing

Blackwell Publishing editorial offices:
Blackwell Publishing Ltd, 9600 Garsington Road, Oxford OX4 2DQ, UK
Tel: +44 (0)1865 776868
Blackwell Publishing Inc., 350 Main Street, Malden, MA 02148-5020, USA
Tel: +1 781 388 8250
Blackwell Publishing Asia Pty Ltd, 550 Swanston Street, Carlton, Victoria 3053, Australia
Tel: +61 (0)3 8359 1011

First published 2008 by Blackwell Publishing Ltd

ISBN: 978-1-4051-7061-1

Library of Congress Cataloging-in-Publication Data

Banks, David.
An introduction to thermogeology: ground source heating and cooling / David Banks.

p. cm.
Includes bibliographical references and index.
ISBN-13: 978-1-4051-7061-1 (hardback : alk. paper)
ISBN-10: 1-4051-7061-1 (hardback : alk. paper)
1. Ground source heat pump systems. I. Title.

TH7638.B36 2008
697′.7–dc22
2007032679

A catalogue record for this title is available from the British Library

Set in 10/13 pt Trump Mediaeval
by Newgen Imaging Systems (P) Ltd., Chennai, India
Printed and bound in Singapore
by C.O.S. Printers Pte Ltd

The publisher's policy is to use permanent paper from mills that operate a sustainable forestry policy, and which has been manufactured from pulp processed using acid-free and elementary chlorine-free practices. Furthermore, the publisher ensures that the text paper and cover board used have met acceptable environmental accreditation standards.

For further information on Blackwell Publishing, visit our website:
www.blackwellpublishing.com

Contents

About the Author

Dave BANKS is a hydrogeologist with 22 years experience of investigating groundwater-related issues. He started his career with the Thames Water Authority in southern England, then moved across the North Sea to the Geological Survey of Norway, where he eventually headed the Section for Geochemistry and Hydrogeology. Since returning to the UK in 1998, he has worked as a consultant from a base in Chesterfield, sandwiched between the gritstone of the Peak District National Park and the abandoned mines of Britain's largest coalfield. He has international experience from locations as diverse as Afghanistan, the Bolivian Altiplano, Somalia, Western Siberia and Huddersfield. During the past few years his attention has turned to the emerging science of thermogeology: He has worked closely with the ground source heat industry and has recently been appointed Senior Research Associate in Thermogeology at the University of Newcastle-upon-Tyne.

In his spare time, Dave enjoys music. With his chum Bjørn Frengstad, he has formed almost half of the sporadically active acoustic lo-fi stunt duo 'The Sedatives'. They have murdered songs by their musical heroes, Jarvis Cocker, Richard Thompson and Kath Williams in a variety of seedy locations.

Acknowledgements

In the late 1990s, I was working for the Norwegian Geological Survey's Section for Hydrogeology and Geochemistry. Despite the Section being choc-a-bloc with brainy research scientists, one of my most innovative colleagues was an engineer who called me, on what seemed a weekly basis, brimming with enthusiasm for some wizard new idea. One day he started telling me all about something called 'grunnvarme' or ground source heat, which was, apparently, very big in Sweden. Initially, it seemed to me to be something akin to perpetual motion – space heating from Norwegian rock at 6°C? – and in violation of the Second Law of Thermodynamics to boot. Nevertheless, he persuaded me that it really did have a sound physical basis. In fact, my chum went on to almost single-handedly sell the concept of ground source heat to a Norwegian market that was on the brink of an energy crisis. A subsequent dry summer that pulled the plug on Norway's cheap hydroelectric supplies and sent prices soaring was the trigger that ground source heat needed to take off. So, first, 'a big thank you' to Helge Skarphagen (for it was he!), who first got me interested in ground source heat.

On my return to England in 1998, I tried to bore anyone who gave the appearance of listening about the virtues of ground source heat (I was by no means the first to try this – John Sumner and Robin Curtis, among others, had been evangelists for the technology much earlier). It was not until around 2003, however, that interest in ground source heat was awakened in Britain and I was lucky enough to fall in with a group of entrepreneurs with an eye for turning it into a business. So, second, many thanks to GeoWarmth of Hexham for the pleasure of working with you, and especially to Dave Spearman, Jonathan Steven, Braid and Charlie Aitken, Nick Smith and John Withers.

I would like to thank the following for taking the time to review the various chapters and for their invaluable comments:

- Professor Paul Younger of the University of Newcastle-upon-Tyne.
- Professor Keith Tovey of the University of East Anglia in Norwich.
- Karl Drage of Geothermal International in Coventry.
- Helge Skarphagen of the Norwegian Water Research Institute (NIVA) in Oslo.
- James Dodds of JDIH Water & Environment Ltd. in Castle Donington.
- Jonathan Steven and Charlie Aitken of GeoWarmth Ltd., Hexham.
- Professor Göran Hellström of Lund Technical University, Sweden.
- Dr Simon Rees of De Montfort University, Leicester.
- Dr Robin Curtis of GeoScience Ltd. in Falmouth.
- Marius Greaves of the Environment Agency, in Hatfield.

The lyrics from *First & Second Law* from *At The Drop of Another Hat* by Flanders & Swann © 1963 are reproduced in Chapter 4 by kind permission of the Estates of Michael Flanders & Donald Swann. The Administrator of The Flanders & Swann Estates, Leon

Berger (leonberger@donaldswann.co.uk) was good enough to facilitate this permission. Additionally, I would like to thank the London Canal Museum for sourcing historical photos of the ice trade, and would wholeheartedly recommend a visit to anyone who has a few hours to spend in the Kings Cross area of London and who wants to view a genuine 'ice well'.

I am grateful to all those who have provided materials, case studies, inspiration, diagrams, photos and more – Pablo Fernández Alonso, Johan Claesson, Richard Freeborn of Kensa Engineering, Bjørn Frengstad of the fabulous 'Sedatives', the International Ground Source Heat Pump Association (IGSHPA), Ben Lawson and Paul Younger of Newcastle University, John Lund of the Geo-Heat Centre in Oregon, Jim Martin, John Parker, Umberto Puppini of ESI Italia, Kevin Rafferty, Randi Kalskin Ramstad of the Geological Survey of Norway, Igor Serëdkin, Chris Underwood of Northumbria University, and many more. I would like to thank John Vaughan of Chesterfield Borough Council, Doug Chapman of Annfield Plain, Les Gibson of Grindon Camping Barn and Lionel Hehir of the Hebburn Eco-Centre for permission to use their buildings as case studies. Janet Giles has been heroic in assisting with administration and printing. Lastly, Ardal O'Hanlon has unwittingly provided me with the alter-ego of *Thermoman*, which I have occasionally assumed while lecturing! Whilst I have tried to contact all copyright holders of materials reproduced in this book, in one or two cases you have proved very elusive! If you are affected by any such omission, please do not hesitate to contact me, via Blackwell, and we will do our utmost to ensure that the situation is rectified for any future editions of this book.

Oh, and by the way, Jenny, I do not know what you have been up to while I have been locked in the attic writing this book, but normal parental service will shortly be resumed!

Dave Banks
Chesterfield, Derbyshire, 2007

1

An Introduction

> Nature has given us illimitable sources of prepared low-grade heat. Will human organisations cooperate to provide the machine to use nature's gift?
>
> John A. Sumner (1976)

Many of you will be familiar with the term *geothermal energy*. It probably conjures mental images of volcanoes or of power stations replete with clouds of steam, deep boreholes, whistling turbines and hot saline water. This book is *not* primarily about such geothermal energy, which is typically high-temperature (or high enthalpy, in techno-speak) energy and which is accessible only in specific geological locations. This book concerns the relatively new science of *thermogeology*. Thermogeology is the study of so-called *ground source heat*: the mundane form of heat that is stored in the ground at normal temperatures. Ground source heat is a much less glamorous topic than high-temperature geothermal energy and its use in space heating is often invisible to those who are not 'in the know'. It is hugely important, however, as it exists and is accessible everywhere. It genuinely offers an attractive and powerful means of delivering CO_2-efficient space heating and cooling.

Let me offer the following definition of thermogeology: the study of the occurrence, movement and exploitation of low-enthalpy heat in the relatively shallow geosphere. By 'relatively shallow', we are typically talking of depths of down to 200 m or so. By 'low enthalpy', we are usually considering temperatures of less than 30°C.[1]

---

[1] Although in conventional geothermal science, anything up to around 90°C is still considered 'low enthalpy'!

1.1 Who should read this book?

This book is designed as an introductory text for the following audience:

- graduate and postgraduate level students;
- civil and geotechnical engineers;
- buildings services and *HVAC*[2] engineers who are new to ground source heat;
- applied geologists, especially hydrogeologists;
- architects;
- planners and regulators;
- energy consultants.

1.2 What will this book do and not do?

This book is not a comprehensive manual for designing ground source heating and cooling systems for buildings: it is rather intended to introduce geologists to the concept of thermogeology. It is also meant to ensure that architects and engineers are aware that there is an important geological dimension to ground source heating schemes. The book aims to cultivate awareness of the possibilities that the geosphere offers for space heating and cooling and also of the limitations that constrain the applications of ground source heat. It aims to equip the reader with a conceptual model of how the ground functions as a heat reservoir and to make him/her aware of the important parameters that will influence the design of systems utilising this reservoir.

While this book will introduce you to design of ground source heat systems and even enable you to contribute to the design process, it is important to realise that a sustainable and successful design needs the integrated skills of a number of sectors: the architect, the buildings services/HVAC engineer, the electromechanical engineer and the thermogeologist. If you are a geologist, you must realise that you are not equipped to design the infrastructure that delivers heat or cooling to a building. If you are a HVAC engineer, you should acknowledge that a geologist can shed light on the 'black hole', which is your ground source heat borehole or trench. In other words, you need to talk to each other and work together. For those who wish to delve into the hugely important 'grey area' where geology interfaces in detail with buildings engineering, to the extent of consideration of pipe materials and diameters, manifolds and heat exchangers, I recommend that you consult one of several manuals or software packages available. In particular, I direct you to the manual of Kavanaugh and Rafferty (1997), despite its use of such unfamiliar units as Btu, feet and °F.

[2] Heating, ventilation and air-conditioning.

1.3 Why should you read this book?

You should read this book because thermogeology is important for the survival of planet Earth! Although specialists may argue about the magnitude of climate change ascribable to greenhouse gases, there is a broad consensus (IPCC, 2007) that the continued emission of fossil carbon (in the form of CO_2) to our atmosphere has the potential to detrimentally alter our planet's climate and ecology. Protocols negotiated via international conferences, such as those at Rio de Janeiro (the so-called Earth-Summit) in 1992 and at Kyoto in 1997, have attempted to commit nations to dramatically reducing their emissions of greenhouse gases [carbon dioxide, methane, nitrous oxide, sulphur hexafluoride, hydrofluorocarbons (HFCs) and perfluorocarbons (PFCs)] during the next decades.

Even if you do not believe in the concept of anthropogenic climate change, recent geopolitical events should have convinced us that it is unwise to be wholly dependent on fossil fuel resources located in unstable parts of the world or within nations whose interests may not coincide with ours. The increasingly efficient use of the fuel resources we do have access to, and the promotion of local energy sources, must be to our long-term benefit.

I would not dare to argue that the usage of ground source heat alone will allow us to meet all these objectives. Indeed, many doubt that we will be able to adequately reduce fossil carbon emissions soon enough to significantly brake the effects of global warming. If we are to make an appreciable impact on net fossil carbon emissions, however, we will undoubtedly need to consider a whole raft of strategies, including

1. A reduction in energy consumption, for example, by more efficient usage of our energy reserves.
2. Utilisation of energy sources not dependent on fossil carbon. The most strategically important of these non-fossil-carbon sources is probably nuclear power (although uranium resources are finite), followed by hydroelectric. Wind, wave, biomass, geothermal and solar power also fall in this category.
3. Alternative disposal routes for fossil carbon dioxide to atmospheric emission, for example, underground sequestration by injection using deep boreholes.

I will argue, however, that utilisation of ground source heat allows us to significantly address issues (1) and (2). Application of *ground source heat pumps* (Chapter 4) allows us to use electrical energy highly efficiently to transport renewable environmental energy into our homes (Box 1.1).

If the environmental argument does not sway you, try this one for size: The regulatory framework in my country is forcing me to install energy-efficient technologies! The Kyoto Protocol is gradually being translated into European and national legislation, such as the British Buildings Regulations, which require not only highly thermally efficient buildings but also low-carbon space heating and cooling technologies. Local

BOX 1.1 Energy, work and power.

Energy is an elusive concept. In its broadest sense, energy is related to the ability to do work. Light energy can be converted, via a photovoltaic cell, to electrical energy, which can be used to power an electrical motor, which can do work. The chemical energy locked up in coal can be converted to heat energy by combustion and then to mechanical energy in a steam engine, allowing work to be done. In fact, William Thomson (Lord Kelvin) demonstrated an equivalence between energy and work. Both are measured in Joules (J).

Work (W) can defined as the product of the force (F) required to move an object and the distance (L) it is moved. In other words,

$$W = F \cdot L$$

Force is measured in Newtons and has a dimensionality $[M][L][T]^{-2}$. Thus, work and energy have the same dimensionality $[M][L]^2[T]^{-2}$ and $1\ J = 1\ kg\ m^2\ s^{-2}$.

Power is defined as the rate of doing work or of transferring energy. The unit of power is the Watt (W), with dimensionality $[M][L]^2[T]^{-3}$.

1 Watt = 1 Joule per second = $1\ J\ s^{-1} = 1\ kg\ m^2\ s^{-3}$.

planning authorities may demand a certain percentage of 'renewable energy' before a new development can be permitted. Ground source heating or cooling may offer an architect a means of satisfying ever more stringent building regulations. It may assist a developer in getting into the good books of the local planning committee.

Finally, the most powerful argument of all: Because you can make money from ground source heat! You may be an entrepreneur who has spotted the subsidies, grants and tax breaks that are available to those who install ground source heating schemes. You may be a consultant wanting to offer a new service to a client. You may be a drilling contractor – it is worth mentioning that, in Norway and the UK, drillers are reporting that they are now earning more from drilling ground source heat boreholes than from their traditional business of drilling water wells. You may be a property developer who has sat down and looked cool and hard at the economics of ground source heat, compared it with conventional systems and concluded that the former makes not only environmental sense but also economic sense.

1.4 Thermogeology and hydrogeology

You do not have to be a hydrogeologist to study thermogeology, but it certainly helps! A practical hydrogeologist often tries to exploit the earth's store of groundwater by drilling wells and using some kind of pump to raise the water to the surface where it can be used. A thermogeologist exploits the earth's heat reservoir by drilling boreholes and using a ground source heat pump to raise the temperature of the heat to a useful

BOX 1.2 Head – a Measure of Groundwater's Potential Energy.

We know intuitively that water tends to flow downhill (from higher to lower elevation). We also know that it tends to flow from high to low pressure. We can also intuitively feel that water elevation and pressure are somehow equivalent. In a swimming pool, water is static: it does not flow from the water surface to the base of the pool. The higher elevation of the water surface is somehow compensated by the greater pressure at the bottom of the pool.
The concept of *head* (h) combines elevation (z) and pressure (P). Pressure (with dimension $[M][L]^{-1}[T]^{-2}$) is converted to an equivalent elevation by dividing it by the water's density (ρ_w: dimension $[M][L]^{-3}$) and the acceleration due to gravity (g: dimension $[L][T]^{-2}$), giving the formula:

$$h = z + \frac{P}{\rho_w g}$$

Groundwater always flows from regions of high head to regions of low head. Head is thus a measure of groundwater's potential energy: it provides the potential energy gradient along which groundwater flows according to Darcy's Law.

level. However, the analogy does not stop here. There is a direct mathematical analogy between groundwater flow and subsurface heat flow.

We all know that water, left to its own devices, flows downhill or from areas of high pressure to low pressure. Strictly speaking, we say that water flows from locations of high *head* to areas of low head (Box 1.2). Head is a mathematical concept that combines both pressure and elevation into a single value. Similarly, we all know that heat tends to flow from hot objects to cold objects. In fact, a formula, known as Fourier's Law, was named after the French physicist Joseph Fourier and permits us to quantify the heat flow conducted through a block of a given material (Figure 1.1):

$$Q = -\lambda A \frac{d\theta}{dx} \tag{1.1}$$

where Q = flow of heat in Joules per second, which equals Watts ($J\ s^{-1} = W$), λ = thermal conductivity of the material ($W\ m^{-1}\ K^{-1}$), A = cross-sectional area of the block of material under consideration (m^2), θ = temperature (°C or K), x = distance coordinate in the direction of decreasing temperature (note that heat flows in the direction of decreasing temperature; hence, the negative sign in the equation) and $d\theta/dx$ = temperature gradient ($K\ m^{-1}$).

The hydrogeologists have a similar law, Darcy's Law, which described the flow of water through a block of porous material, such as sand:

$$Z = -KA \frac{dh}{dx} \tag{1.2}$$

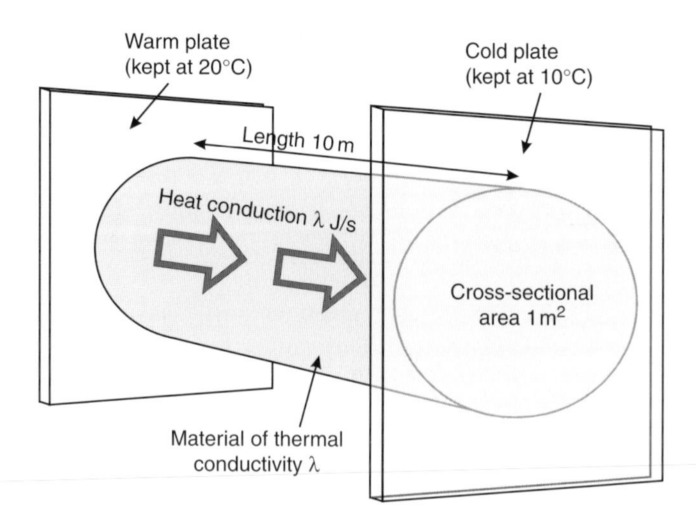

Figure 1.1 The principle of Fourier's Law. Consider an insulated bar of material of cross-sectional area 1 m^2 and length 10 m. If one end is kept at 20° and the other end at 10°, the temperature gradient is 10 K per 10 m, or 1 K m^{-1}. Fourier's Law predicts that heat will be conducted from the warm end to the cool end at a rate of λ J s^{-1}, where λ is the thermal conductivity of the material (in W m^{-1} K^{-1}). We assume that no heat is lost by convection or radiation.

where Z = flow of water (m^3 s^{-1}); K = hydraulic conductivity of the material (m s^{-1}), often referred to as the permeability of the material; A = cross-sectional area of the block of material under consideration (m^2); h = head (m); x = distance coordinate in the direction of decreasing head (m); and dh/dx = head gradient (dimensionless).

A hydrogeologist is interested in quantifying the properties of the ground to ascertain whether it is a favourable target for drilling a water well (Misstear *et al.*, 2006). Two properties are of relevance. First, the hydraulic conductivity (or permeability) is an intrinsic property of the rock or sediment, which describes how good that material is at allowing groundwater to flow through it. Second, the storage coefficient describes how much groundwater is released from pore spaces or fractures in a unit volume of rock, for a 1 m decline in groundwater head. A body of rock that has sufficient groundwater storage and sufficient permeability to permit economic abstraction of groundwater is called an *aquifer* (from the Latin 'water' + 'bearing').

In thermogeology, we again deal with two parameters describing how good a body of rock is at storing and conducting heat. These are the *volumetric heat capacity* (S_{VC}) and the *thermal conductivity* (λ). The former describes how much heat is released from a unit volume of rock as a result of a 1 K decline in temperature, while the latter is defined by Fourier's Law (Equation 1.1). We could define an *aestifer* as a body of rock with adequate thermal conductivity and volumetric heat capacity to permit the economic extraction of heat (from the Latin *aestus*, meaning 'heat' or 'summer'). In reality, however, all rocks can be economically exploited (depending on the scale of the system required – see Chapter 4) for their heat content, rendering the definition rather superfluous.

Table 1.1 The key analogies between the sciences of hydrogeology and thermogeology.

	Hydrogeology	**Thermogeology**
What are we studying?	Groundwater flow	Subsurface heat flow
Key physical law	Darcy's Law $Z = -KA\frac{\mathrm{d}h}{\mathrm{d}x}$	Fourier's Law (conduction only) $Q = -\lambda A\frac{\mathrm{d}\theta}{\mathrm{d}x}$
Flow	Z = groundwater flow ($\mathrm{m^3\ s^{-1}}$)	Q = heat flow = ($\mathrm{J\ s^{-1}}$ or W)
Intrinsic property of conduction	K = hydraulic conductivity ($\mathrm{m\ s^{-1}}$)	λ = thermal conductivity ($\mathrm{W\ m^{-1}\ K^{-1}}$)
Measure of potential energy	h = groundwater head (m)	θ = temperature (°C or K)
Measure of storage	S = groundwater storage (related to porosity)	S_{VC} or S_C = specific heat capacity ($\mathrm{J\ m^{-3}\ K^{-1}}$ or $\mathrm{J\ kg^{-1}\ K^{-1}}$)
Exploitable unit of rock	Aquifer (Lat. *aqua*: water)	Aestifer (Lat. *aestus*: heat)
Tool of exploitation	Well and pump	Borehole or trench and heat pump
Measure of well/borehole efficiency	Well loss	Borehole thermal resistance

Table 1.1 summarises the key analogies between thermogeology and hydrogeology, to which we will return later in the book.

2

Geothermal Energy

> It has pressed on my mind, that essential principles of Thermo-dynamics have been overlooked by … geologists
>
> William Thomson, Lord Kelvin (1862)

2.1 Geothermal energy and ground source heat

Let us clear up this business of 'geothermics' once and for all, because high-temperature geothermal energy will not be covered later in this book. We have already stated that, in the context of this publication, we will use the terms *geothermal energy* and *geothermics* to describe the high-temperature energy that

- is derived from the heat flux from the earth's deep interior;
- one finds either in very deep boreholes or in certain specific locations in the earth's crust (or both).

We will reserve the terms *ground source heat* and *thermogeology* to describe the low-enthalpy heat that

- occurs ubiquitously at 'normal' temperatures in the relatively shallow subsurface;
- may contain a component of genuine geothermal energy from the deep-earth heat flux, but will usually be dominated by solar energy that has been absorbed and stored in the subsurface.

The distinction between geothermal energy and ground source heat may rile specialists in the former discipline, so let me explain my reasoning. I have selected the term *thermogeology* because I believe that the practical utilisation of ground source heat has reached a stage where it has developed a theoretical substructure distinct from that applied to high temperature geothermics. Thermogeology is a highly suitable word for this theoretical framework because it invites an analogy with the science of hydrogeology. Hydrogeology is the study of the occurrence, movement and exploitation of water in the geosphere (in other words, the study of groundwater). The comparison appears fortuitous, but once we start looking more closely, the analogy is rigorous and the two sciences enjoy pleasingly parallel theoretical frameworks (see Table 1.1).

2.2 Lord Kelvin's conducting, cooling earth

Since deep mining commenced in the sixteenth and seventeenth centuries, it was known that the earth became warmer with increasing depth, while in 1740, the first geothermometric measurements were taken by de Gensanne in a French mine (Prestwich, 1885; Dickson and Fanelli, 2004). In other words, it gradually became clear that there is a geothermal gradient. Fourier's Law tells us that, if there is a geothermal gradient and if rocks have some finite ability to conduct heat, then the earth must be conducting heat from its interior to its exterior.

$$Q = \lambda A \frac{d\theta}{dz} \quad \text{Fourier's Law} \tag{2.1}$$

where Q = heat flow (W), A = cross-sectional area (m^2), θ = temperature (°C or K), z = depth coordinate and λ = thermal conductivity (W m^{-1} K^{-1}) of rocks.

From here it is a short leap to the deduction that the earth is losing heat and cooling down. It is exactly this chain of logic that William Thomson (later Lord Kelvin – Box 2.1) used in the middle of the nineteenth century to deduce the age of the earth, staking his claim to be the world's first thermogeologist (Thomson, 1864, 1868). At that time, Thomson and many other geologists suspected that the earth had been born as a globe of molten rock and had subsequently cooled. They believed that the observed geothermal gradient was due to the residual heat in the earth's interior gradually leaking away into space through a solid lithosphere. As heat was lost, the thickness of the lithosphere increased at the expense of the molten interior. Thomson combined Fourier's Law with the one-dimensional equation for heat diffusion by conduction (which we will meet again in Chapter 3):

$$\frac{\partial^2 \theta}{\partial z^2} = \frac{S_{VC}}{\lambda} \frac{\partial \theta}{\partial t} \tag{2.2}$$

where z = depth below the earth's surface (m), $\partial\theta/\partial z$ = geothermal gradient, S_{VC} = specific volumetric heat capacity of rocks (J m^{-3} K^{-1}) and t = time (s).

Thomson tried to work out the age at which the earth's crust had formed, by making assumptions about the initial temperature of the earth's interior (around 7000°F hotter than current surface temperature, or somewhat over 4100 K; Thomson, 1864; Ingersoll *et al.*, 1954; Lienhard and Lienhard, 2006) and a reasonable estimate of the thermal diffusivity of rocks. In 1862, he was able to conclude that the age of the earth,

BOX 2.1 William Thomson, Lord Kelvin.

William Thomson was born in Belfast, Ireland, in 1824 to James Thomson, an engineering professor (O'Connor and Robertson, 2003). The family subsequently moved to Scotland after James was appointed professor of mathematics at Glasgow University. The young William started as a student at the same university at the tender age of 10; while by his mid-teens he was writing essays on the earth and studying Joseph Fourier's theories of heat. After further study and research in Cambridge and Paris, he was appointed professor of natural philosophy at Glasgow in 1846.

Thomson is probably best remembered for the hugely important theoretical underpinning that he provided for the science of thermodynamics. Amongst other things, he proposed (in 1848) the absolute temperature scale (the Kelvin scale) and developed the principle of the equivalence between mechanical work, energy and heat. He seems to have been the first person to propose the notion and lay the theoretical foundations for the heat pump, where mechanical energy is used to transfer heat from a low-temperature environment to a high-temperature one.

Later in his career, Thomson was created Baron Kelvin of Largs by the British Government. This title provides us with the name of the SI unit of temperature, possibly the only SI unit to be ultimately named after a Glaswegian river!

Thomson can also lay claim to being the world's first thermogeologist and was able to estimate the age of the earth from the earth's heat flux. His estimate of around 100 million years was wrong (due to the lack, at that time, of any concept of heat generation by radioactive isotopes in the earth), but his techniques were fundamentally correct.

Kelvin died in his eighties in 1907 near Largs in Scotland. It is time that his reputation is rehabilitated and that he is recognised not only as one of the fathers of the heat pump but also as the founder of the science of thermogeology.

that is, the time it would take to cool down to the current observed temperature and geothermal gradient, was somewhere between 20 and 400 million years (he later homed in on 100 million years, and eventually settled on an age at the younger end of the range; Lewis, 2000).

We will not worry about the mathematics here, but combining Equations 2.1 and 2.2, followed by integration, yields the following expression (Ingersoll *et al.*, 1954; Clark, 2006):

$$t_{earth} = \frac{S_{VC} \cdot (\theta_i - \theta_s)^2}{\lambda \cdot \pi \cdot \psi^2} \tag{2.3}$$

where ψ is the current geothermal gradient near the surface, t_{earth} is the age of the earth's crust (in seconds), θ_s is the surface temperature and θ_i is the temperature of the earth's molten interior. You can try the calculation yourself by choosing 'guesstimates' of input values (see Table 3.1, p. 35, for typical values of λ and S_{VC}, and try a value of $0.02\ K\ m^{-1}$ for ψ).

Thomson's estimate placed him at odds with conservative Christians, who accepted a young earth, based on tendentious genealogical calculations from the Bible. It also won him little popularity with some contemporary geologists and biologists, who thought that the lower of Thomson's age estimates were rather short to account for the observed stratigraphy of the earth and the evolution of life. Thomson's estimate ultimately proved to be a gross underestimate: we currently reckon the earth to be around 4.5 billion years old. As a result, Thomson is sometimes ridiculed by modern geologists. But his calculations were fundamentally correct, given the knowledge and conceptual model he had at the time. We now know that the earth's interior is kept hot by the continuous decay of radionuclides, chiefly isotopes of uranium (^{238}U and ^{235}U), potassium (^{40}K) and thorium (^{232}Th); hence, it cools far slower than Thomson's prediction. But Thomson could not know this: radioactivity was only discovered by Antoine Henri Becquerel in 1896, and radioactive elements were only isolated by Marie and Pierre Curie in 1902.

2.3 Geothermal gradient, heat flux and the structure of the earth

Thomson assumed, not unreasonably, that the transport of heat through the earth's lithosphere was dominated by conduction, and that it was spatially homogeneous. In other words, he assumed that the geothermal gradient and heat flux were uniform over the earth's surface. In the later part of the nineteenth century, workers such as Joseph Prestwich and J.D. Everett (and other colleagues, including my thermogeological predecessor at the University of Newcastle, George A.L. Lebour) focused on quantifying the magnitude of the gradient from measurements in English coal mines, Cornish tin mines and drilled boreholes (Everett *et al.*, 1876, 1877, 1880, 1882; Everett 1882; Prestwich, 1886). Indeed, Prestwich (1885) deduced a value of 0.037°C m^{-1} from determinations in coal mines and a value of 0.042°C m^{-1} from the Cornish mines. In fact, we now know that geothermal gradient varies considerably between different locations, although typical values are in the range 2–3.5°C per 100 m (0.02–0.035 K m^{-1}). The typical geothermal heat flux is of the order 60–100 mW m^{-2}, with a global average estimated at 87 mW m^{-2} (Pollack *et al.*, 1993; Dickson and Fanelli, 2004). Using Fourier's Law (see above), we can try these values for size to derive a typical thermal conductivity of the earth's subsurface of

$$\lambda = \frac{0.087\,\mathrm{W\,m^{-2}}}{0.0275\,\mathrm{K\,m^{-1}}} = 3.2\,\mathrm{W\,m^{-2}\,K^{-1}}$$

It has also become clear that the earth has a somewhat more complicated internal structure than Kelvin's conceptual model presupposed. In terms of geochemistry and mineralogy, the earth's structure can be considered to comprise (Figure 2.1)

i. a solid inner core of metallic iron–nickel, of radius 1370 km;
ii. a molten outer core of iron–nickel, of thickness 2100 km;

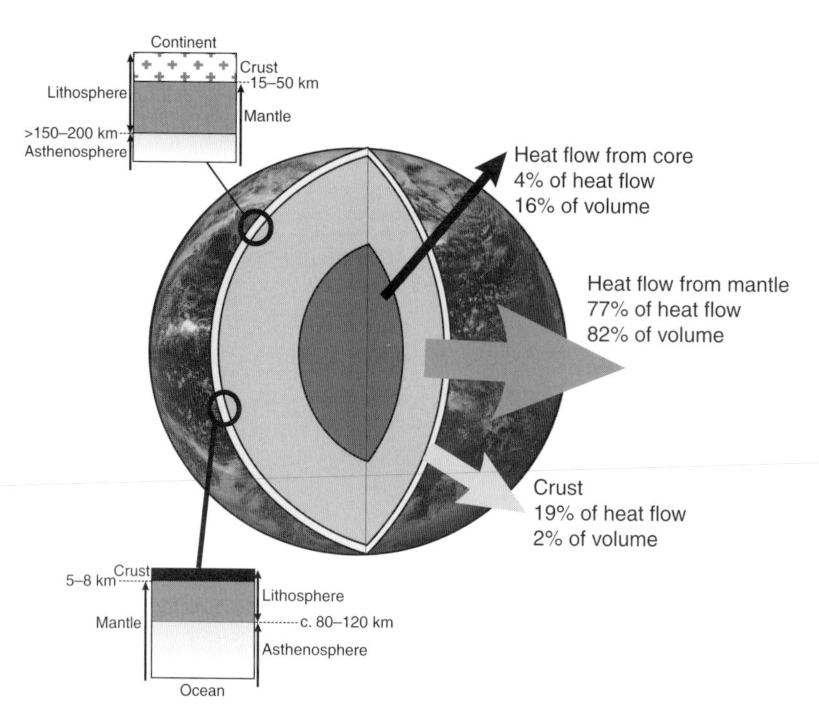

Figure 2.1 Schematic diagram of the structure of the earth, showing the percentage of geogenic heat derived from the core, mantle and crust (numbers adjacent to arrows), compared with the volume of these three portions of the earth. The figure also shows the typical structure of the lithosphere in continental and oceanic plates.

iii. a mantle of ultrabasic Fe- and Mg-rich composition, of thickness 2900 km. The uppermost part of the mantle is, for example, dominantly composed of the minerals olivine and pyroxene, the constituents of a rock called peridotite;
iv. a very thin crust (the boundary between crust and mantle is called the *moho*).

The oceanic crust is wholly different from the crust beneath the continents. The former is very thin (only some 5–8 km) and is predominantly composed of basic minerals and rocks (e.g. gabbro, dolerite, and basalt). Continental crust is somewhat thicker (15–50 km, and even more under mountain belts; Smith, 1981) and less dense than oceanic crust. It is geochemically more acidic and 'sialic' (i.e. rich in *si*licon and *al*uminium) and contains minerals such as quartz and feldspar, which are the familiar constituents of the granites, gneisses and sedimentary rocks that we encounter during our land-based geology field trips.

The earth's radius is about 6370 km. Thus, its circumference is around 40 000 km and its surface area is around 510 million km^2 (5×10^{14} m^2). As the average geothermal heat flux from the earth is known, it can be estimated that the total heat flow from the earth is around 44 TW (Dickson and Fanelli, 2004). The proportions of geogenic heat flux from the core, mantle and crust of the earth are shown in Figure 2.1. The heat derived from the core is small, relative to the core's volume. This is due to the core being composed largely of metallic iron and nickel, impoverished in the heat-generating radioactive nuclides. The crust (and especially the continental crust) is responsible for more than its volumetrically 'fair' share of heat production, at around 19%, being relatively rich in radioactive uranium, potassium and thorium minerals.

2.4 Internal heat generation in the crust

Wheildon and Rollin (1986) point out that the earth's geothermal gradient changes with depth, due to radiogenic heat generation within the crust itself. If Å is the heat production per unit volume of the earth's crust (W m^{-3}) and q is the geothermal heat flux, then

$$\frac{\partial q}{\partial z} = \frac{\partial}{\partial z}\left[\lambda \frac{\partial \theta}{\partial z}\right] = -\text{Å} - S_{VCf} v_D \frac{\partial \theta}{\partial z} + S_{VC} \frac{\partial \theta}{\partial t} \tag{2.4}$$

The second term on the right-hand side relates to heat transport by convection, where S_{VCf} is the volumetric heat capacity of the convecting fluid (e.g. groundwater, gas or magma) and v_D is its vertical fluid flux rate (positive upwards). The third term on the right-hand side represents the change in heat stored in the rock with time, where S_{VC} relates to the volumetric heat capacity of the rock. Thus, if convectional heat transfer is negligible and if we consider a steady-state situation:

$$\frac{\partial q}{\partial z} = \frac{\partial}{\partial z}\left[\lambda \frac{\partial \theta}{\partial z}\right] = -\text{Å} \tag{2.5}$$

We have already derived Fourier's Law (Equation 2.1), which relates heat flow to geothermal gradient, with the caveat that the geothermal gradient may vary with depth due to internal production of heat. We now also have Equation 2.5, which is a form of Poisson's equation that relates the change in average geothermal gradient to internal heat production. This sounds temptingly simple, but we should also remember that internal heat production will be depth-dependent! In fact, Å will typically decrease with increasing depth, as the crust becomes less sialic in nature and with a lesser content of radioactive minerals.

Equation 2.4 predicts that we would expect the highest geothermal heat fluxes in regions with high radiogenic heat production in the upper crust or with strong upward convection of hot fluids from depth. For example, if we consider Figure 3.7, p. 51, (showing the geothermal heat flux in the UK), the highest heat fluxes are from areas underlain by granites (south-west England and Weardale). In the granitic terrain of Devon and Cornwall (south-west England), internal radiogenic heat production (Å) reaches 5 μW m^{-3}, while heat fluxes (q) in excess of 100 mW m^{-2} are observed. Other anomalies, such as those in central England (around the Peak District) are more likely to be the result of deep convection of groundwater (see Brassington, 2007).

2.5 The convecting earth?

While Kelvin's conceptual model involved a static conducting globe, we now know (thanks to the plate tectonic paradigm shift of the 1960s) that the earth is not a rigid sphere. Over long periods of geological time and at the temperature and pressure conditions prevailing in the earth's mantle, we can envisage rocks behaving more like fluids than solids. It is widely believed by many geologists that the earth's mantle, at some scale, is subject to convection processes. Put very simply, just as a saucepan full of milk, heated from below, will begin to form roiling convection cells, the earth's interior is in constant, slow fluid motion.

We can think of the earth's tectonic plates as a kind of stiff, low-density 'scum' (or *lithosphere*) of rock floating on a deeper, fluidly deforming *asthenosphere*. The boundary between the lithosphere and asthenosphere does not coincide with the crust/mantle boundary (or moho). Rather, the lithosphere comprises the crust and a rigid portion of the underlying mantle, while the asthenosphere lies wholly in the mantle (Figure 2.1). Below the oceanic crust, the lithosphere may be around 80–120 km thick [although at mid-oceanic ridges (Figure 2.2) it may only be several km thick]. Below continents, the lithosphere is believed to be considerably thicker, exceeding 200 km in places.

It is widely believed that the motion of the lithosphere's tectonic plates is in some way coupled to convection cells within the mantle/asthenosphere. Tectonic plates move away from each other at mid-ocean ridges, where the lithosphere is thin and the asthenosphere rises and diverges. At subduction zones and compressive plate margins, chunks of lithosphere override each other (Figure 2.2). In fairness, most geologists agree that there are a number of 'driving forces' behind the motion of tectonic plates, including gravitational forces acting on descending slabs of lithosphere at subduction zones. Furthermore, it is also recognised that parts of the lower crust also undergo significant fluid deformation on large scales (Westaway *et al.*, 2002). Thus, mantle convection is, at best, only part of a complex picture.

Thus, far from being a uniform, gently cooling globe, the earth is a heterogeneous (at least in its upper portions), convecting sphere. The outer shell of the earth is composed of materials of varying thermal properties and is in slow, constant motion. Volcanic and seismic activities are concentrated along tectonic plate margins (Figure 2.3). Moreover, the geothermal heat flux at these margins can average 300 mW m^{-2} (Boyle, 2004), and it should be no surprise that the earth's major geothermal resources are also concentrated along these zones.

2.6 Geothermal anomalies

In most locations on earth, direct use of true geothermal energy is not an especially realistic option. With a geothermal gradient of 0.02°C m^{-1}, we would need to drill 1.75 km to reach a temperature of 45°C (about the minimum necessary for space

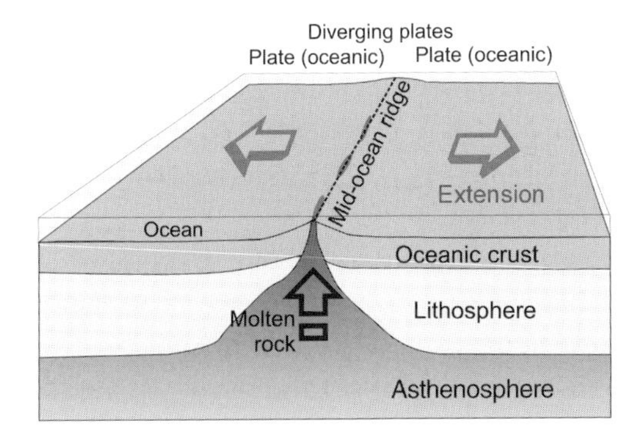

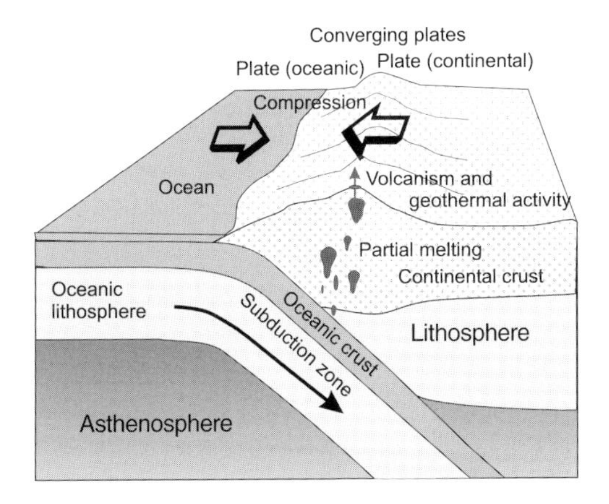

Figure 2.2 A simplified cross section of the earth's lithosphere showing both divergent (top) and convergent (bottom) plate margins. The rising, partially molten asthenosphere and the thinning of both the crust and the lithosphere at oceanic ridges result in a strongly elevated geothermal gradient and volcanic activity. At subduction zones, the presence of water in the descending oceanic lithospheric slab, coupled with prevailing temperature and pressure conditions, gives rise to partial melting along the slab. This creates bodies of magma that rise through the overlying lithosphere and eventually give rise to localised volcanism and geothermal fields in the island groups or mountain ranges located above the subduction zone.

heating). Alternatively, we could look at things another way: to utilise sustainably the earth's geothermal heat flux to heat a small house, with a peak heat demand of 10 kW, we would need to capture the entire flux (say, 87 mW m^{-2}) over an area of 115 000 m^2 (11.5 ha). Both a 1.75 km deep hole and an 11.5 ha heat-capture field per house are rather unrealistic propositions for the average householder!

Fortunately, the earth's geothermal heat flux and temperature gradient are not uniformly distributed and there do exist anomalous areas of the earth's surface where the heat flux is much larger than average and/or we encounter high temperatures at shallow depth. We can call these anomalies potential geothermal fields, and they can be due to a variety of geological factors.

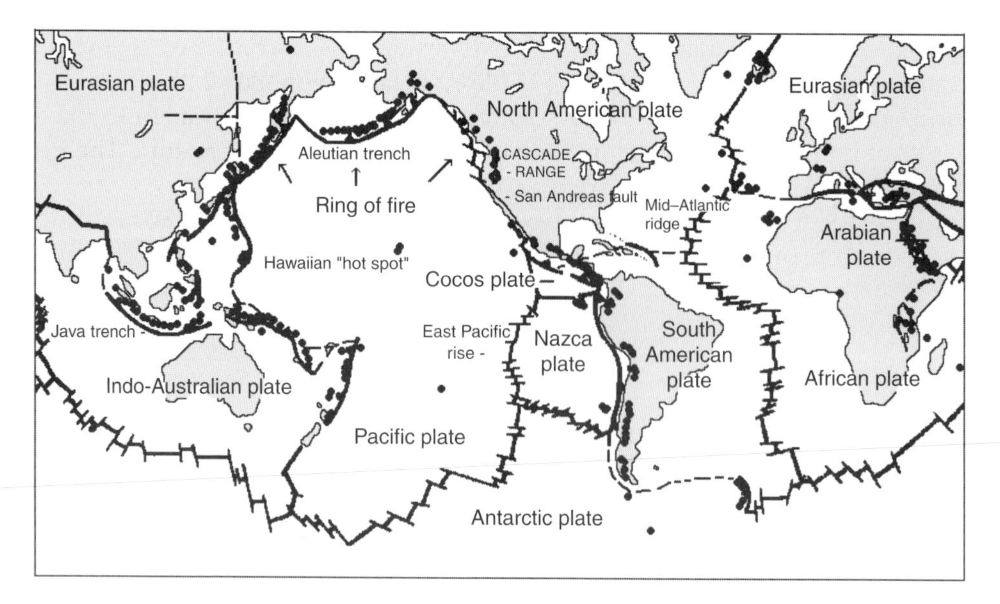

Figure 2.3 Simplified plate tectonic map of the world, showing locations of active volcanoes as dots. These tend to fall along plate boundaries. *Public domain material produced by United States Geological Survey (USGS)/Cascades Volcano Observatory and accessed from http://vulcan.wr.usgs.gov.*

High-temperature geothermal fields are usually related to plate tectonic features (see Figure 2.3). They typically occur at one of three tectonic locations and are often associated with current or historic vulcanism.

1. Extensional plate margins: typically mid-oceanic ridges (e.g. Iceland and the Azores), or proto-rifts such as the Great Rift Valley of central and eastern Africa, and the Rhine Graben. At such extensional margins, the crust and lithosphere are rather thin and are being 'ripped' apart. The geothermal gradient is very high and the asthenosphere may only be a few km deep. Geochemically basic magma intrudes into the extensional cracks and fissures related to rifting and may overspill at the surface as volcanoes. The geothermal fields around Iceland's capital, Reykjavík, are examples of systems drawing their energy from the presence of magma at shallow depth in an extensional rifting regime (Franzson *et al.*, 1997).
2. Convergent plate margins: the presence of water in a subducting slab of oceanic crust (Figure 2.3), coupled with the particular pressure and temperature conditions at depth, can lead to partial melting along the slab. This generates bodies of magma that rise slowly through the overthrust lithospheric slab. If these 'diapirs' of magma reach the surface, the water-rich molten rock can explode as a violent volcanic eruption, such as that of Krakatoa in 1883. The accumulated volumes of magmatic material, coupled with the tectonic forces associated with subduction, usually give rise to linear mountain belts (such as the Andes) or island arcs (such as the Aleutians or

Japan) above the plate margin. These margins are usually associated with geothermal or volcanic loci, such as those in the Mediterranean region (Box 2.2).

3. Below some tectonic plates, localised plumes of warm material rise from the deep mantle at hot spots seemingly unrelated to the broader tectonic picture. The mechanisms of these 'mantle plumes' are still poorly understood, but these volcanic and geothermal loci can persist for geologically extended periods. The Hawaiian island chain was formed by successive volcanic eruption centres as the Pacific Plate drifted slowly across the location of a mantle plume. The Yellowstone 'supervolcano' and associated geothermal field is another example of 'plume' activity.

BOX 2.2 Larderello – the history of geothermal energy in a nutshell.

The Larderello site is situated in southern Tuscany, Italy, and is usually regarded as the great-granddaddy of geothermal energy schemes. The area is renowned for its phreatic volcanic activity, that is, periodic eruptions of steam. The last such major eruption was from Lago (Lake) Vecchienna in 1282 AD, when ash and blocks of rock were also disgorged. The geothermal activity is believed to be related to the presence of a cooling body (or pluton) of granite at relatively shallow depth beneath a cover of metamorphic and sedimentary rocks (GVP, 2006).

Geothermal waters because they are usually deeply derived (a long time for hydrochemical evolution) and because they are warm (enhanced chemical kinetics) often have high and unusual mineral contents (Albu *et al.*, 1997). The hot waters at Larderello were historically renowned for their mineral content: the Romans used the sulphur-rich waters for bathing. Later, the waters were extracted from shallow boreholes and used to produce the element boron, which they contained in abundance. The site was not known as Larderello at that time, but as Montecerboli: it was renamed after the Frenchman François de Larderel, who in 1827 first used the geothermal steam to assist in extracting boron from cauldrons of 'volcanic' mud. This drew attention to the locality's thermal, as well as hydrochemical potential. Shortly afterwards, geothermal steam was being exploited to perform mechanical work at the boron works. In 1904, it was first used to attempt to generate electricity, followed by the construction of a power plant in 1911. The site remained the world's sole geothermal electricity producer until 1958, when New Zealand opened its first plant. It was not until 1910–1940 that the geothermal heat from the site was actually used for space heating (Dickson and Fanelli, 2001). The development of this geothermal site provides an interesting perspective on historical priorities: first mineral production, then mechanical work and finally space heating. It should serve as a reminder that it is relatively recently that the era of consequence-free, dirt-cheap fossil fuel has drawn to a close. It is only now that we are beginning to prioritise the need for sustainable, affordable, low-carbon space heating.

Away from these specific tectonic settings, more modest geothermal anomalies (either positive or negative) are related to the earth's dynamic behaviour over geological timescales, to heterogeneities within the crust, or to the effects of fluid flow in transporting heat from one location to another. For example, they may be related to

4. Variations in thermal conductivity of rocks. Assuming that we have a constant flux of heat from the earth's interior, Fourier's Law implies that, in order to conduct this constant flux, a layer of rock with a low thermal conductivity must possess a high geothermal temperature gradient (Figure 2.4). We thus expect to find anomalously high temperatures beneath thick layers of rock with low thermal conductivity. The low-temperature Paris (Boyle, 2004) and Southampton (Box 2.3) geothermal systems are examples of reservoirs with an anomalously high temperature due to an overlying blanket of low-conductivity mudstones or limestones.
5. The fact that some rock bodies have internal heat production (e.g. radioactive decay of uranium and potassium in granites, or chemical oxidation of sulphides in mine waste). An excellent target rock for hot dry rock geothermal systems (Section 2.12) is thus a granite with a high internal heat production, overlain by a thickness of low-conductivity sediments, thus ensuring a high geothermal gradient and high temperature at relatively shallow depth (Box 2.4).
6. Groundwater flow can transport heat rapidly by advection from one location to another (Figure 2.5). Geothermal anomalies may thus occur where faulting allows deep warm groundwater to flow up towards the surface, carrying a cargo of heat

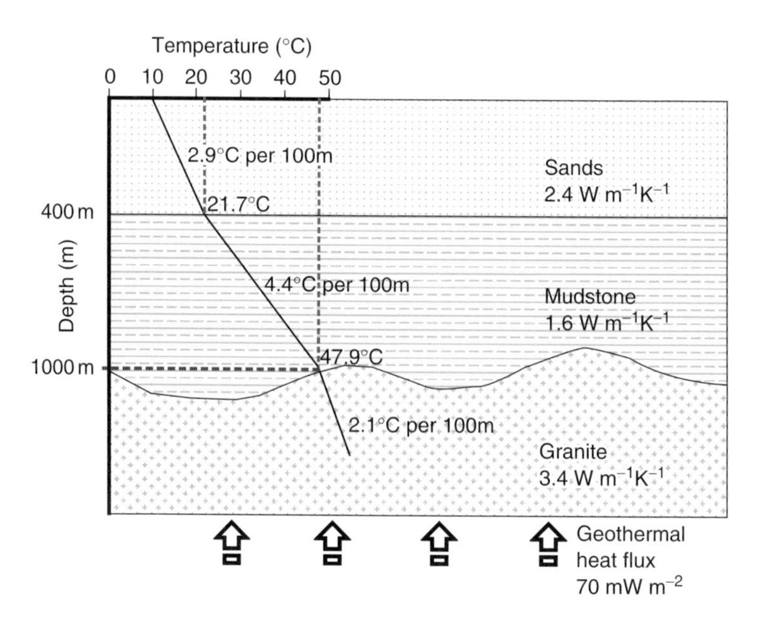

Figure 2.4 Schematic cross section through a three layer 'sandwich' of different rock types. In order to maintain a constant geothermal heat flux of, say, 70 mW m^{-2}, the geothermal gradient in the lower conductivity mudstone layer must be higher than in the sand layer. Therefore, temperatures at the top of the granite are higher than they otherwise would have been, given the initial geothermal gradient in the top (sand) layer.

BOX 2.3 The Southampton geothermal system.

A quick glance at the map of Figure 3.7 (p. 51) reveals that the city of Southampton, on the English south coast, is not associated with any especially high geothermal heat flux. The fact that it is England's first and most famous functional geothermal heating system demonstrates that one does not necessarily require extraordinary geological conditions – one merely needs

- a thick sedimentary sequence of relatively low thermal conductivity (and thus high geothermal gradient), such that elevated temperatures are found at modest depth;
- an aquifer horizon at that depth.

Exploratory drilling at Southampton commenced in the early 1980s, with the first 1.8 km deep production well being commissioned in 1987. Southampton lies in a Tertiary synclinal structure and the borehole penetrated Tertiary and Mesozoic clays, sandstones and limestones before encountering, at a depth of some 1730 m, the geothermal reservoir rock – the 70 m thick Triassic Sherwood Sandstone (Smith, 2000) – which possesses both a relatively high porosity (groundwater storage) and transmissivity. The Sherwood Sandstone was found to contain brine at 76°C, with a static water level around 100 m below ground level. Brine is pumped from the well to the surface (the design yield was 10–15 L s^{-1}), where it passes through a heat exchanger, with associated heat pumps. Heat is thus transferred to a carrier fluid (water) which provides heat, via a network of insulated mains, to a number of properties in Southampton city centre, including homes, hotels, a college campus, a store, a stadium and numerous offices. The cooled brine (around 28°C) is discharged to the estuary of the River Test. The heat yield of the geothermal source was originally 1 MW_t by direct heat exchange to the carrier fluid, although this has now been increased, by the use of heat pumps, to nearer 2 MW_t (Boyle, 2004).

In fact, the geothermal source is now integrated with a combined heat and power plant (CHP), including a high-efficiency 5.7 MW generator, supplying the district heating and cooling scheme. The CHP provides heat to the circulating carrier fluid. It also sources the electricity required to run the system's pumps and the heat to run the system's absorption heat pumps. Furthermore, fossil fuel boilers can be activated to support the peak heating load of the scheme (which is around 12 MW_t).

Heat pumps are also used to support a separate insulated network of pipes supplying customers with chilled water for air-conditioning. Cooling needs have increased dramatically since the system's conception to such a level that it is planned, during summer months, to use nighttime electricity to produce ice by means of heat pumps. This ice will then be used to provide chilled water during daytime – a simple but elegant means of storing 'coolth' produced at times of surplus electrical capacity (Energie-Cités, 2001).

While the geothermal borehole now only provides around 10–20% of the total peak heating load supplied by the integrated scheme, the scheme's total impact is impressive, with an annual carbon dioxide emission saving of at least 10 000 tonnes, compared with conventional technologies (Energie-Cités, 2001; EST, 2005).

BOX 2.4 The Weardale exploration borehole.

Figure 3.7 (p. 51) shows that the region of Britain with the highest geothermal heat flux is that underlain by the Hercynian granite batholiths of Devon and Cornwall, which have an internal radioactive heat production of up to 5 $\mu W\ m^{-3}$. The next most prominent feature stretches west of Sunderland from Weardale to the Lake District and does not clearly correspond with any surface geological outcrop. The anomaly is, however, known to be underlain by Caledonian (lower Devonian) granites with an estimated radiogenic heat production of 3.3–5.2 $\mu W\ m^{-3}$ (Wheildon and Rollin, 1986). The granites are overlain by a thickness of several hundred metres of Carboniferous sedimentary rocks of low thermal conductivity. We have seen (Section 2.6) that we should expect high temperatures at shallow depths where such 'hot' radioactive rocks are overlain by an 'insulating' sedimentary cover. Indeed, a team at Newcastle University found that thermochemical signatures of waters in flooded mines in the region provided tentative indications of elevated temperatures (Younger, 2000).

In 2003, a proposal was made to revive a disused cement works site at Eastgate in Weardale as a 'renewable energy village'. Hydrochemical evidence from nearby fluorite mines suggested a temperature anomaly associated with the Slitt Vein – a mineralised fault zone in the Carboniferous sedimentary rocks presumed to pass beneath Eastgate (Manning *et al.*, 2007). In summer 2004, five 50–60 m inclined boreholes were drilled to locate the Slitt Vein at Eastgate beneath a cover of Quaternary superficial deposits.

In August 2004, a deep exploration borehole commenced above the subcrop of the Slitt Vein, at a diameter that would allow it to be commissioned as a production well, should thermal water be encountered. Drilling proceeded through the Lower Carboniferous sedimentary rocks (including 67 m of the dolerite Whin Sill), until the granite was encountered at 272 m depth. Drilling proceeded into the granite until, at 410 m, a major fracture was encountered (the drilling bit appeared to drop through a void of some 50 cm aperture). Water entered the borehole from this fracture and its level eventually stabilised around 10 m below the ground level. The potential short-term yield of this horizon exceeded 16 $L\ s^{-1}$, comprising a hypersaline sodium–calcium–chloride brine (Paul Younger, *pers. comm.*; Manning *et al.*, 2007).

Drilling continued and, despite high bit attrition and corrosion rates, eventually terminated at 995 m. A bottom-hole temperature of 46°C was measured (compared with a predicted temperature of around 29°C assuming an average geothermal gradient of 2°C). The water yielded by the borehole as a whole is, of course, dominated by the major fracture at 410 m, and hence has a lower temperature of around 27°C – hot enough to support a saline kurbad (should that be desirable), but inadequate for direct heating purposes. Alternative options for achieving a higher temperature include

- Hydraulic fracturing of the lower sections of the borehole to allow groundwater flow in the deepest, hottest part of the granite – a type of 'enhanced geothermal system' (see Section 2.12).

BOX 2.4 *(Continued)*

- Operating the Eastgate borehole as a form of 'standing column well' system (see Chapter 11). By doing this, one is sacrificing advectional heat transfer for conductive heat transfer and, while it may be possible to achieve a higher temperature, it will be at the expense of flow volume and thus of overall heat production rate.
- Utilise heat pumps to raise the temperature of 27°C to a useful space heating value, with a high degree of efficiency.

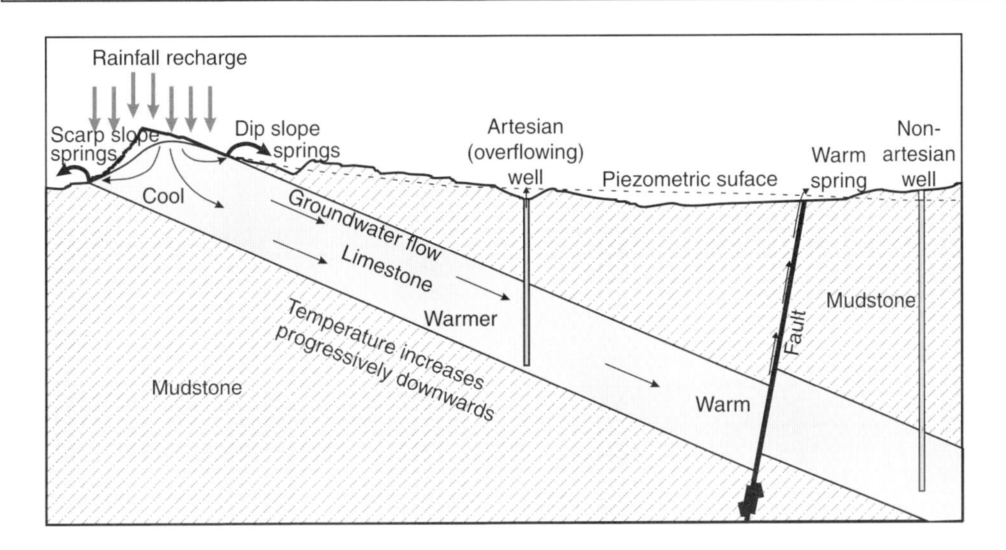

Figure 2.5 Schematic cross section through a groundwater system. Recharge falling on the limestone aquifer outcrop slowly flows down-dip, equilibrating with progressively higher temperatures with increasing depth. Small quantities of water are able to exit the aquifer system via a 'short-circuiting' fault. The ascent along a high permeability fault may be so rapid that the water does not substantially cool during its re-ascent, emerging as a warm spring. The grey shaded strata are water-saturated limestone.

(geothermal short-circuiting). The British hot springs at Bath, Buxton and Matlock are all related to faulting that allows deep groundwater from Carboniferous limestone strata to flow to the surface (Banks, 1997; Brassington, 2007).

7. Geothermal anomalies can also occur due to the fact that earth (and its climate) is a dynamic system. The subsidence of thick sedimentary basins can 'rapidly' carry cold sediments downwards. Conversely, isostatic rebound (e.g. following the last glaciation) can raise the elevation of rocks at a rate of several centimetres per year. Furthermore, climatic cooling during the Pleistocene glaciation in the UK and northern Europe is believed to have depressed the temperature of sediments and rocks down to at least 300 m depth (Wheildon and Rollin, 1986; Šafanda and Rajver, 2001). In the slight twitches and anomalies of geothermal gradients measured in boreholes, specialists can make deductions about the past climate and about rates

of crustal uplift and subsidence. For deep geothermal exploration, we must also be aware of the potential violation of one of the assumptions behind Equation 2.5: that the geological environment is in a steady state. In fact, in northern Europe, it is not – it is still recovering (thermally and isostatically) from the Pleistocene glaciations. Wheildon and Rollin (1986) suggest that ignoring this perspective may cause us to significantly underestimate our geothermal heat flows.

2.7 Types of geothermal system

Geothermal energy systems can be classified into low-, intermediate- and high-enthalpy systems (Figure 2.6): here, the term 'enthalpy' is closely related to the temperature of the system. Various authors disagree about the boundaries between these classifications and they are, frankly, of little practical value. It is possibly better to classify geothermal systems based on their potential for use or on the characteristics of the fluids they produce (Dickson and Fanelli, 2004).

2.7.1 Water- and vapour-dominated geothermal systems

The fluid produced by wells drilled into water-dominated systems is mostly liquid water as the pressure-controlling phase, with some steam present, for example, as bubbles. The temperatures of these systems may be well above 100°C – remember that water only boils at 100°C at 1 atm pressure. In subsurface pressure conditions, water can exist as a liquid at much higher temperatures, only boiling ('flashing') when pressure is released during transit to or on arrival at the surface. Thus, water-dominated systems may produce hot water, mixtures of water and steam, wet steam or even dry steam (see Figure 2.7).

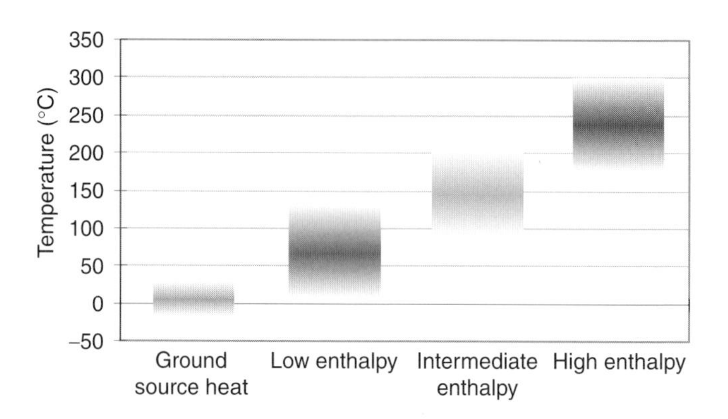

Figure 2.6 Classification of geothermal systems according to temperature (based on suggestions made by Dickson and Fanelli, 2004).

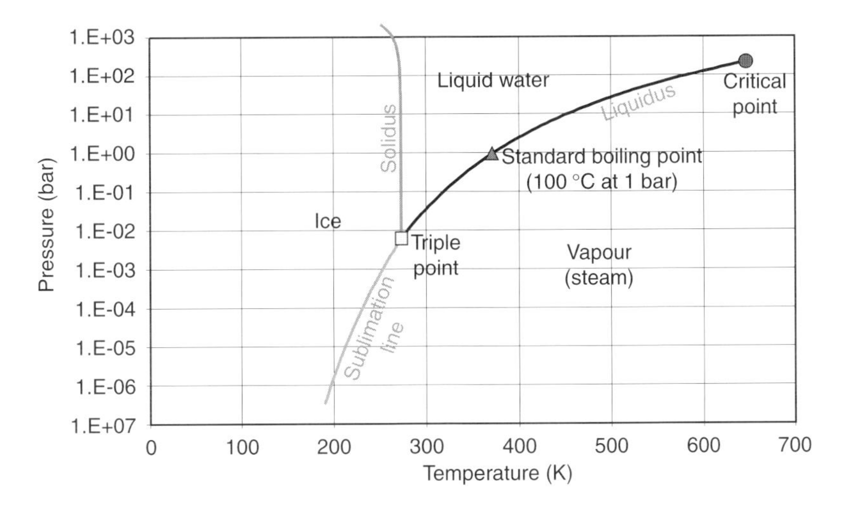

Figure 2.7 A simplified phase diagram for water, showing the phases present at various pressure and temperature conditions. The liquidus is the line dividing the liquid and steam fields, the solidus is the line dividing the ice and liquid fields. Wet steam is steam that co-exists with water at the temperature of volatilisation for a given pressure (i.e. a mixture of steam and water on the liquidus); dry steam is steam without any water phase present, but still lying on the liquidus; superheated steam is steam whose temperature is above the temperature of volatilisation for a given pressure (i.e. it plots below the liquidus).

Vapour-dominated systems are rarer. Here water and steam may be present in the reservoir rocks, but steam will be the pressure-controlling phase. These systems typically produce dry or superheated steam.

2.8 Use of geothermal energy by steam turbines

The use of steam turbines to directly generate electricity (Figure 2.8) typically requires geothermal systems producing fluids at temperatures over 150°C. In systems producing mixtures of water and steam or wet steam, a separator may be required to remove the water component from the steam. The efficiency of the turbine can be improved by using a condensing unit to condense the steam following passage through the turbine.

2.9 Binary systems

At lower temperatures, it is still possible to generate electricity from geothermal fluids. Here, however, we may not be able to use steam directly as the working fluid. We must use a heat exchanger to transfer heat energy to a secondary fluid that has a lower temperature of vaporisation (boiling point), such as n-pentane or butane (Boyle, 2004). It is this secondary fluid, which, having volatilised, performs mechanical work

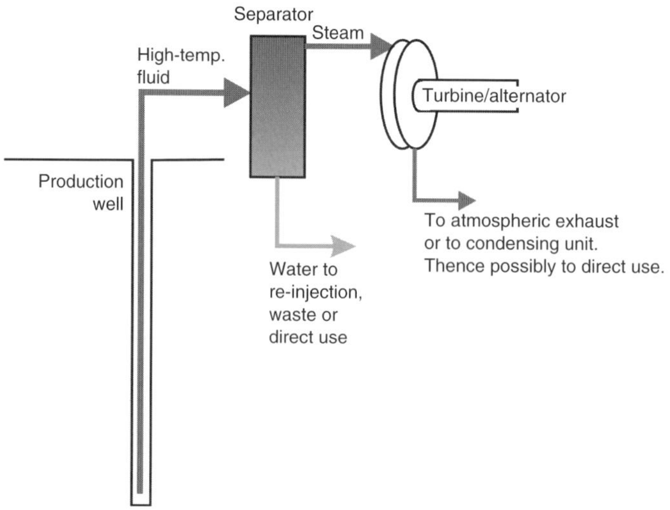

Figure 2.8 The use of a high-temperature geothermal fluid to power a steam turbine.

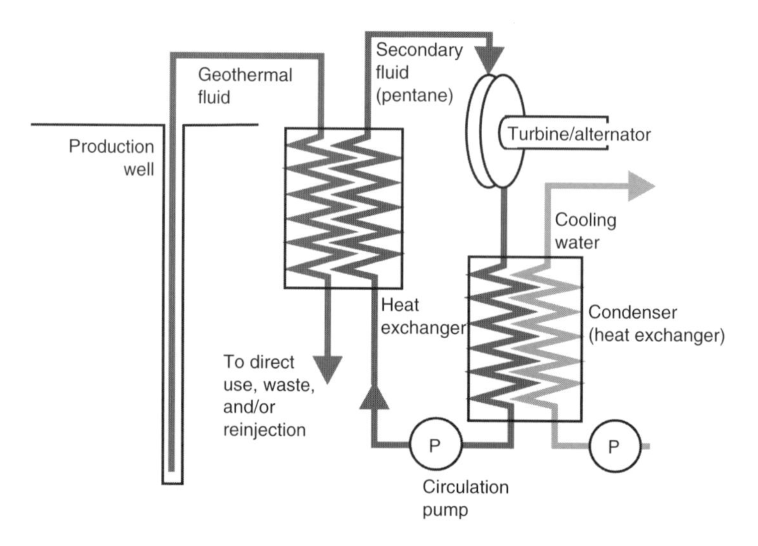

Figure 2.9 The use of an intermediate-temperature geothermal fluid to power a binary system.

in rotating a generating turbine (Figure 2.9), using what is often referred to as an organic Rankine cycle (ORC). Binary systems can generate electricity from geothermal fluids of temperatures down to 85°C (or, theoretically, even lower). Recent developments such as the ammonia-water-based Kalina cycle seem set to improve efficiencies at lower temperatures (Boyle, 2004; Dickson and Fanelli, 2004).

2.10 Direct use

Lower temperature geothermal fluids can also be used for direct uses (Lienau, 1998), which include

- space heating;
- industrial uses;
- swimming pools;
- agriculture and aquaculture (i.e. fish farming).

Heat pumps (Chapter 3) may or may not be employed to deliver heating to these uses. Usually, the geothermal fluid will transfer its heat, via a heat exchanger (Figure 2.10), to a delivery fluid, whose chemical composition is controlled. In this way, problems with chemical incrustation or corrosion are limited to the heat exchanger and will not affect the remainder of the heating circuit.

2.11 Cascading use

Of course, in reality, the energy of many geothermal systems is not extracted by only one means of exploitation, but by several successive 'cascading' applications.

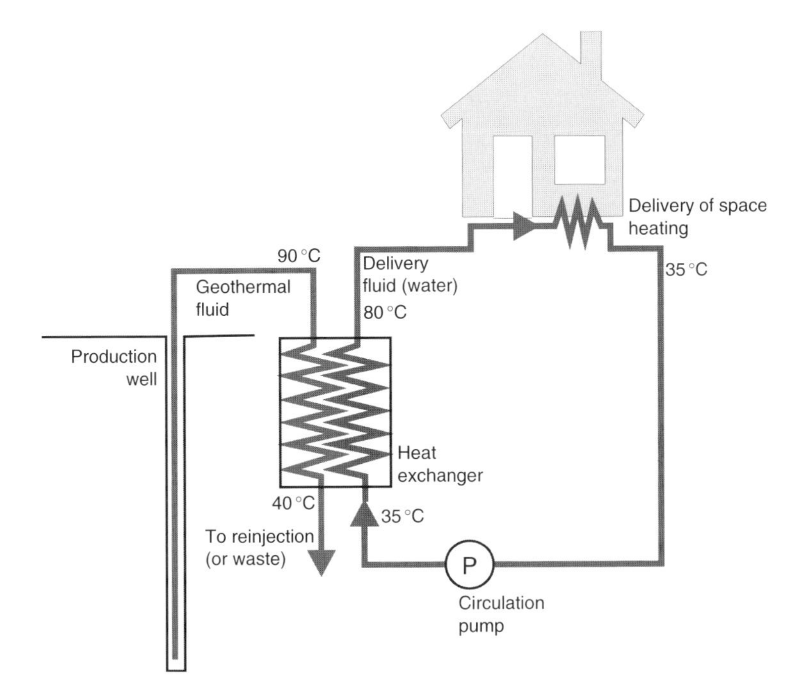

Figure 2.10 'Direct use' of a lower temperature geothermal fluid to deliver space heating.

For example, a high-enthalpy geothermal system with high steam content may drive steam turbines in two steps: first using the primary steam content and second by 'flashing' the water in the separator to drive another turbine (a so-called dual-flash turbine system). Thereafter, the heat in the 'waste' steam from the turbines may be extracted via heat exchangers to a secondary fluid, where it can be used directly: first, maybe, for a high-temperature industrial application and thereafter for residential space heating purposes.

Figures 2.11a and b show schematic diagrams for the Nesjavellir high-enthalpy geothermal field in Iceland and for the other fields contributing to the Reykjavik district-heating scheme (Lund, 2005). Note that the Nesjavellir geothermal wells produce over 1600 L s^{-1} of water at over 200°C and 14 bar. In 2005, the capacity for electricity generation by turbine as Nesjavellir was 120 MW_e. The waste steam from the turbines is condensed and the heat exchanged to cold groundwater from a separate wellfield, in turn raising its temperature to around 85°C. Following de-aeration to remove excess oxygen, this hot groundwater is piped around 30 km to Reykjavik, where it contributes over 200 MW_t to the capital city's geothermal district heating scheme (which is also fed by several other geothermal fields and whose total capacity is over 800 MW_t – see Figure 2.11b).

2.12 Hot dry rock systems (a.k.a. 'enhanced geothermal systems')

In the discussions above, we have blithely assumed that if we drill into a geothermal reservoir, we will find a geothermal fluid (water and/or steam) that we can extract and utilise. However, some geothermal reservoirs have rather low permeability and we cannot extract large volumes of natural fluid from them. These are known as 'hot dry rock' resources, and may comprise poorly fractured hard rocks such as granites. These are attractive as they may be associated with a higher than average heat flux and geothermal gradient, due to the fact that they often contain elevated concentrations of radioactive isotopes of uranium, potassium and thorium. Such granites may represent significant geothermal anomalies, especially if buried underneath thicknesses of insulating sediments of low thermal conductivity (see Figure 2.4 and Box 2.4).

So, how do we extract heat from such a poorly permeable block of hot granite sitting in the deep subsurface? Projects such as the Cornish Hot Dry Rock project in the Carnmenellis granite of south-west England and the European trial at Soultz-sous-Forêts (Alsace) have discovered that we can inject cool water deep into the formation and circulate it through artificially created fractures in the granite, before re-abstracting it at a higher temperature (Figure 2.12). These fractures thus effectively act as heat exchange surfaces between the rock and the water. We can create artificial fractures by using explosives down a borehole, but we can create much more controlled fractures by utilising a technique known as hydraulic fracturing or hydrofraccing (Less and Andersen, 1994; Banks and Robins, 2002; Misstear *et al.*, 2006). Here, water is injected down a borehole, through a packer, at very high pressure. In fact, the pressure is so high that it exceeds the *in situ* geological stress acting on the rock mass and the rock's own

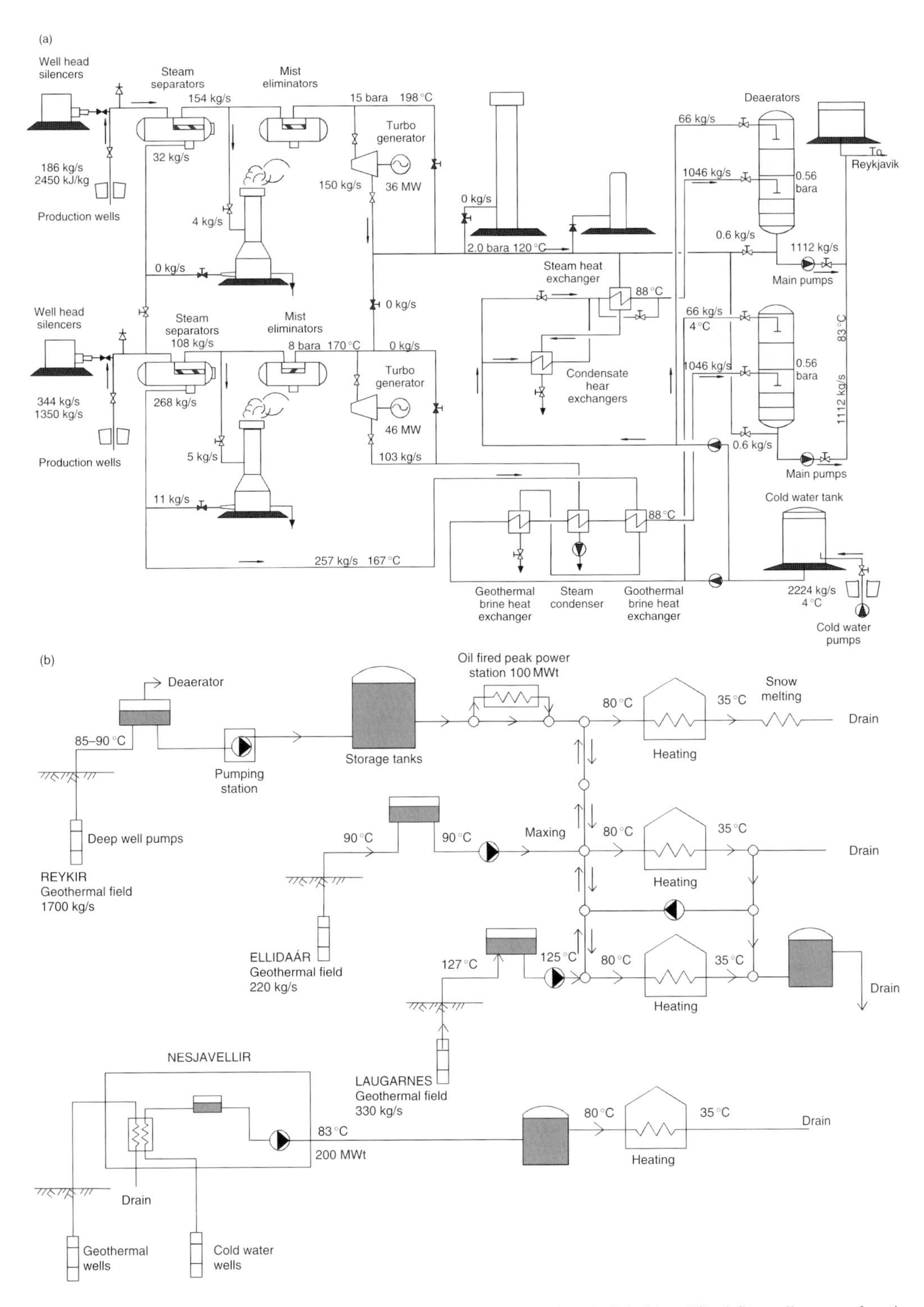

Figure 2.11 (a) Flow diagram of the Nesjavellir plant, Iceland. (b) Simplified flow diagram for the district heating in Reykjavik. *Both diagrams after Lund (2005) and reproduced by kind permission of Dr John W Lund and the GeoHeat Center, Klamath Falls, Oregon.*

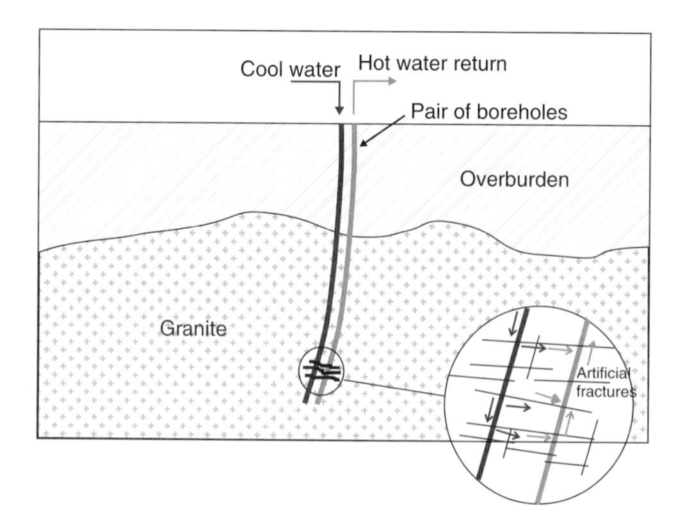

Figure 2.12 A hot dry rock geothermal exploitation system.

inherent tensile strength and eventually creates a new fracture (or opens existing planes of weakness or joints in the rock). As an example of the dimensions we are considering, the test wells at Soultz-sous-Forêts (Alsace) were initially drilled to around 3.5–3.8 km (corresponding to a temperature of $\approx$165°C) and, following hydrofraccing of the granite, a flow of 25 L s^{-1} was circulated across the 450 m separation between the downflow and return wells. The wells were subsequently deepened to 5 km (Rabemanana *et al.*, 2003).

Although hot dry rock projects at Carnmenellis, Los Alamos, Japan and (most recently) at Soultz-sous-Forêts, where a binary plant has recently been commissioned, have demonstrated that the technology is feasible, the combination of (1) the need for a specific geological environment (which may not be in close proximity to major demand centres) and (2) the capital expenditure on research, investigation and plant, have conspired to make 'hot dry rock' applications unattractive in the current energy climate. However, in a rapidly changing global energy market where low-carbon energy sources can increasingly be sold at a premium, the day may soon arrive when other binary plants similar to that at Soultz-sous-Forêts make economic sense.

2.13 The 'sustainability' of geothermal energy and its environmental impact

When discussing the sustainability of our usage of any natural resource, it is, first, wise to be clear about our definition of the word 'sustainable' and, second, to have a clear perspective of the timescale we are considering. One of the classic definitions of sustainability is that of Gro Harlem Brundtland's commission: sustainable development meets the needs of the present without compromising the ability of future generations to meet their own needs. Stefansson and Axelsson (2003) offer a more complex definition, tailored to geothermal exploitation:

> For each geothermal system, and for each mode of production, there exists a certain level of maximum energy production, E_0, below which it will be possible to maintain constant energy production from the system for a very long time (100–300 years). If the production rate is greater than E_0 it cannot be maintained for this length of time. Geothermal energy production below, or equal to E_0, is termed *sustainable production*, while production greater than E_0 is termed *excessive production*.

The non-sustainability of our exploitation of a geothermal system may manifest itself in one of two ways (or, of course, both):

i. The temperature of the system may fall to an unusable level, because we are extracting a greater heat flux than can be naturally replenished on the time scale of the operation.
ii. The supply of geothermal fluid (whether it be water or steam) may diminish, as we are abstracting the fluid at rate greater than its natural replenishment.

Thus, to assess the sustainability of a geothermal operation we need to have a very clear understanding of both the heat budget and the water budget of system, and the boundary conditions of these systems. When constructing our heat budget, we should remember that, in many geothermal systems, there are at least two, and maybe three, mechanisms of heat recharge to a subsurface geothermal system (Stefansson and Axelsson, 2003):

- advection of magma;
- advection of groundwater (or geothermal fluid);
- conduction.

Figure 2.13 shows an example of an approximate heat budget, on a national scale, for Iceland. As regards our fluid budget, we should understand that as we extract hot fluid from a geothermal reservoir, we may deplete the fluid resource of the system, and if there is no natural recharge of groundwater, downhole heads or pressures will drop to unusable levels. If the reservoir is open to groundwater recharge, on the other hand, abstraction may induce the additional recharge of cold groundwater to the system, maintaining fluid reserves but ultimately depleting the temperature of the reservoir and resulting in the breakthrough of cooler fluid to production wells. It has been demonstrated, both by theoretical studies (e.g. Gringarten, 1978) and empirical example, that the practice of re-injecting 'waste' geothermal fluids back into the reservoir will not only serve to maintain fluid pressures and reservoir lifetime, but can also serve to maximise the total heat extracted from the reservoir. Indeed, after years of consistently falling steam productivity at the Geysers field in California caused by paucity of natural recharge, the progressive introduction of re-injection of wastewater

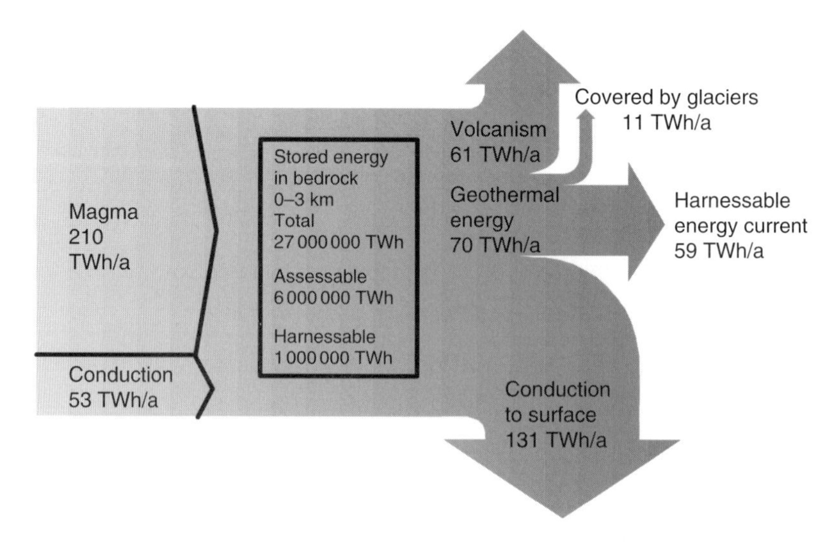

Figure 2.13 The terrestrial energy current in Iceland (after Stefansson and Axelsson, 2003). *Reproduced by kind permission of Orkustofnun and Íslenskar orkurannsóknir.*

throughout the 1980s is believed to have assisted in braking the decline by around the year 1995 (Stefansson and Axelsson, 2003).

Boyle (2004) provides some simple calculations for the Italian Tuscan geothermal fields, the Imperial Valley of California and the Krafla field of Iceland, demonstrating that the rate at which we extract heat in many geothermal operations significantly exceeds the areal rate of replenishment and argues that our exploitation of these geothermal fields is ultimately unsustainable. Rybach (2003a) also considers the sustainability of geothermal fields and indicates that we are typically considering periods of decades or centuries for the longevity of high-enthalpy geothermal fields and comparable periods for their recovery to natural conditions following exploitation. Stefansson and Axelsson (2003) also present examples of geothermal operations that can be described as unsustainable on a scale of decades to centuries, including the Geysers field of California and the Hamar field of central Iceland. They also, on a somewhat more optimistic note, present data from the Laugarnes field in Iceland, appearing to demonstrate some form of stability (=sustainability?) in production.

Finally, when considering the issue of sustainability, we would do well to remember timescale. Even though a reservoir (i.e. a set of geographically and hydrogeologically related extraction and injection wells) may turn out to have a finite lifetime of decades or centuries, the earth's tectonic and heat flow processes have a timescale of millions of years. If we have to abandon a geothermal field after 100 years, the ultimate heat source (geothermal heat flux and magmatism associated with plate tectonics) remains. Conduction, magmatic convection and groundwater convection continue to supply heat and we can usually ultimately expect our abandoned geothermal field to recover towards its pre-exploitation temperatures on a similar timescale to that of its productive life (i.e. typically decades to centuries; Rybach, 2003a). If we are lucky enough to live in such a geologically active country as Iceland, we may be able to simply shift the centre of our operation to a new set of wells in a different geological location some tens of kilometres away and continue production while we await the recovery of our original reservoir. Such a nation may therefore be able to indefinitely sustain a national energy policy founded on geothermal energy, via a sound understanding and holistic planning of its resources of heat and geothermal fluids.

In the discussion above, we have defined 'sustainability' rather narrowly and constrained it to discussions of whether production of fluids and heat can be maintained over protracted periods. In popular perception, the discussion of sustainability is bound with the concept of pollution and environmental impact. Clearly, geothermal power stations and district heating schemes can potentially have adverse (and occasionally, beneficial) environmental impacts. Both Dickson and Fanelli (2004) and Boyle (2004) discuss these in detail, but we can here briefly list the main factors:

- Noise: from production wells and turbines. This is probably worst during the drilling, development and testing phase.
- Smell: from emission of gases such as H_2S ('rotten eggs').
- Emissions of other gases, including gases such as CO_2, from the produced geothermal fluids. According to Boyle (2004), emissions of CO_2 from geothermal operations range from 0.004 to 0.74 kg CO_2 per kWh, with an average of 0.12 kg CO_2 per kWh. From Table 4.1 (p. 65), we can see that this is far from insignificant, but is much less than that from conventional electricity generation.
- Subsidence and microseismicity (especially where re-injection of fluid is not practiced or is inadequate).
- Saline wastewaters: although these may be re-injected or run to waste in the sea. They may even be 're-branded' as tourist amenities, such as the Blue Lagoon at the Svartsengi geothermal field in Iceland (Franzson *et al.*, 1997).

In summary, therefore, although the exploitation of geothermal energy is not without some environmental drawbacks, these are widely considered to be significantly less than with conventional fossil fuel technology (Boyle, 2004).

2.14 And if we do not live in Iceland?

Meanwhile, those of us who live in such distinctly un-geothermal provinces as Henley-on-Thames and Scunthorpe are left pondering the question: What use do I have for an understanding of thermogeology? The remainder of this book is dedicated to exactly those poor souls, for the rocks beneath the Thames Valley and the Humber Estuary are blessed with the ubiquitous low-temperature reserves of heat that we will term *ground source heat*. Although Kelvin realised it 150 years ago, it has taken the rest of us over a century to recognise that we can utilise this low-temperature heat for space heating. In fact, as a strategic 'green' energy resource, it is almost certainly far more significant for the planet's future than the geothermal energy considered in this chapter.

So, let us leave geothermal energy and geothermics behind us. Onwards! To the infinitely more exciting realm of ground source heat and thermogeology ....

3

The Subsurface as a Heat Storage Reservoir

> Yonder the harvest of cold months laid up,
> Gives a fresh coolness to the Royal Cup;
> There Ice like Christal, firm and never lost,
> Tempers hot July with December's frost …
> … Strange! that extreames should thus preserve the Snow
> High on the Alpes and in deep Caves below!

Edmund Waller (1606–1687)
on the occasional of a new royal icehouse at St James's Park, London,
cited by Buxbaum (2002)

In its most basic form, a household storage heater is a box full of bricks that is heated by electric elements when electricity is cheap, usually at night. The heat is then slowly released during the day to keep the house warm. The storage heater uses bricks because such silicate-based media not only have a very high capacity to store heat, but also have a rather modest *thermal conductivity* – they release heat relatively slowly.

The earth's shallow subsurface can be regarded as a huge storage heater. It is warmed up by solar energy during the summer. We can access and extract that heat during winter. Most rocks are silicate-based and, like bricks, they have a huge potential to store heat. Their thermal conductivity is modest: not so high that the stored heat dissipates immediately, but not so low that we cannot draw it out of the ground through well-designed heat exchangers (typically installed in boreholes or trenches).

You have probably noticed that we are introducing two fundamental thermogeological properties here: thermal conductivity and storage. Thermal conductivity describes

BOX 3.1 Jean Baptiste Joseph Fourier.

Fourier was born in 1768 into a large tailor family in Auxerre, France (O'Connor and Robertson, 1997). By the age of 13, his mathematical talents had become clear, but in his studies he was torn between following a career as a priest and indulging his love for maths. The latter eventually won. In 1790, he took a post as teacher at his former college, the École Royale Militaire of Auxerre. After a brush with Revolutionary politics, which nearly terminated his life at the guillotine's blade, he went on in 1795 to the École Normale, followed by the Central School of Public Works (*École Centrale des Travaux Publiques*), where he benefited from close contact with the mathematicians, Joseph-Louis Lagrange and Pierre-Simon Laplace.

Fourier's subsequent career included positions as Professor at the École Polytechnique and adviser to Napoleon during his invasion of Egypt. Later (at Napoleon's bidding) he 'accepted' the Prefecture of the Department of Isère in Grenoble, where he accomplished a number of civil engineering works. At Grenoble, he obviously had some time to pursue theoretical research; between 1804 and 1807 he completed his work *On the Propagation of Heat in Solid Bodies*, which introduced the world to his mathematical technique of expressing mathematical functions as what we now call 'Fourier series'. Fourier's theory was not universally popular: there was a tendency amongst some scientists to regard heat as a substance called 'caloric' and expend much effort discussing what this substance was. According to Greco (2002), Fourier's great insight was to avoid this speculation about what heat was and to focus on what it did. In other words, he treated heat flow as a mathematical process, rather than attempting to squeeze it into an ill-fitting physical conceptualisation.

Fourier died in 1830 in Paris.

a material's ability to transfer heat by conduction, as described by Fourier's Law (Box 3.1). We have already met this property in Equation 1.1. Already in the latter half of the nineteenth century, scientists were beginning to measure the thermal conductivity of rocks: indeed William Thomson required a value of thermal conductivity as input to his model of the cooling earth (Section 2.2, p. 9). In the 1860s, Thomson, J.D. Everett and A.J. Ångström had started making determinations of the thermal diffusivity/conductivity of sediments and soils (Everett, 1860; Thomson, 1868; Rambaut, 1900). Professors A.S. Herschel and G.A. Lebour of the University of Durham College of Science (later to become the University of Newcastle-upon-Tyne) were, in the 1870s, able to present a series of determinations of thermal conductivities of rocks (Herschel and Lebour, 1877; Herschel *et al.*, 1879; Prestwich, 1885, 1886; Barratt, 1914).

The property describing storage of heat is called specific or volumetric heat capacity. It is new to us, so let us consider it in more detail.

3.1 Specific heat capacity: the ability to store heat

The ability of a medium (solid, liquid or gas) to store heat is termed its *specific heat capacity* (S_C). This is the amount of heat locked up in the medium for every degree Kelvin of temperature. It is measured in Joules per Kelvin per kilogram (J $K^{-1}kg^{-1}$). The specific heat capacity of water is particularly high at around 4180 J K^{-1} kg^{-1} at around 20°C; that of most rocks is around 800 J K^{-1} kg^{-1} (see Table 3.1). This means

Table 3.1 The thermal conductivity and volumetric heat capacity of selected rocks and minerals.

	Thermal conductivity (W m^{-1} K^{-1})	Volumetric heat capacity (MJ m^{-3} K^{-1})
Rocks and sediments		
Coal	*0.3*	*1.8*
Limestone	1.5–3.0 (*2.8, massive limestone*)	1.9–2.4 (*2.3*)
Shale	1.5–3.5 (*2.1*)	*2.3*
Wet clay	0.9–2.2 (*1.6*)	*2.4*
Basalt	1.3–2.3 (*1.7*)	*2.4*–2.6
Diorite	1.7–3.0 (*2.6*)	*2.9*–3.3
Sandstone	2.0–6.5 (*2.3*)	*2.0*–2.1
Gneiss	2.5–4.5 (*2.9*)	2.1–2.6 (*2.1*)
Arkose	2.3–3.7 (*2.9*)	*2.0*
Granite	3.0–4.0 (*3.4*)	1.6–3.1 (*2.4*)
Quartzite	5.5–7.5 (*6.0*)	1.9–2.7 (*2.1*)
Minerals		
Plagioclase	1.5–2.3	1.64–2.21
Mica	2.0–2.3	2.2–2.3
K-feldspar	2.3–2.5	1.6–1.8
Olivine	3.1–5.1	2.0–3.6
Quartz	7.7	1.9–2.0
Calcite	3.6	2.24
Pyrite	19.2–23.2	2.58
Galena	2.3–2.8	1.59
Haematite	11.3–12.4	3.19
Diamond	545	–
Halite	5.9–6.5	1.98
Other		
Air	0.024	1.29×10^{-3} at 1 atm.
Glass	0.8–1.3	1.6–1.9
Concrete	0.8 (*1.6*)	*1.8*
Ice	1.7–2.0 (*2.2*)	*1.9*
Water	*0.6*	4.18
Copper	390	3.5
Freon-12* at 7°C (liquid)	0.073	1.3
Oak	0.1–0.4	1.4
Polypropene	0.17–0.20	1.7
Expanded polystyrene	0.035	–

* Dichlorodifluoromethane (CCl_2F_2).

Data from *inter alia* Halliday and Resnick (1978), Sundberg (1991), Clauser and Huenges (1995), Eskilson *et al.* (2000), Banks and Robins (2002), Banks *et al.* (2004), Waples and Waples (2004a), and Lienhard and Lienhard (2006). Italics show recommended values cited by Eskilson *et al.* (2000).

that if a mass (m) of 1 kg rock cools down (by an amount $\Delta\theta$) from 13°C to 11°C then two degrees' worth of heat is lost, 1600 J:

$$\text{Heat lost} = m \times S_C \times \Delta\theta = 1\,\text{kg} \times 800\,\text{J}\,\text{K}^{-1}\,\text{kg}^{-1} \times 2\,\text{K} \qquad (3.1)$$

We can also express specific heat capacity as Joules per degree Kelvin per unit volume. This is termed the volumetric heat capacity (S_{VC}). For water, $S_{VC} \approx 4180\ \text{J K}^{-1}\ \text{L}^{-1}$ at around 15–20°C (as its density ≈ 1 kg L^{-1}), whereas most rocks have values of S_{VC} in the range 2.0–2.4 MJ $\text{K}^{-1}\ \text{m}^{-3}$. So, from every cubic metre of rock, we can release up to 10 megaJoules (MJ) of energy, simply by dropping its temperature by 4 K. Conversely, we need to put a similar amount of energy back into our cubic metre of rock to raise it temperature by 4 K. The heat energy in the material is ultimately stored as molecular vibrational or kinetic energy; the hotter the material, the faster the molecules of a fluid whiz around, or the molecules of a crystal vibrate.

Volumetric heat capacity varies somewhat with temperature, partly (but not wholly) due to changes in density of the material. Figure 3.1 shows how this affects the thermal properties of water at varying temperature, while Figure 8.4 (p. 190) shows the same effect for a solution of ethylene glycol antifreeze.

We should remember that heat may also be stored or released from a substance due, not just to change in temperature, but to change in phase. This stored heat is called *latent heat* (Box 3.2).

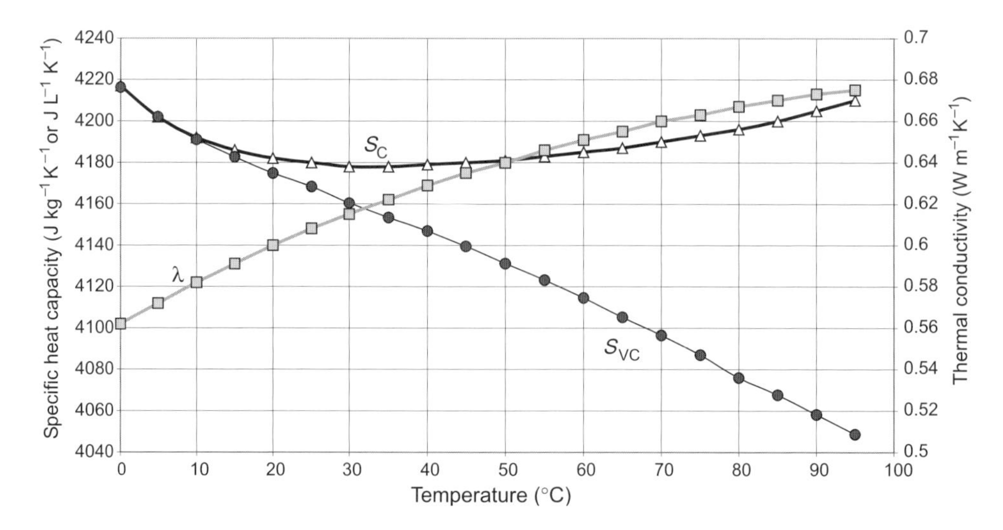

Figure 3.1 The temperature dependence of the thermal properties of water. In this diagram, S_C and S_{VC} are the specific heat capacity and volumetric specific heat capacity, while λ is the thermal conductivity. Values on the graph are derived or calculated from data provided by Eskilson *et al.* (2000).

BOX 3.2 Latent heat.

Some energy is stored in materials simply as a property of their phase, that is, a gas, liquid or solid. For example, steam at 100°C is far more energetic than liquid water at the same temperature. To convert a liquid to a gas (i.e. to move it across the liquidus in Figure 2.7, p. 23), we have to inject an additional amount of energy, without actually raising the temperature at all. For the water–steam transition, this *latent heat of vaporisation* is 2.272 MJ kg^{-1}. The uptake and loss of heat during volatilization and condensation is one of the features that makes the compression–expansion cycle at the heart of refrigerators and heat pumps function (see Chapter 4).

Similarly, when a solid melts to a liquid, it requires an uptake of heat to effect the phase transition, without any temperature rise resulting. To convert ice at 0°C to water at 0°C requires 335 kJ kg^{-1}: the *latent heat of fusion*.

3.2 Movement of heat

It is not enough to know that rocks, sediments and groundwater in the earth's subsurface store heat, we also need to understand how the subsurface absorbs and loses heat, and how we can induce heat to move to places where we can extract it. Heat is transferred by three main mechanisms:

- conduction (we have already met this in Fourier's Law, Equation 1.1);
- convection;
- radiation.

In the shallow subsurface environment, conduction through minerals or pore fluids and convection via groundwater are probably the two most important mechanisms of heat flow. In some cases, radiation may also be important, so let us consider that, too.

3.2.1 Conduction

Heat conduction describes the process by which heat diffuses through a solid, liquid or gas by processes of molecular interaction. Put crudely, if we warm up one end of a chunk of granite, the molecules at that end start vibrating more strongly. These vibrating atoms or molecules cause their neighbours to start vibrating as well, such that the heat energy (and temperature) gradually diffuses throughout the piece of rock. It is this process that Fourier's Law describes by means of Equation 1.1.

The thermal conductivity describes how good the medium is at conducting heat: copper is very good, rocks are less good and plastics are generally poor. Note (Table 3.1) that the thermal conductivities of rocks and other geological materials tend to fall within a rather narrow range, typically between 1 and 3 W m^{-1} K^{-1}. Note also that, of the common rock-forming minerals, quartz has the highest thermal conductivity, at around 7 W m^{-1} K^{-1}. Thus, the thermal conductivity of rocks and sediments depends to a large extent on their quartz content.

Although we tend to treat thermal conductivity as a constant, it is, in fact temperature-dependent to a minor degree in most materials (e.g. see Figure 3.1). In thermogeology, we are working within a relatively narrow temperature range and can neglect this effect in most cases. Thermal conductivity is also dependent on a material's phase: note (Table 3.1) that the thermal conductivity of ice is much higher than that of water (its specific heat capacity is lower, however).

3.2.2 *Convection*

Fluids store heat: for example, water stores around 4180 J L^{-1} for every °C of temperature. Thus, if we move hot water from a boiler house to a bathroom, we are also moving heat. Heat transport that occurs by virtue of the motion of a fluid is termed convection. If we pump hot water from one place to another, we term the heat transport forced convection or *advection*, because the heat is flowing due to a force externally imposed on the carrier fluid. Isaac Newton proposed a simple formula for the rate at which heat is transferred to, or away from, bodies in a moving stream of fluid. This formula (Newton's Law of Cooling) can be stated in the following form:

$$q^* = \bar{h} \cdot (\theta_{\text{body}} - \theta_{\text{fl}}) \tag{3.2}$$

where

- q^* is the heat transfer from body to fluid in W m^{-2} of surface area.
- $\bar{h}$ = local coefficient of heat transfer (W m^{-2} K^{-1}), which will depend on the nature of the fluid, its rate of flow, the body's surface properties, and so on.
- θ_{body} and θ_{fl} are the temperatures of the body and the fluid, respectively.

Newton's Law of Cooling is really a working approximation rather than a law. It functions pretty well for forced convection situations where the temperature difference between the body and the fluid is not too large.

Convectional heat transfer can also take place from a hot body in a fluid that is initially static, with no externally imposed forces. Imagine switching on an electric bar fire in your lounge: the air near the bar heats up (by conduction and radiation) and it expands slightly (see Box 3.3). It thus becomes less dense than the surrounding air and starts to rise, being displaced by denser cold air. This new cold air soon warms up and rises and, before you know it, a convection cell has started within the room. The bar fire provides the heat source. The heat may ultimately be lost from the air to a ceiling, or window or external wall, but heat transport with fluid (air) flow has taken place. We call this 'free convection'.

Newton's Law of Cooling is even more difficult to apply to situations involving free convection, as the heat transfer coefficient will vary as a function of the temperature

BOX 3.3 Boyle's Law and the Universal Gas Constant.

The Irish-born scientist, Robert Boyle (1627–1691), spent much time studying gases. In an appendix, written in 1662, to his work *New Experiments Physio-Mechanicall, Touching the Spring of the Air and its Effects*, he quantified the relationship between their pressure and volume (O'Connor and Robertson, 2000). We intuitively understand that if we pressurise a given quantity of gas, we reduce its volume. Boyle found that, for a given mass of ideal gas (and most gases behave relatively ideally) at a constant temperature, which has an original volume V_1 and pressure P_1, but which is pressurised to a pressure P_2:

$$P_1 \cdot V_1 = P_2 \cdot V_2 = \text{constant}$$

where V_2 is the final volume of the gas at P_2.

Later, the temperature of the gas was found to fit neatly into this relationship.

$$\frac{P_1 \cdot V_1}{T_1} = \frac{P_2 \cdot V_2}{T_2}$$

where T_1 and T_2 are the initial and final temperatures (expressed as degrees Kelvin). In other words, when a gas is pressurised, it has a tendency to heat up (think of a bike pump heating up as you pump up your tyres). When pressure is released, it has a tendency to cool down. You may have used a bottled gas stove when cooking baked beans on a camping trip. When you open the valve on the pressurised gas cylinder, allowing gas to flow to the burner, you are slowly decreasing the pressure in the cylinder. The remaining fluid cools and you may notice drops of condensation on the cylinder. (Note, however, that neither of these examples are strictly appropriate, as the mass of gas in the tyre and the cylinder change during these 'experiments'.)

In fact, it turns out that we can specify a universal gas constant from which we can calculate the volume of a given amount of gas at any temperature and pressure:

$$n = \frac{M}{m} = \frac{P \cdot V}{R \cdot T}$$

This is the so-called Ideal Gas Law, where n = the amount of gas (in moles – a mole is simply a chemical quantity that allows us to compare different gases):

M = mass of gas (kg)
m = molar mass of gas (kg mole^{-1})
P = pressure of gas (Pa = N m^{-2} = kg m^{-1} s^{-2})
V = volume of gas (m^3)
T = absolute temperature of gas (K)
R = the universal gas constant = 8.3145 kg m^2 s^{-2} K^{-1} mole^{-1}

continued

BOX 3.3 *(Continued)*

Note (Box 1.1, p. 4) that we have already seen that 1 Joule = 1 kg m^2 s^{-2}. Thus,

$$R = 8.3145 \text{ J K}^{-1} \text{ mole}^{-1}$$

The Gas Law helps us to predict what happens to a gas when we heat it. If we heat a closed vessel of gas, the volume is fixed and constant. Thus, as the temperature rises, the pressure increases to ensure that the quantity PV/RT remains constant. If, however, we just heat a region of air in a room, the pressure is constant (atmospheric) and the region of air must expand slightly (become less dense) for PV/RT to remain constant. This decreasing density causes that region of air to become more buoyant than the surrounding air and to rise: the beginning of free convection.

Note that the equations in this box only hold true for gases (and ideal gases, at that!).

difference. In fact, Lienhard and Lienhard (2006) suggest that, for free convection:

$$\overline{h} \propto (\theta_{\text{body}} - \theta_{\text{fl}})^n \tag{3.3}$$

where n is a power, often in the range 0.25–0.35, but sometimes reaching as high as 2 in boiling liquids.

In Section 2.5 (p. 14), we have already considered the huge convection cells of fluid mantle rock that are related to global plate tectonic processes and continental movements. In the applied science of thermogeology, however, it is not this type of subsurface convection that interests us. Rather, we need to know about the forced convection (advection) of heat that occurs with groundwater flow. Groundwater in the shallow subsurface flows from areas of high head to low head (Box 1.2, p. 5), usually ultimately driven by gravitational forces. In doing so, it carries with it a huge cargo of heat (around 4.2 kJ K^{-1} L^{-1}). If we sink a well and start pumping it, we locally reduce the groundwater head, forming a cone of depression (Figure 3.2), causing groundwater to flow towards the well, where it can be abstracted. The water can be used for drinking, for household or industrial use – but we can also extract the advected cargo of heat and utilise that, too (see Box 3.4). By the way, in thermogeology, we may also need to consider free convection cells that can establish themselves in groundwater (or even soil gas) within or around water wells or heat extraction boreholes.

3.2.3 Radiation

All bodies radiate energy in the form of electromagnetic radiation: stars, humans, lakes and the earth's cool surface. The hotter the body, the more energy it radiates. Indeed, Stefan (via experimental work in 1879) and Boltzmann (via theoretical consideration, in 1884) stated that, for an ideally radiating body (a so-called *black body*), the energy

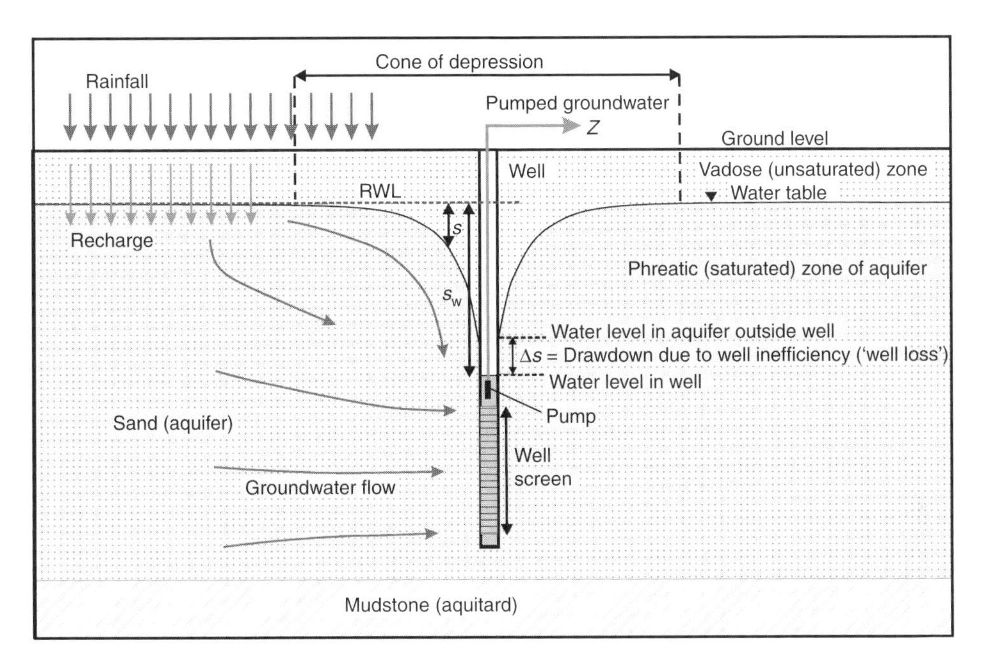

Figure 3.2 Conceptual model of a well abstracting groundwater (at a rate Z) from an unconfined aquifer. The abstracted water also carries a cargo of heat (advection) that can be utilised. Drawdown (s) is defined as the difference between rest water level in the aquifer (RWL) and the groundwater level during pumping. The drawdown in the pumped well (s_W) is the sum of the drawdown in the aquifer and additional drawdown (Δs) due to the hydraulic inefficiency of the well.

BOX 3.4 Case study: Gardermoen International Airport.

Norway's main international airport, at Gardermoen, is also located on the nation's largest aquifer: the Øvre Romerike glaciofluvial sand and gravel aquifer (Odling *et al.*, 1994), containing huge reserves of groundwater at around 5–6°C (Sæther *et al.*, 1992). Before Oslo's airport was shifted from Fornebu to Gardermoen in 1998, it was decided that the massive new airport deserved a ground source heating scheme, based on pumped groundwater. During most of the year, water is pumped from nine 45–50 m deep 'warm' abstraction wells (spaced at 50 m and each with a yield of up to 70 L s^{-1}) and heat is extracted from the groundwater flow (via a heat exchanger) by a heat pump array. The resulting chilled water is used to satisfy any cooling needs in other parts of the airport complex and is then returned to nine similarly spaced 'cold' re-injection wells, some 150 m away from the abstraction wells. In summer, the polarity of the wells is reversed and water is abstracted from the 'cold' wells to perform passive and, if necessary, active cooling, before the water (with its cargo of waste heat) is re-injected to the warm wells. The maximum installed capacity of the system is around 8 MW heating and 6 MW cooling (SINTEF, 2007). The payback period for the additional capital cost of the ground source heating and cooling installation, compared with conventional technology, was reckoned to be around 2 years (Eggen and Vangsnes, 2005).

radiated (E_b) is proportional to the fourth power of the absolute temperature (θ, in Kelvin):

$$E_b = \sigma\theta^4 \tag{3.4}$$

where σ is the Stefan–Boltzmann constant of 5.67×10^{-8} W m^{-2} K^{-4}.

Hot bodies radiate electromagnetic energy through a wide spectrum of wavelengths: the sun's radiation is, for example, dominated by ultraviolet, visible and infrared wavelengths. However, in general, the hotter the body, the shorter the modal wavelength (i.e. the wavelength of peak intensity) in the radiation spectrum – this is known as Wien's Law. Thus, the hottest stars are bluish in hue, the cooler ones red. We know that a horseshoe in the blacksmith's furnace glows 'red hot' – it is also radiating heat in the visible spectrum.

At more familiar temperatures, such as that of our bodies or the ground that we stand on, heat is primarily radiated in the invisible, infrared part of the spectrum. The earth's surface radiates such heat energy, and an infrared camera (maybe satellite-mounted) is able to sense such radiation and compile temperature maps of the earth's surface.

Very cold bodies radiate electromagnetic energy at much longer wavelengths. The universe ('space') can be regarded as having a background temperature of some 3 K above absolute zero. Throughout the universe, we can detect radiation (the so-called cosmic background) with a wavelength of some 1.9 mm (corresponding to 3 K) in the microwave region of the spectrum (even longer wavelength than visible light and infrared).

The heat that we feel on our skin or which is absorbed by the soil on a sunny day is infrared radiation emitted by the sun (along with visible light). Thus, radiation of heat and absorption of radiated energy from the sun and atmosphere are important heat transfer mechanisms at the surface of the earth (Box 3.5). Particularly when we consider the heat budget of bodies such as lakes and ponds (Chapter 9), the radiation of heat energy may be an important component in their heat budget.

3.3 The temperature of the ground

Rocks and sediments have high values of volumetric heat capacity (S_{VC}) but modest values of thermal conductivity (λ). Thus, they have rather low values of thermal diffusivity (Box 3.6). Heat pulses do not propagate very fast or far throughout the subsurface of the earth (at least, in the absence of advection by groundwater).

In summer, the surface of the earth heats up due to intensified solar radiation and elevated air temperatures. This heating effect propagates a few metres down into the earth's subsurface, but not beyond. In fact, below a few metres depth, the temperature of the subsurface is remarkably stable, at a value approximating to the long-term annual average surface temperature. Figure 3.3 shows a hypothetical example for typical Swedish ground conditions (Rosén *et al.*, 2001), where throughout the course of a year, the surface temperature varies by around 20°C. At 6 m depth, the amplitude of seasonal

BOX 3.5 Insolation and atmospheric radiation.

Insolation describes the (dominantly shortwave) solar radiation arriving at the earth's surface. The maximum solar irradiance, incident on a plane perpendicular to the sun (i.e. directly facing the sun), just outside the earth's atmosphere, is estimated to be 1366 W m^{-2}. This would be the theoretical maximum insolation on a perfectly clear day, at noon at a location on the equator, at the equinox, neglecting any effects of atmospheric absorption, reflection or back-scattering!

If the incoming solar irradiance on the earth's cross section is 1366 W $m^{-2} \times \pi r_e^2$, where r_e is the radius of the earth (6.37×10^6 m), the total incoming radiation on the side of the earth facing the sun is 1.74×10^{17} W. This amount, averaged over the earth's entire surface area ($4\pi r_e^2$), gives a mean insolation (outside the atmosphere) of 342 W m^{-2}. Of course, the actual amount of insolation at a given location will depend on latitude and time of day.

However, solar radiation will be absorbed and back-scattered during its passage through the earth's atmosphere. Thus, the shortwave insolation at the earth's surface (q_{sw}) will be somewhat less (around half, as a global average) than that outside the atmosphere. In fact, Connelly (2005) estimates annual mean insolation rates of around 300 W m^{-2}in tropical regions, around 200 W m^{-2} in temperate zones, and as little as 100 W m^{-2} in northern Eurasia and North America. Moreover, the actual instantaneous insolation rate varies from season to season and throughout the day (Table 3.2). Only a relatively small proportion (<6%) of the incoming short-wave radiation is harvested by plants to drive photosynthesis (Linacre and Geerts, 1997).

We must also remember that incoming radiation serves to warm up the earth's atmosphere and, as a warm body, the atmosphere and clouds re-radiate this thermal energy as long-wave (infrared) radiation. This long-wave atmospheric radiation (q_{lw}) must be added to the short wave solar radiation to give the total incoming radiation incident on the earth's surface. Formulae to estimate the long-wave radiation include

$$q_{lw} = \varepsilon_a \sigma \left(\theta_a^0\right)^4 \ \mathrm{W\,m^{-2}} \qquad \text{(Kapetsky and Nath, 1997)}$$

$$q_{lw} = 208 + 6\theta_{sc} \ \mathrm{W\,m^{-2}} \qquad \text{(Linacre, 1992)}$$

where ε_a is the atmospheric emissivity, σ is the Stefan–Boltzmann constant, θ_a^0 is the absolute effective atmospheric temperature (K) and θ_{sc} is the mean daily screen temperature in °C. For a 15°C screen temperature, we can see that a q_{lw} of around 300 W m^{-2} might be typical. Atmospheric long-wave radiation increases with cloud cover and Linacre (1992) suggests that his calculated flux should be multiplied by a factor ($1 + 0.0034\mathrm{Cl}^2$) where Cl is cloud coverage in oktas (eights of the sky).

It appears that atmospheric long-wave radiation is typically of a similar order of magnitude to short-wave insolation, but we must also remember that the ground

continued

BOX 3.5 *(Continued)*

itself re-radiates long-wave 'back radiation' (q_{back}) in relation to its temperature (Linacre, 1992; Hostetler, 1995):

$$q_{back} = \varepsilon\sigma\left(\theta^0_{sur}\right)^4 \text{ W m}^{-2}$$

where ε = surface emissivity ($\approx$0.97), σ = Stefan–Boltzmann constant and θ^0_{sur} = surface temperature in K. The back radiation is typically similar in magnitude to the incoming long-wave atmospheric radiation: a surface temperature of 12°C yields a q_{back} of 363 W m^{-2}. However, the *net* long-wave radiation budget ($q_{back} - q_{lw}$) is typically of the order of 20–90 W m^{-2} *from* the ground, for a screen temperature of 15°C and depending on cloudiness (Linacre, 1992).

We can now define a radiation budget for the earth's surface. The net incoming radiation (R_n) is given by

$$R_n = (1 - \alpha_{sw})q_{sw} + (1 - \alpha_{lw})q_{lw} - q_{back}$$

where α_{sw} and α_{lw} are the short-wave and long-wave albedos (reflectivity) of the earth's surface. The latter is usually very small. Linacre and Geerts (1997) have constructed this radiation budget for an entire clear day in a meadow in Oregon, USA (44°N). The short-wave insolation (q_{sw}) is estimated as 336 W m^{-2}, of which 24% (81 W m^{-2}) is reflected from the meadow's surface. Incoming atmospheric radiation (q_{lw}) is 289 W m^{-2} and outgoing back-radiation (q_{back}) from the ground is 376 W m^{-2}. Thus, the net incoming radiation (R_n) is 168 W m^{-2}. (Note that, during night-time, there would be no incoming short-wave radiation and there would be a net loss of long-wave radiation, leading to a cooling of the ground). The annual mean net incoming radiation typically exceeds 100 W m^{-2} in the tropics, with 40–80 W m^{-2} being typical for temperate Europe and <40 W m^{-2} for northern climes (Linacre and Geerts, 1997).

What becomes of the incoming net radiation? It heats up the ground, and is ultimately lost by evapotranspirative heat losses (q_{evap}) of latent heat in water vapour or convective losses (q_{conv}) of sensible heat (i.e. the ground heats up the adjacent air). We can say (neglecting any geothermal heat flux) that

$$R_n = q_{evap} + q_{conv} + G$$

where G is change in stored heat in the ground per m^2 (which should be negligible in the long term) *or* any heat actively removed from the ground by a ground source heating scheme.

We can begin to see that the magnitude of net incoming radiation is at least two orders of magnitude greater than the typical geothermal heat flux (<0.1 W m^{-2}). Ground source heat systems are thus not truly geothermal energy systems, they

BOX 3.5 (*Continued*)

merely utilise the earth's surface as a huge solar collector and storage. We can also begin to understand why so many of the design criteria (which we will meet in later chapters) boil down to specific energy extraction rates of the order of 10–20 W m^{-2} of earth's surface – this approximates to the amount of useful solar and atmospheric radiation we can 'harvest' with a ground source heat scheme (and remember that part of this radiation is reflected from the earth's surface or re-radiated by it).

In fact, we could construct an energy budget for the ground–atmosphere interface in an area where we are planning a ground source heat scheme (as we will do for a lake in Chapter 9), where the components would include short-wave insolation, long-wave atmospheric radiation, long-wave back-radiation, reflection, evapotranspiration and convective heat transfer. We can intuit that a ground source heating scheme that lowers the temperature of the ground, albeit slightly, will tip the energy budget in favour of heat transfer from atmosphere to the ground (e.g. by decreasing q_{back} and also by impacting convective and evaporative heat transfer).

Table 3.2 Average monthly and annual rates of insolation at selected European locations, after data provided by Whitlock *et al.* (2000). Figures are cited as kWh m^{-2} day^{-1} (and *W m*$^{-2}$).

	December	July	Average annual
Oslo	0.19 (*8*)	4.84 (*202*)–June	2.27 (*95*)
Edinburgh	0.32 (*13*)	4.34 (*181*)–June	2.26 (*94*)
London	0.52 (*22*)	4.74 (*198*)	2.61 (*109*)
Athens	1.63 (*68*)	7.61 (*317*)	4.56 (*190*)
Malaga	2.14 (*89*)	7.64 (*318*)	5.16 (*215*)

BOX 3.6 Thermal diffusivity.

The ratio of thermal conductivity to volumetric specific heat capacity is known as thermal diffusivity (α), which has units m^2 s^{-1}.

$$\alpha = \frac{\lambda}{S_{VC}} = \frac{\lambda}{\rho S_C}$$

where λ = thermal conductivity (W m^{-1} K^{-1}), S_C = specific heat capacity by mass (J K^{-1} kg^{-1}), S_{VC} = specific heat capacity by volume (J K^{-1} m^{-3}) and ρ = density (kg m^{-3}).

The thermal diffusivity represents the rate and extent to which a heat signal or heat pulse is propagated throughout a medium.

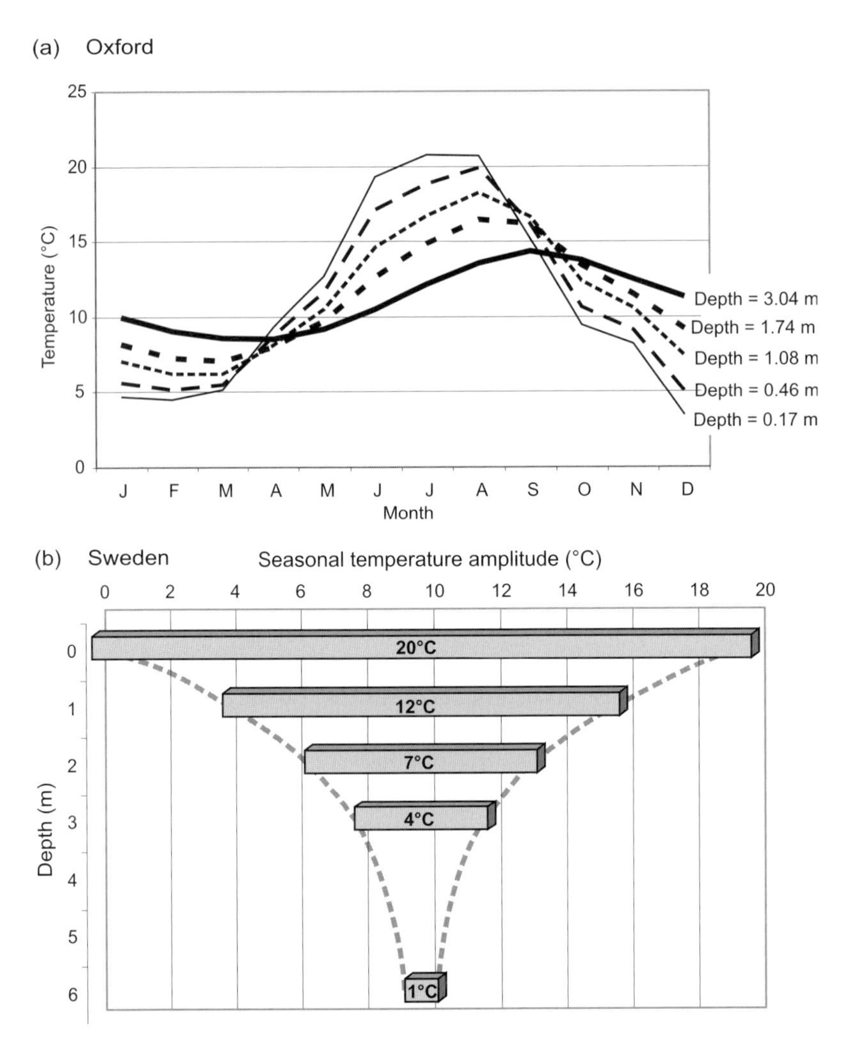

Figure 3.3 (a) The seasonal temperature fluctuation in the subsurface at various depths, as observed at Radcliffe, Oxford, UK, in 1899 by Rambaut (1900). Not only does the amplitude decrease with depth, the temperature maxima and minima become progressively delayed in time. (b) The amplitude of subsurface temperature fluctuation with depth, resulting from an annual temperature variation of 20°C at the surface, based on typical Swedish conditions, using data cited by Rosén *et al.* (2001).

temperature variation is no more than 1°C. The seasonal temperature signal at various depths formed the basis for some of the earliest determinations of the thermal diffusivity of soils by the likes of Ångström, Thomson and Everett (see above). For a full treatment of how seasonal and diurnal temperature cycles propagate into the ground, chapter 5 of Ingersoll *et al.* (1954) is difficult to beat.

In much of the UK, the annual average surface temperature is in the range 9–12°C and the subsurface temperature reflects this (Box 3.7). In other words, the earth's subsurface temperature is warmer than the air temperature in winter, but cooler in summer. The

BOX 3.7 Annual average air temperature, soil temperature and subsurface temperature: the effect of winter snow cover and freezing.

In the chapter, we have talked rather evasively about the annual average 'surface' temperature. When we start looking in detail, we find that the annual average air temperature usually differs slightly from the annual average soil temperature (and is typically a little lower). There are many factors that influence the magnitude of any such difference: for example, the radiative budget and the efficiency of heat transfer between soil and atmosphere, the aspect of the terrain (south-sloping or north-sloping; shaded?), the vegetation cover, the soil moisture at different times of the year and snow cover patterns.

Snow is reflective, so if snow cover exists at times and latitudes where days may be sunny, it may serve to reflect incoming solar radiation and decrease absorption of solar energy. On the other hand, porous snow has a low thermal conductivity: it will insulate the underlying soil from extreme winter air temperatures. In northern latitudes, this effect dominates: in Sweden, Rosén *et al.* (2001) claim that every 100 days with snow cover increases the annual average soil temperature by 1.5°C, relative to average annual air temperature.

Look at the climate data in Figure 3.4 for the town of Taiga (near Tomsk) and the city of Kemerovo in southern West Siberia. The annual average air temperature in both cases is a little below zero: we might even expect to see permafrost beginning to form here – but we do not. This part of Siberia is free from permafrost. Indeed, the temperature observed in the shallow subsurface (e.g. the groundwater in shallow wells) is relatively high, typically +3–5°C (Parnachev *et al.*, 1999; Banks *et al.*, 2004b). We can speculate that at least two factors may be important for the observed differential between annual average air and subsoil temperatures:

i. The fact that a substantial snow cover persists for several months of the winter, insulating the soil from the extremely low winter temperatures.
ii. The majority of the precipitation occurs in summer. A small portion of this 'warm' precipitation may infiltrate into the subsoil, elevating subsurface temperatures. The rather small amount of winter precipitation will not infiltrate the subsurface – it will remain frozen at the surface as snow or ice. When the accumulated snow cover eventually melts in around March–April, it will infiltrate the subsoil at 0°C.

earth thus provides a convenient source of heat in winter and a source of cooling (i.e. a sink to 'dump' waste heat in summer) (Figure 3.4).

Figure 3.5 shows how the low thermal diffusivity of the earth's subsurface subdues the amplitude of the annual air temperature 'signal' and also delays it in time. The diagram is derived from a waterworks at Elverum in inland Norway, which has both

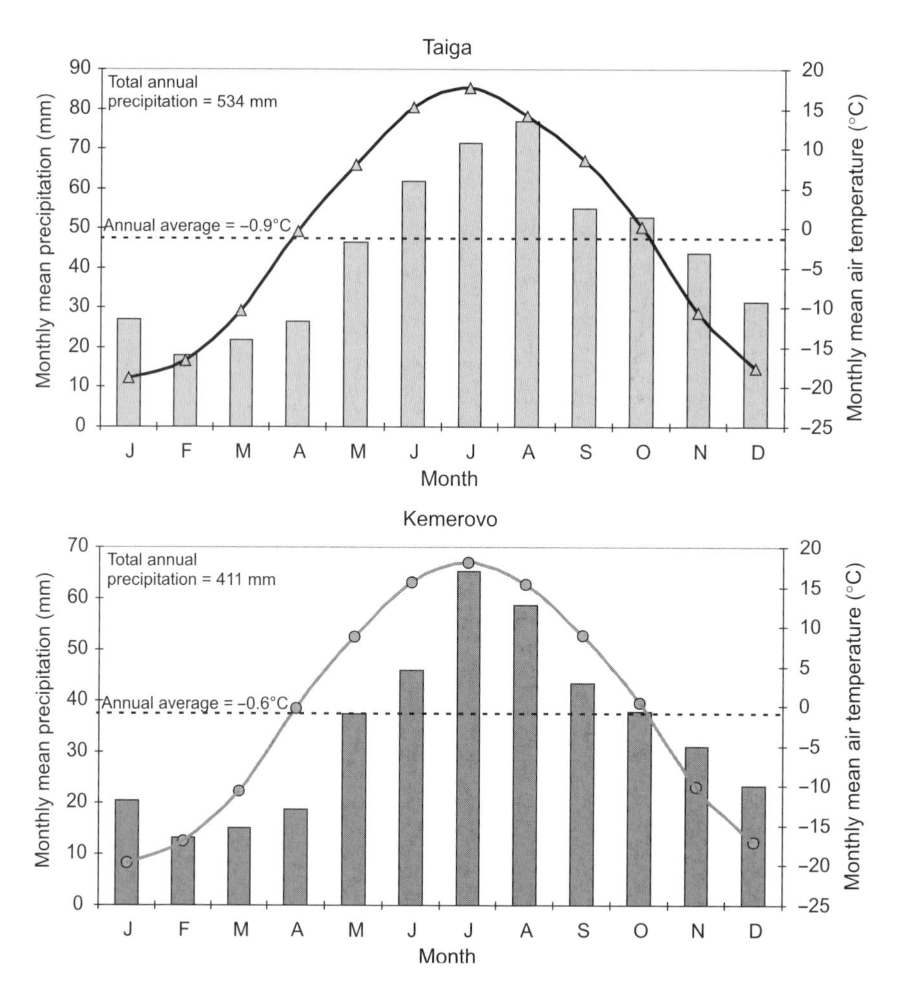

Figure 3.4 Typical annual precipitation histogram and monthly mean temperature profile for the southern Siberian town of Taiga (near Tomsk) and the city of Kemerovo, Russian Federation [based on data cited by Belyanin *et al.* (eds.), 1969].

a river water intake (from the River Glomma) and some groundwater wells. Furthermore, a weather station monitors air temperature throughout the year. It can be seen that the air temperature is subject to rapid fluctuation, reaching a minimum of −24°C in winter and the positively balmy heights of +20°C in summer. The river water temperature shows a more subdued curve, with lower 'amplitude': the river is frozen (0°C) in winter and does not rise above 15°C in summer. Furthermore, the temperature maximum (in August) is delayed relative to the maximum air temperature. The groundwater trace shows almost no temperature signal: it has a constant temperature of 5°C, just a little above the mean annual air temperature of ≈3°C (The Norwegian Meteorological Institute, climate data for station 6600 Elverum, 1961–1990; see also Box 3.7).

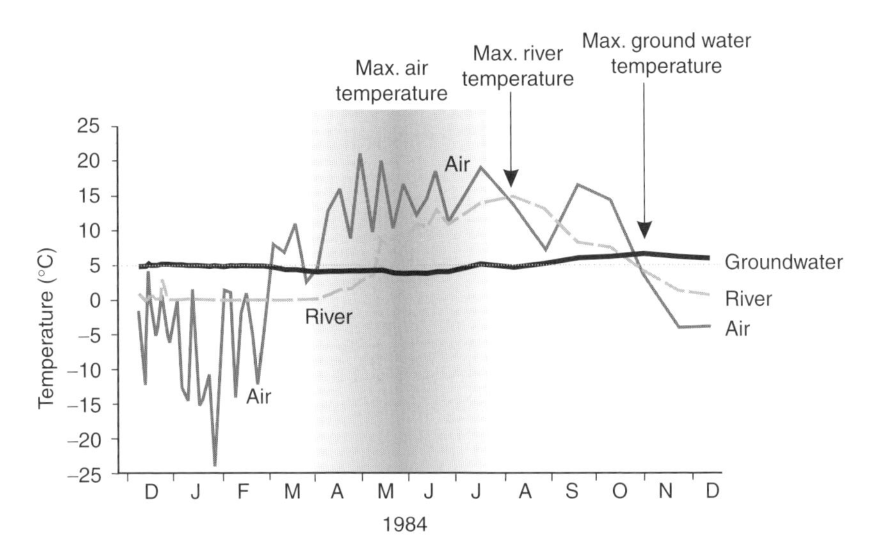

Figure 3.5 Graph showing fluctuation of temperature of the air, the River Glomma and shallow groundwater, at Elverum, inland Norway, during 1984. *Based on an original diagram by Randi Kalskin Ramstad and reproduced by kind permission of Norges geologiske undersøkelse, Trondheim, Norway.*

The imaginative eye may, however, discern a very faint amplitude, with a maximum as late as November.

The temperature of the shallow subsurface (below the zone of seasonal fluctuation) varies according to climatic zone: in the UK we have seen that it is 10–12°C, in Norway 4–6°C. In parts of northern and eastern Siberia, the annual average air temperature is well below zero, and so is the subsurface temperature – in fact, the ground freezes, forming permafrost. In parts of Greece, on the other hand, ground temperatures above 20°C may be common (Katsoyiannis *et al.*, 2007). In Chapter 13, we will also see how the presence of cities can affect subsurface temperatures.

3.4 Geothermal gradient

The temperature of the shallow subsurface is thus largely controlled by the annual average air temperature, and the heat that we extract via ground source heating schemes is dominantly ultimately derived from solar energy absorbed by the earth's surface. The earth's surface, in fact, acts as a huge solar collector (Box 3.5).

However, there is also a minor component of true geothermal heat flux, derived from the earth's interior and migrating towards the earth's surface. This manifests itself as a geothermal temperature gradient, superimposed on the annual average surface temperature (Figure 3.6). The earth's geothermal gradient, outside of anomalous or volcanically active areas, is normally in the range 0.01–0.03°C m^{-1}, or around 1–3°C per 100 m. This represents a geothermal heat flux of some 40–100 mW m^{-2} (Figure 3.7).

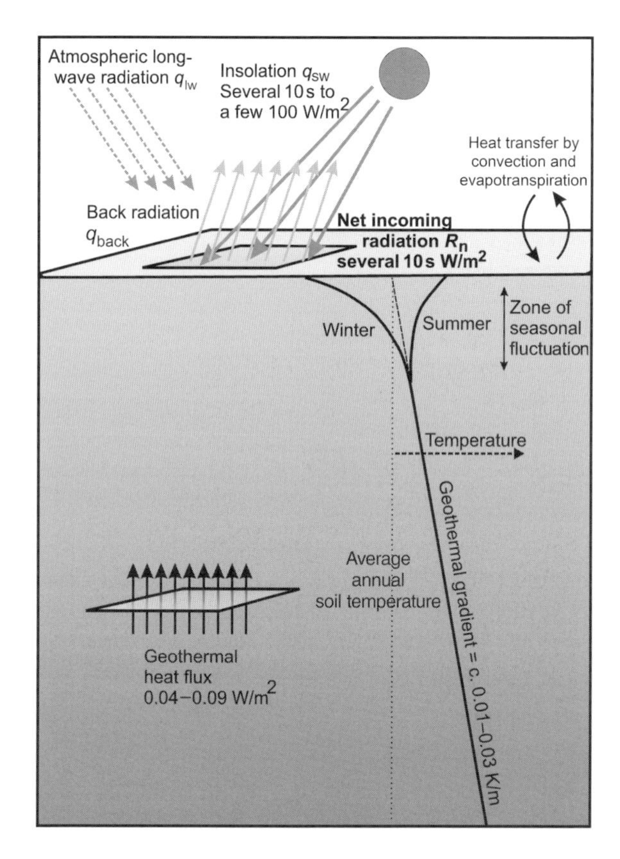

Figure 3.6 Schematic block diagram showing the downward increase in temperature in the earth due to the geothermal gradient, the seasonal zone of fluctuation in temperature and the relative magnitudes of geothermal heat flux and insolation.

In the UK, the measured geothermal gradient ranges from about 0.015 to 0.04 K m^{-1}, while the average gradient (for onshore areas down to a depth of around 4 km) is believed to be around 0.02 K m^{-1} (Wheildon and Rollin, 1986). Thus, if we drill a borehole to 100 m depth in a part of England where the annual average surface temperature is 10°C and the geothermal gradient is 0.02 K m^{-1}, we would expect the temperature at the base of the borehole to be around 12°C, with an average temperature over the depth of the borehole of 11°C (see Figure 3.8).

The British Geological Survey has compiled maps of geothermal heat flux in mW m^{-2} for the whole of the UK (Figure 3.7). Note particularly the high geothermal heat fluxes in Cornwall and Devon, where granites with a high content of radioactive elements outcrop or subcrop at relatively shallow depths. Note also the high fluxes in the Weardale-Lake District zone (west of Newcastle), where granite also occurs at shallow depth (see Box 2.4, p. 20).

The mean geothermal heat flux measured in the UK is calculated to be 69±28 W m^{-2} (comparable with the mean for continental Europe of 64 W m^{-2}). This value is believed by Wheildon and Rollin (1986) to be something of an overestimate, due to bias in

Figure 3.7 The geothermal heat flux in the UK, in mW m^{-2}. Reproduced by permission of the British Geological Survey © NERC. All rights reserved. IPR/90-20DR.

the measurements towards geological formations with high heat flow. An attempt to compensate the data set for bias has resulted in a revised estimate of mean geothermal heat flux in the UK of 54 ± 12 W m^{-2}.

We can use Fourier's Law to relate geothermal heat flux (Q) to estimate the average thermal conductivity (λ) of the subsurface:

$$\lambda = \frac{Q}{A \cdot (\mathrm{d}\theta/\mathrm{d}z)} \tag{3.5}$$

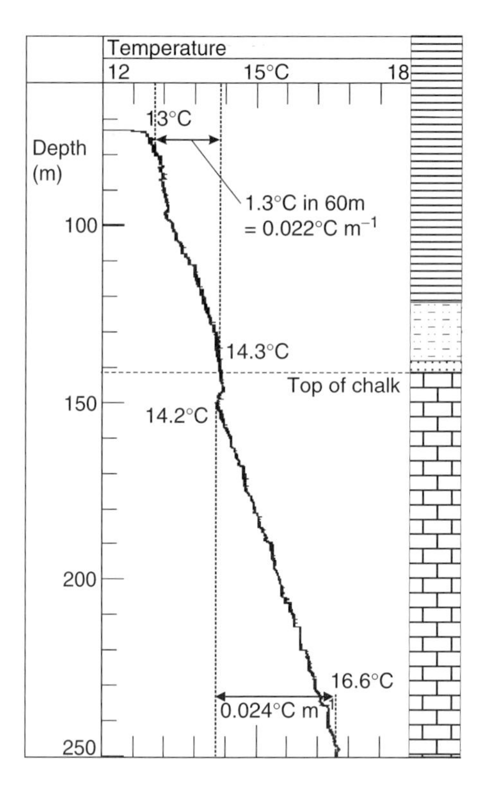

Figure 3.8 A groundwater temperature log of a water well in Surrey, southern England. The yield of the well was very poor (≈0.5 L s^{-1}), so there is very little groundwater throughflow to disturb the geothermal gradient. The borehole passes through 141 m of dominantly Tertiary clays and silts (and occasional sands) before encountering the Chalk at 141 m. The geothermal gradient is 0.024°C m^{-1} in the Chalk and 0.022°C m^{-1} in the Tertiary section. *Public domain information, provided by and reproduced with the permission of the Environment Agency of England and Wales (Thames Region).*

where A is a unit cross-sectional area of the earth's surface (m^2) and $d\theta/dz$ is the geothermal temperature (θ) gradient with depth (z). If a typical geothermal gradient is 0.02 K m^{-1} and a typical heat flux (Q/A) is 0.06 W m^{-2}, this yields a thermal conductivity of around 3 W m^{-1} K^{-1}, corresponding well with figures for the more crystalline and/or lithified lithologies in Table 3.1. Thermal conductivities of some of the main British geological units are shown in Table 3.3.

3.5 Geochemical energy

So far, we have seen that the temperature of the shallow subsurface of the ground is controlled by the annual average air and soil temperature (solar energy), and modified

Table 3.3 Thermal conductivity data for selected lithologies in the UK, based on laboratory measurements made on samples extracted from boreholes. The thermal conductivity data are believed to represent water-saturated samples. Abstracted from Rollin (1987). Reproduced by permission of the British Geological Survey © NERC. All rights reserved. IPR/90-20DR. N = number of determinations.

Formation	Lithology	N	Thermal conductivity ($W\ m^{-1}\ K^{-1}$)
London Clay (Palaeogene)	Sandy mudstone	5	2.45 ± 0.07
Reading Beds (Palaeogene)	Sandy mudstone	4	2.33 ± 0.04
	Mudstone	10	1.63 ± 0.11
Chalk (Cretaceous)	Limestone	41	1.79 ± 0.54
Upper Greensand (Cretaceous)	Sandstone	18	2.66 ± 0.19
Gault (Cretaceous)	Sandy mudstone	32	2.32 ± 0.04
	Mudstone	4	1.67 ± 0.11
Kimmeridge Clay (Jurassic)	Mudstone	58	1.51 ± 0.09
Oxford Clay (Jurassic)	Mudstone	27	1.56 ± 0.09
Mercia Mudstone (Triassic)	Mudstone	225	1.88 ± 0.03
Sherwood Sandstone (Permo-Triassic)	Sandstone	64	3.41 ± 0.09
Magnesian Limestone (Permian)	Limestone	12	3.32 ± 0.17
Westphalian (Coal Measures)	Sandstone	37	3.31 ± 0.62
	Siltstone	12	2.22 ± 0.29
	Mudstone	25	1.49 ± 0.41
	Coal	8	0.31 ± 0.08
Namurian (Millstone Grit)	Sandstone	7	3.75 ± 0.16
Lower Carboniferous limestone	Limestone	14	3.14 ± 0.13
Upper Old Red Sandstone (Devonian)	Sandstone	27	3.26 ± 0.11
Silurian slates near Selkirk	Slate	67	3.33 ± 0.05
Hercynian granites	Granite	895	3.30 ± 0.18
Basalt	Basalt	17	1.80 ± 0.11

by a geothermal temperature gradient (geothermal energy). In most cases, these are the two main natural energy sources that we need to consider in a ground source heat budget. In some special cases, there may also be a component of geochemical energy affecting the heat budget of the subsurface.

Some minerals weather very rapidly in the presence of air and moisture. One such group of minerals are the sulphide minerals, such as pyrite (FeS_2), marcasite (FeS_2), sphalerite (ZnS) and galena (PbS). These minerals commonly occur in many types of metal ore deposit and are commonly found in and around coal deposits. When these minerals are exposed to air and to water (in underground mines or in mine wastes deposited at the surface), they oxidise and release a cocktail of acid, dissolved metals and sulphate that is commonly known as acid mine drainage.

$$ZnS + 2O_2 = Zn^{2+} + SO_4^{2-} \tag{3.6}$$

Sphalerite + oxygen = Dissolved metal + sulphate

$$2FeS_2 + 2H_2O + 7O_2 = 2Fe^{2+} + 4SO_4^{2-} + 4H^+ \tag{3.7}$$

Pyrite + water + oxygen = Dissolved metal + sulphate + acid

This is a major potential environmental pollution issue in many mining areas. It can also present a hazard, as these sulphide oxidation reactions are typically exothermic; that is, heat energy is released during oxidation. Thus, mine spoil tips can be very hot inside, sometimes in excess of 50°C. In coal stores or coal mine waste tips, the heat released by pyrite and marcasite oxidation can even lead to spontaneous combustion of the wastes. In Norwegian metals mines, sulphide oxidation can lead to tropical down-mine temperatures, and the phenomenon of *kisbrann* ('sulphide-fire') was well-known. Indeed, each mole of pyrite oxidised by Equation 3.7 releases 1400–1500 kJ mol^{-1} of heat energy, while the sphalerite reaction (Equation 3.6) releases over 1700 kJ mol^{-1} (Banks *et al.*, 2004).

Banks *et al.* (2004) have speculated that, if we could extract this geochemical energy from mines or mine waste dumps (maybe using heat pumps: Chapter 4) and use it for space heating, it would be an elegant means of converting an environmental liability into a sustainable energy resource.

Other exothermic reactions releasing potentially usable geochemical energy include organic degradation reactions such as might take place in landfills or manure heaps or other accumulations of organic waste. These, too, could be conceived as sources of usable energy. Indeed, Sæther *et al.* (1992) document a landfill, beneath and down-gradient of which groundwater temperatures are 1–2°C warmer than normal.

3.6 The heat energy budget of our subsurface reservoir

Let us consider a block of the subsurface as a heat reservoir, or *aestifer* (Figure 3.9). There will be natural components contributing to the heat budget of the block. We can refer to these as the 'boundary conditions' of our conceptual model.

i. There will be heat entering the aestifer from the geothermal heat flux. We can regard this as a constant flux. In other words, the base of our aestifer can be conceptualised as a constant flux boundary.
ii. There may be groundwater flow passing through our aestifer, carrying with it a load of heat. In many cases, this groundwater flow can be regarded as constant over the long term and we may be able to regard the heat being advected into the model (from the left in Figure 3.9) as another constant flux boundary.
iii. The aestifer may be gaining or losing heat from the ground surface. Under natural conditions, there will, in the long term, be a net loss from the surface equal to the geothermal heat flux. If we start extracting heat from the ground, however, we will cool down the aestifer and may start to induce a flux of atmospheric (solar) energy from the surface into the ground. The long-term direction and magnitude of the heat energy flux to/from the surface depends on the magnitude of the temperature difference between the ground and the annual average surface temperature. The annual average surface temperature, at the ground–atmosphere interface, can be regarded as approximately constant in the long term. We can thus describe the top of our aestifer as a constant temperature boundary.

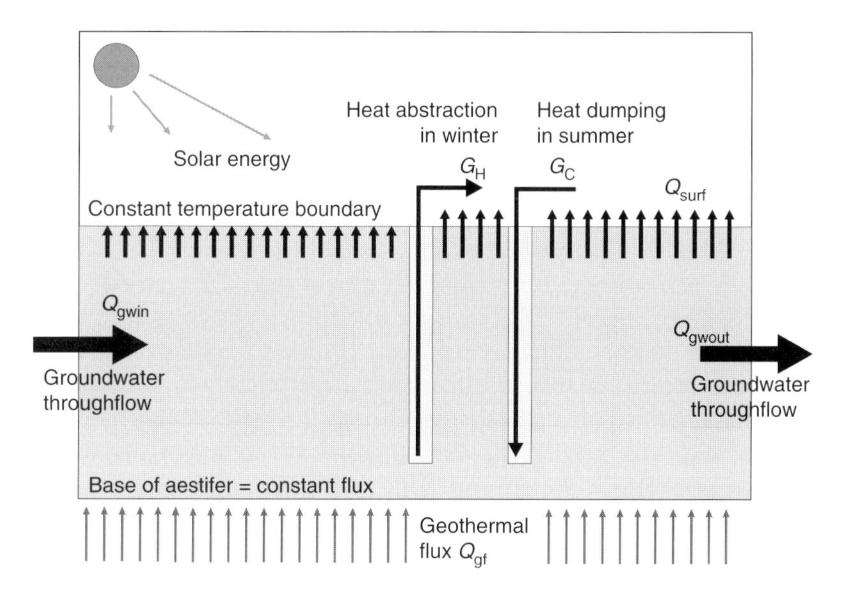

Figure 3.9 A block of the subsurface (an *aestifer*) showing possible elements of its heat budget.

If there is no geochemical (or significant internal radiogenic) component to the heat budget, then we now need to consider the amount of heat that we are planning to extract from the ground (in a ground source heating scheme) or reject to the ground (in a cooling scheme). Of course, throughout a year, we may extract heat (G_H) in winter and dump it (G_C) to the ground in summer. If $G_H > G_C$ over the course of a year, we refer to our ground source heat scheme as a 'net heating' scheme in the long term and there will be a tendency for the ground to cool down. If $G_C > G_H$, we are dealing with a 'net cooling' scheme, imparting a tendency for the ground to heat up. In some particularly happy circumstances, the heat rejected in summer balances the heat extracted in winter, resulting in minimal disturbance to the long-term heat budget of the aestifer and thus to its temperature.

The final component of the heat budget to consider is the heat stored in the ground. An increase in the heat stored in the ground is represented by ($V_{aest} \times \Delta\theta \times S_{VC}$), where V_{aest} is the volume of aestifer under consideration, $\Delta\theta$ is the average temperature change of the aestifer and S_{VC} is its specific volumetric heat capacity. Under natural conditions, over the long term:

$$Q_{gwout} + Q_{surf} = Q_{gwin} + Q_{gf} \tag{3.8}$$

An equilibrium is presumed to exist and the ground temperature does not change with time. However, if we start to interfere and extract ground source heat to warm up an office block (a net heating scheme)

$$G_H + Q_{gwout} + Q_{surf} > G_C + Q_{gwin} + Q_{gf} \tag{3.9}$$

then the heat stored in the aestifer ($V_{aest} \times \Delta\theta \times S_{VC}$) will be depleted and the temperature will fall. Hopefully, this fall in temperature will induce

- Q_{surf} to decline and eventually become negative; that is, a flux of (ultimately solar) heat is induced from the surface. Q_{surf} becomes more negative in response to a greater temperature gradient between surface and subsurface in accordance with Fourier's Law.
- The temperature of exiting groundwater to fall and thus Q_{gout} to decrease.

In this way, we would hope that, in response to our extraction of ground source heat, a new equilibrium condition would ultimately be established, such that subsurface temperatures would eventually stabilise at a new, lower (but still acceptable) level. We can get it wrong, however! If G_H is too great, the temperature might continue falling until we start to freeze the ground. This might be undesirable for many reasons: geotechnical, environmental or operational.

Thus, ground source heat extraction can be genuinely sustainable, if we have a good understanding of our aestifer's heat budget. However, it is not automatically sustainable: overoptimistic ground source heat schemes may have an in finite design life. (Of course, we may deliberately design our scheme with a finite lifetime, if we are dealing with a temporary construction or event that needs to be heated.)

Conversely, if we are dealing with a net cooling scheme:

$$G_H + Q_{gwout} + Q_{surf} < G_C + Q_{gwin} + Q_{gf} \quad (3.10)$$

the ground will begin to heat up, Q_{surf} (i.e. heat loss from subsurface to atmosphere) will ultimately increase and Q_{gout} may increase. Eventually, a new equilibrium may establish itself at a higher temperature and our heat rejection (net cooling) operation can be thought of as 'sustainable'. Here, if we get it wrong, temperatures in the ground may increase beyond the design limits of our scheme or our heat pump. The scheme may become progressively more inefficient and eventually inoperable.

4

What Is a Heat Pump?

> Heat won't pass from a cooler to a hotter.
> You can try it if you like but you'd far better not-a
> ... (that's entropy, man!)
>
> Michael Flanders and Donald Swann

A lot of heat is stored away in the earth's subsurface, even at normal temperatures and shallow depths. We have seen, in Chapter 3, that in the UK, the rocks, sediments and groundwater beneath our feet are typically at temperatures of 9–13°C down to depths of around 100 m (i.e. within the range of most drilling rigs). But how can we use this energy at such a low temperature? How can we heat our homes using a medium with a temperature of 12°C, when most of us enjoy a room temperature in the region of 20°C? The simple answer is – we cannot, using natural temperature gradients alone. Try as we might, we cannot get heat to flow naturally from rocks at 12°C to a living room at 20°C.

We can, however, envisage other scenarios. Let us imagine that we live in Scandinavia, where the temperature drops below 0°C for some months of the year and where there may be persistent heavy snowfall. There is thus often a need for de-icing pavements, roads or parking surfaces. In Scandinavia, the subsurface temperature of rocks and groundwater may be in the region of 4–7°C (a little higher than the annual average air temperature – Box 3.7, p. 51). We could conceive of drilling a small well and pumping up groundwater at, say, 6°C in order to circulate it in a network of pipes beneath the pavement to keep it ice-free. We are working with a favourable temperature gradient – heat flows happily from 6°C to 0°C! This is termed 'free' heating or passive heating.

In the same way, we could imagine circulating cool groundwater around a large British office building in the summertime, through a network of pipes, beams and panels. The office owner wants to keep staff at a comfortable 20°C. If the network of panels and beams and so on is large and efficient enough, we can envisage that heat will

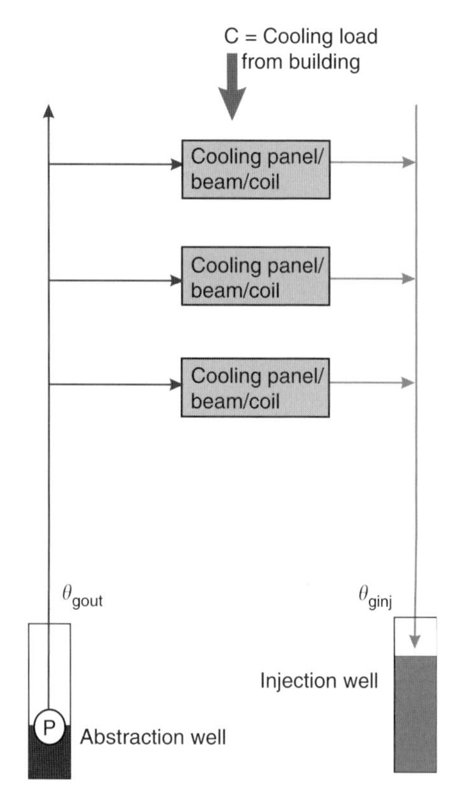

Figure 4.1 Using groundwater to perform 'passive' cooling in a British office block during the summer. θ_{gout} is the temperature of the abstracted groundwater (maybe 11°C) and θ_{ginj} is that of the waste groundwater (maybe 14°C), which in this diagram is re-injected to the aquifer via an injection well.

flow naturally from the office interior, be absorbed by the circulating groundwater (at 12°C) and be carried away, keeping the building cool. This is termed passive cooling or 'free' cooling, and is one of the most environmentally friendly forms of space cooling available (Figure 4.1).

But the natural temperature of the subsurface and groundwater in the UK (10–13°C) places limits on the use of passive heating and cooling. We cannot use these techniques to heat a space above ≈13°C or cool it below 10°C. How can we overcome this limitation? Answer: we pump heat from a lower temperature to a higher temperature using a device called a heat pump. But before we consider the heat pump, let us look at its opposite: the heat engine.

4.1 Engines

One can define an engine as a device that converts potential energy to work (Boxes 1.1 (p. 4) and 4.1). To do work, we need a potential energy gradient. This may be a

BOX 4.1 Kilowatts and kilowatt-hours.

A kilowatt is a unit of power: a rate of energy generation, transfer or consumption. 1 kilowatt (1 kW) is equal to 1000 Joules per second (1000 J s^{-1}).

A kilowatt-hour (kWh) is a unit of energy. It is the amount of energy generated/consumed by an appliance rated at 1 kW left running for 1 hr. In other words:

$$1 \text{ kWh} = 1000 \text{ J s}^{-1} \times 3600 \text{ s} = 3\,600\,000 \text{ J or } 3.6 \text{ MJ}$$

Similarly

$$1 \text{ MWh} = 3.6 \text{ GJ}$$

$$1 \text{ GWh} = 3.6 \text{ TJ and so on.}$$

chemical energy gradient, a gravitational potential energy gradient, a head gradient or a temperature gradient. Let us take the example of head of water. If we have two reservoirs of water with a difference in head between them (Box 1.2, p. 5), we can allow water to flow from a higher head to a lower head and perform work, perhaps by employing a water wheel. Similarly, if we have two heat reservoirs at different temperatures, we can perform mechanical work. The net effect is a flow of heat from the high-temperature source to a low-temperature exhaust.

A steam engine is a heat engine: the high-temperature source is steam at a temperature in excess of 100°C, while the low-temperature exhaust is often a water condensate. Internal combustion engines are also heat engines: the source is a hot gas (from fuel combustion), from which the heat is degraded (while doing mechanical work) to a lower-temperature exhaust.

Sadi Carnot studied the theory of ideal heat engines and was able to demonstrate that, for such an engine, the work performed was determined by the difference in temperature between the source/inlet (at θ_1) and the exhaust (at θ_2). Lord Kelvin, in 1851, further demonstrated that the maximum efficiency (E_{max}) of an ideal heat engine, defined as the ratio between work delivered (W) and heat input at high temperature to the engine (H_{in}) could be described by the following formula (Sumner, 1948):

$$E_{max} = \frac{W}{H_{in}} = \frac{(\theta_1 - \theta_2)}{\theta_1} \tag{4.1}$$

Thus, to take Sumner's (1948) example: if we have a heat source at temperature $\theta_1 =$ 82°C (=355 K) and a heat sink at 27°C (=300 K), the maximum efficiency of an ideal heat engine would be 55 K/355 K = 0.15 or 15%. In other words, no more than 15% of the heat flow could be converted to useful work.

4.2 Pumps

We have established that there is a close analogy between hydrogeology and thermogeology. That groundwater exists deep in the earth has long been known, but it has not always been accessible. Prior to 1863, the villagers of the English Chiltern village of Stoke Row (Banks, *in press*) had to walk to distant springs and rivers to collect their household water, despite the fact that abundant groundwater existed in the Chalk aquifer deep beneath their feet. They had not the money to reach it, however, until the Maharajah of Benares (modern Varanasi, India) visited the village and was so shocked at their plight that he paid for a 113-m deep well to be sunk to access the water table (Figure 4.2). The villagers used a hand-powered pump to bring the water to the surface: in reality, this was a bucket on a rope, which presumably required considerable muscle power to operate. Nevertheless, the well and its 'pump' allowed the villagers to access a new resource, bringing it from a previously inaccessible depth (low *head*) to the surface where it was accessible.

For ancient miners, groundwater was a curse rather than a benefit (Younger, 2004). It flooded their mines, wet their socks and dampened their lunchtime snap boxes. More

Figure 4.2 The Maharajah's Well at Stoke Row in the English Chiltern Hills (*photo by Dave Banks*).

importantly, it prevented them from mining to deep levels, accessing the reserves of coal and metals that were presumed to exist. Miners had to make do with tedious low-capacity hand-powered or animal-powered pumps (or in some cases, water-powered pumps) until a certain Thomas Newcomen (1663–1729) developed a steam-powered, atmospheric condensing piston engine for pumping water, in the period between 1705 and 1725. This was truly one of the world's great inventions and one of the earliest efficient 'machines' (at least in our modern understanding of the term), predating James Watt's steam engine by over 40 years and revolutionising the mining industry (Figure 4.3a).

Nowadays, industrialised European nations seldom use hand-powered or steam-powered pumps to raise water from a low head to a high head. We tend to use petrol- or diesel-powered pumps or, most commonly, electrical pumps to do this job (Figure 4.3b).

A pump is thus the opposite of an engine. In a water engine, water flows down a head gradient allowing mechanical work to be 'extracted' from the system. In a water pump, electrical/mechanical work is done to move water up the head gradient (against its natural tendency), that is, from a locus of low head (low elevation or low pressure) to a locus of high head (high elevation or pressure). It allows us to transfer water from

Figure 4.3 (a) The last *in situ* example of a Newcomen Beam Engine, at Elsecar Colliery, South Yorkshire, England; the inset shows the cylinder and piston (*photos by Dave Banks*). The device is both an engine and pump: the potential energy locked in coal, and the heat it produces, is converted to mechanical work. This work is used to raise water from a low head (i.e. low potential energy: deep in the mine) to high head (a drainage adit near the surface). (b) A small modern electrical submersible pump (*photo by Bjørn Frengstad*).

a location where it is no use to anyone (deep in an aquifer) to a useable elevation (the surface or, better still, a water tower).

4.3 Heat pumps

A trivial, but wholly correct, definition of a heat pump is that it is 'a device that pumps heat'! With a water pump, we can expend energy and perform mechanical work to get water to flow uphill or from low pressure to high pressure. With a heat pump, we can persuade heat to flow from a low-temperature environment to a high-temperature one – but we must perform mechanical work and expend energy to do so! A heat pump is the opposite of a heat engine. A heat pump raises the temperature of the available heat from an unusable level to a usable one (Box 4.2).

We are all familiar with water pumps, but we have conceptual difficulties with heat pumps. Nevertheless, despite our theoretical misgivings, we all happily put our trust in them. Most readers will have invested in one already – it is called a fridge. Our domestic refrigerator pumps heat from a low-temperature environment (our chilled salad compartment) into our kitchen. This will be evident if you put on some protective rubber gloves and venture – carefully – into the hidden world of lost sausages and stray herrings behind your fridge. It is warm back there! In fact, there is most likely a metal radiator grid on the back of the fridge pumping heat from the fridge interior out into the kitchen.

But any heat pump requires an energy input and the fridge is no exception. Most domestic fridges require an input of some few hundred watts of electrical energy that performs mechanical work by powering a compressor.

4.4 The rude mechanics of the heat pump

In this book, we will not dwell too much on exactly how a heat pump works, but it is best to have at least a conceptual understanding. Heat pumps (including your fridge) transfer heat by means of circulating a refrigerant fluid around a compression–expansion cycle (Figure 4.4). Consider the inside of your fridge, which should be at around 4°C. There are four parts to the refrigerant cycle:

i. Within a network of pipes in the fridge (formally, a heat exchanger known as the evaporator), the refrigerant fluid is circulating at a low (sub-zero) temperature (θ_D). The refrigerant fluid is chosen such that it boils (under the pressure conditions in the refrigerant circuit) at a temperature below 0°C (i.e. below the target temperature of, say, 4°C in your chiller cabinet). As it boils, it absorbs a large amount of latent heat of vaporisation from the fridge's interior.
ii. The refrigerant fluid, now a vapour at temperature θ_A, then passes through a compressor, powered by the electrical energy input to the fridge. We all recognise that when you compress a gas, the temperature rises – think about pumping up

BOX 4.2 The heat pump at the Eco-Centre, Tyneside, northern England.

Figure 4.5 shows a schematic example of a real application of the type discussed in Section 4.7, installed at the Eco-Centre in Tyneside, northern England. Here, groundwater is pumped at a rate of around 3 L s^{-1} from a 60 m deep well drilled into Carboniferous sandstone strata. The temperature of the groundwater is around 10°C. The heat pump extracts some 63 kW of heat, leaving a chilled groundwater at 5°C to be rejected into the Tyne estuary.

The heat pump itself delivers a nominal 88 kW of heat effect. It runs on a refrigerant cycle utilising R407C (a mixture of the fluorinated hydrocarbons CH_2F_2, CHF_2CF_3 and CH_2FCF_3), powered by twin reciprocating compressors running off a 415 V, three-phase electricity supply. As the groundwater is saline (around 25 000 ppm), the heat pump is designed for marine use: the evaporator is of stainless steel and is corrosion-resistant.

In a neat twist, the Eco-Centre has a large wind turbine, which generates electricity for sale to the National Grid. The Centre then re-purchases electricity on a nighttime tariff to power the heat pump: the National Grid is effectively being used as a 'buffer' for the wind generated electricity. Thus the heat pump could be regarded as being run by a green electricity source – a truly zero-carbon space-heating solution.

We have seen that ground source heat pumps are at their most efficient when they deliver heat to a low-temperature central heating system at a constant rate over prolonged periods. The Eco-Centre's heat pump delivers heat at 45°C to an underfloor warm water central heating system flowing at around 2.1 L s^{-1}. The flow return's design temperature is 35°C. The building has a high thermal mass (i.e. it takes a long time to heat up or cool down in response to heat inputs). Thus, the heat pump is typically run during winter on cheap electricity at night. The building is thus warm when office hours commence and the accumulated heat is retained throughout the day.

The ratio of heat delivered to electricity consumed (the coefficient of performance) was designed to be around 3.5. Accumulated experiences over the lifetime of the heat pump suggest that the actual figure is in the region of 3. One might expect a slightly better figure from a 'state of the art system', but we should remember that the Eco-Centre's system was one of the earlier installations in Britain, being commissioned in 1996.

a bicycle tyre – remember how hot the pump can get. The pressurised gas thus emerges from the compressor at a high temperature θ_B ($\theta_B > \theta_R$, where θ_R is room temperature).

iii. The refrigerant passes through another heat exchanger (the condenser: the radiator grid on the back of your fridge); heat flows from the refrigerant vapour to your kitchen, and the vapour starts to condense back to a liquid, shedding more

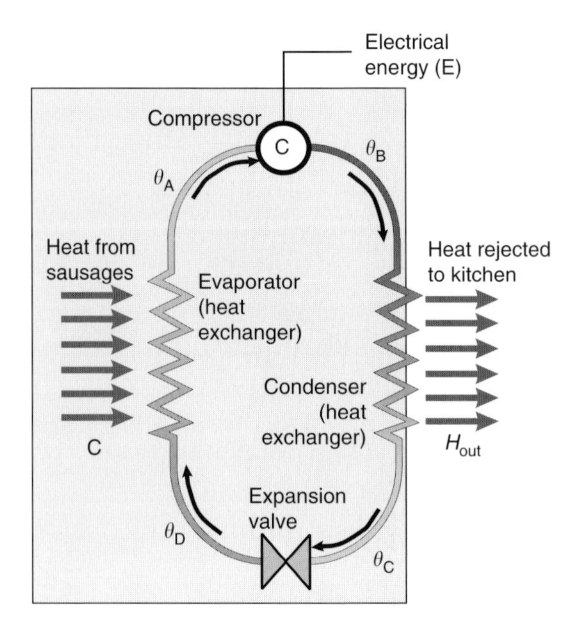

Figure 4.4 A schematic diagram showing how your refrigerator works.

latent heat as it does so. After passing through the condenser, the resultant, still pressurised, fluid now has a temperature θ_C.

iv. The refrigerant completes the cycle by passing through an expansion valve. We should be familiar with the idea (Box 3.3, p. 39) that expanding fluids tend to cool down – think of the condensation that forms on a butane cylinder as gas expands out of it to power your camping stove, or of how an aerosol container cools a little on use. As the refrigerant fluid passes through the expansion valve, the temperature drops significantly, back to θ_D.

A heat pump, used for heating a building, performs exactly the same cycle, although the pressures and vaporisation/condensation temperatures may differ from those of a fridge. The refrigerant fluid within a heat pump can be of many types: it should, however, be thermally stable, have a suitable specific heat capacity, have a volatility/boiling point tailored to the operating temperature and pressure of the heat pump and be environmentally benign. John Sumner's pioneering Norwich Heat Pump (Sumner, 1948 – see Box 5.2, p. 92) utilised sulphur dioxide (SO_2), which has a boiling point of −1°C at 22 psi: ideal for use with near-freezing winter river water as an environmental heat source.

The earliest refrigerants used on a large scale were ammonia, sulphur dioxide and chloromethane, but these were either flammable or toxic (Heap, 1979; Table 4.1). The chlorofluorocarbons (CFCs), including Freon (a DuPont trading name), were developed in the late 1920s as low toxicity, non-flammable alternatives and R12 became commonly used in refrigeration equipment in the 1940s. CFCs were common in domestic fridges and heat pumps until relatively recently. In the late 1970s and 1980s

Table 4.1 The properties of some refrigerants (data derived from Calm and Hourahan, 2001). Boiling point is cited at atmospheric pressure; flammability = lower flammability limit (% in air by volume), HET = indicative human exposure threshold/limit in air (ppm), ODP = indicative ozone depletion potential (Freon 11 = 1), GWP = global warming potential integrated over 100 years (CO_2 = 1). Note that the human health thresholds are indicative only and should not be used in any formal risk assessment.

Number	Name		Boiling point (°C)	Flammability (%)	HET (ppm)	ODP	GWP
R11	Trichlorofluoromethane (Freon 11)	CCl_3F	23.7	None	1000	1	4600
R12	Dichlorodifluoromethane (Freon 12)	CCl_2F_2	−29.8	None	1000	0.82	10600
R32	Difluoromethane	CH_2F_2	−51.7	12.7	1000	0	550
R40	Chloromethane	CH_3Cl	−24.2	8.0	50	0.02	16
R125	Pentafluoroethane	CHF_2CF_3	−48.1	None	1000	0	3400
R134a	Tetrafluoroethane	CF_3CH_2F	−26.1	None	1000	0	1300
R290	Propane	C_3H_8	−42.2	2.1	2500	0	≈20
R407c	Mixed refrigerant: 23% R32 + 25% R125 + 52% R134a		−43.8	None	—	0	1700
R717	Ammonia	NH_3	−33.3	15	25	0	<1
R744	Carbon dioxide	CO_2	−78.4	None	5000	0	1
R764	Sulphur dioxide	SO_2	−10	None	2	0	–

it became increasingly clear that the eventual release of CFCs (following disposal of the fridge) was resulting in destruction of the ozone layer. Nowadays, possible refrigerants include fluorinated hydrocarbons (such as R407c – see Table 4.1), hydrocarbons and ammonia (UNEP, 2002).

Note that the compressor is a mechanical device. It is commonly powered by electricity, but does not have to be. Indeed, refrigerators and heat pumps can be designed whose compressors are powered by diesel, steam, manual effort or even water power. We can even design heat pumps that do not use mechanical compressors. The absorption heat pump (Box 4.3) functions in an analogous cycle to the 'conventional' version, but the compressor is replaced with a chemical absorption reaction and the electrical energy source is replaced with a heat source (e.g. a gas heater).

BOX 4.3 The absorption heat pump.

The absorption heat pump or refrigerator is not hugely different from a vapour-compression–expansion unit. The absorption cycle uses a heat source, rather than mechanical or electrical energy, as its energy input. The compressor is replaced by an 'absorber' – a reservoir of absorbent medium, in which the cycling refrigerant has a high solubility. The 'classic' combination, utilised by Carré in 1858–1859, was water as the absorbent and ammonia as the refrigerant. The ammonia may be mixed with a low-solubility carrier gas, such as hydrogen, which does not take part in the refrigeration process, but which essentially regulates pressure in the system. Other combinations could be used: historically, water (refrigerant) and sulphuric acid (absorbent) were employed, or, in more recent times, water (refrigerant) and lithium bromide (absorbent).

In an ammonia-based system, liquid ammonia volatilises in the evaporator of the heat pump, but the resulting ammonia gas is highly soluble in the water of the absorber. This results in very low partial pressures of ammonia at the evaporator and an increased tendency of the ammonia to volatilise at correspondingly low temperatures.

The ammonia-rich water in the absorbent is then heated by the heat source (steam or a gas flame, or 'waste' heat from another source) in the 'generator' unit to expel the dissolved ammonia from the water in a 'separator', resulting in a high temperature and ammonia pressure at the condenser. The cycle is completed, as in a conventional heat pump, by an expansion valve.

Historically, absorption refrigerators were very important in the large-scale production of ice. However, the toxicity of ammonia, the improvement of vapour-compression machines and the increasingly reliable supply of electricity has led to a decrease in their usage. Their application may still be favoured, however, in remote regions where electricity is unreliable or unavailable. They may also still be attractive where low evaporator temperatures are desired or where waste heat is available to drive the absorption–distillation unit.

We may ask, 'What becomes of the electrical (or other) energy used to power the heat pump's compressor?' In the action of the compressor, this electrical energy input is converted, partly to sound energy (the hum of the compressor), but mostly to heat energy, which is absorbed in the refrigerant and must be discharged at the *condenser*. This allows us to answer the tantalising question – Is it sensible to leave the door of the fridge open on a hot summer's day to cool your kitchen? Although opening the fridge door might bring some temporary relief, as cool air flows out into the kitchen, the long-term answer is 'no'. The fridge's heat pump will simply be transferring energy from the kitchen in front of the fridge, via the open door, to the kitchen at the back of the fridge. And with each cycle of the heat pump, say, 100 W of electrical energy are being converted to heat energy and being added to the discharged heat load. This tempting practice will steadily lead to your kitchen heating up!

4.5 Heat pumps for space heating

Take a look at Figure 4.4, which shows a schematic diagram of a fridge. A quantity of heat (C, the cooling load, in J s^{-1} or W) is absorbed from the chiller compartment and transferred via a heat pump and an external radiator grid to your kitchen. The electrical power required for the compressor is E and the total rejected heat is H_{out}. We can say (neglecting the small loss of power as acoustic noise – your humming fridge)

$$H_{out} = C + E \tag{4.2}$$

The temperature in the fridge is 4°C (although the temperature of the refrigerant in the evaporator will be much colder than this) and the temperature of the rejected heat may be, say, 30–40°C (the temperature of the refrigerant in the evaporator will be hotter than this).

A refrigerator is a heat pump that is extracting heat from the sausages that we place in the chiller compartment, and is using it to heat our kitchen. In fact, we can regard our fridge as a 'sausage-sourced' heat pump! It should now be possible to understand that we can take heat from any source that is thermally coupled to the evaporator, and use it to heat our house. We have already seen that the environment around us contains huge reserves of low-grade heat. By using a heat pump, we can take heat from sewage, from rivers or from the sea. By circulating outside air over the evaporator, we can extract heat from the atmosphere and use it to heat our home. This is the principle of an air-sourced heat pump. By somehow coupling the ground to the evaporator, we can also extract heat from the geological environment – a *ground source heat pump* (GSHP) (Figure 4.5).

Let us assume that we are able to extract a flux of heat (Q_{env}) from some environmental reservoir with a temperature θ_{env}: this may be the ambient outdoor air on a cool spring day or a flow of groundwater pumped from a well (Box 4.2). A heat pump transfers this heat energy to the interior of our house and, on the way, it is upgraded to

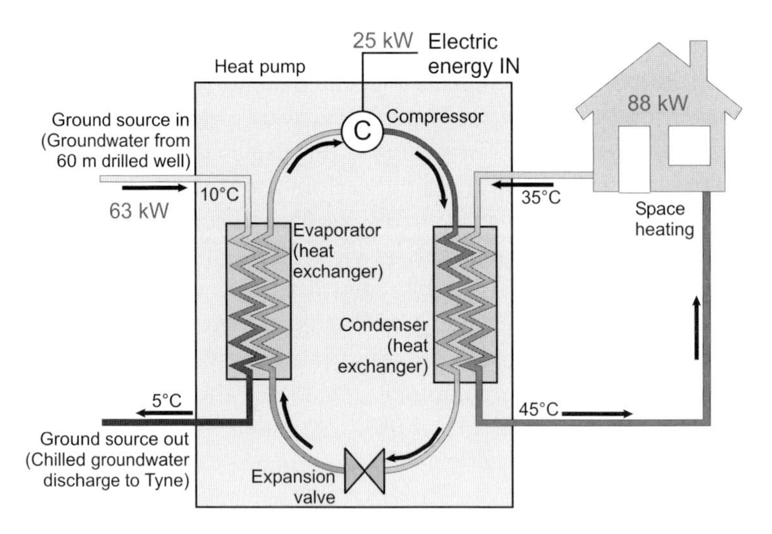

Figure 4.5 A schematic diagram of a GSHP, based on the design for the Eco-Centre at Hebburn, Tyneside (Box 4.2).

a temperature θ_{in} that is adequate to support our domestic heating system. This value of θ_{in} may be

- over 60°C if we have a ropey old conventional hot water central heating system;
- 45–55°C if we have a more modern low-temperature central heating system, with a high radiator surface area in our house;
- 30–45°C, if we have underfloor waterborne central heating;
- 25–30°C if we use warm air circulation as our means of heating.

The total heating effect H is given by

$$H = Q_{env} + E \tag{4.3}$$

where E is the electrical energy required to power the heat pump.

4.6 The efficiency of heat pumps

We have already established that there is a theoretical limit on the efficiency of heat engines (Equation 4.1). An ideal heat pump is simply the reverse of an ideal heat engine. This, if we define the efficiency of a heat pump as the ratio of heat delivered (H) at elevated temperature (θ_1) to work performed by the compressor (W), then from Equation 4.1 we can see that the theoretical maximum heat pump efficiency, using an idealised Carnot cycle (Heap, 1979), is

$$E_{max} = \frac{H}{W} = \frac{\theta_1}{(\theta_1 - \theta_2)} \tag{4.4}$$

In modern heat pumps, the compressor is powered by electricity, so W is replaced by E (the electrical power input). The efficiency of the heat pump is usually referred to as its coefficient of performance (COP_H), where

$$COP_H = \frac{H}{E} \tag{4.5}$$

From Equation 4.4, we see that there is a theoretical maximum for COP_H, which depends on the temperature at the delivery side (θ_1) and the source side (θ_2). If $\theta_1 = 35°C$ (308 K) and $\theta_2 = 5°C$ (278 K), the maximum theoretical efficiency of an ideal heat pump would be around 1000% (i.e. the maximum COP_H would be 10). We also see that the efficiency of the heat pump decreases with increasing delivery temperature and decreasing source temperature. In other words, a given heat pump does not have a fixed COP_H: this will depend on the operating conditions and temperatures.

In practice, the real COP_H of a heat pump will be much lower than the ideal, for several reasons (Heap, 1979):

- The evaporator temperature (θ_2) is usually significantly below the environmental source temperature in order to ensure a kinetically rapid transfer of heat from the environment to the refrigerant (the ideal Equation 4.4 assumes that the evaporator temperature is very similar to the environmental source). Similarly, the condenser temperature (θ_1) will be higher than the temperature of the space to be heated.
- Real vapour compression heat pumps do not use the ideal Carnot cycle of vapour compression but often a cycle called the Rankine cycle, which is more practical but slightly less efficient.
- Compression inefficiencies and other inefficiencies in the system.

For most space heating GSHPs, we would hope for a COP_H of at least 3, and probably approaching 4, under operational conditions. Air-sourced heat pumps will generally have a lower COP_H. Let us consider a small house, with a peak heating load H of 6 kW, heated by an air-sourced heat pump with a $COP_H = 3$, then the electrical energy input (E) required will be given by Equation 3.5: $E = 6$ kW/3 $= 2$ kW. Furthermore, from Equation 3.3, we can calculate that the heat energy derived from the outside air $H_{env} = 6$ kW -2 kW $= 4$ kW. Thus, with our air-sourced heat pump, we are expending 2 kW of electricity to transfer 4 kW of free environmental energy from the outside air to our house. Of the 6 kW total heating effect delivered, 2 kW is electrically derived and 4 kW is renewable, environmental energy 'conjured' from thin air.

We should note, from Equation 4.4, that the efficiency, or COP_H, of the heat pump is greatest when the difference between θ_2 (the environmental source) and θ_1 (the heat delivery temperature) is minimised. We would expect very low heat pump efficiencies if we were trying to extract heat from cold winter's air at $-5°C$ and deliver it to an old high-temperature radiator central heating system at 65°C. It would be far more efficient to deliver heat to a low-temperature heating system, such as warm air circulation or underfloor central heating. It would be even better if we could extract heat from an environmental source that retains a relatively high temperature of, say, 10°C even in winter. This is where GSHPs score over air-sourced heat pumps.

4.7 Ground source heat pumps

A GSHP is a heat pump where the source of environmental energy is the ground, or a medium thermally coupled to the ground. By this last phrase, we mean a medium in intimate thermal contact with the ground, such as groundwater. Some people might even class heat pumps based on deep lakes, ponds or river intakes as ground-sourced (or ground-coupled) heat pumps.

Let us consider a simple GSHP, based on groundwater being pumped from a well. We have already seen that, in the UK, groundwater might be expected to be at a temperature of around 11°C. We can thus pump groundwater from the ground at a rate Z such that it passes into our heat pump's evaporator. The heat pump will extract a heat energy flux (G) from the groundwater and its temperature will drop. A typical magnitude for this temperature drop ($\Delta\theta$) might be around 5°C, leaving us with a cool groundwater at a temperature of 11°C − 5°C = 6°C to dispose of (more of this in Chapter 7). The heat extracted is upgraded in the heat pump to a temperature θ_{in}, and is used to support a domestic heating system. Again, assuming that energy loss due to acoustic noise is negligible and that all extracted heat and heat of compression is efficiently delivered to a point of use, the total heating effect (H) is given by

$$H \approx G + E \tag{4.6}$$

where G is the ground source heat (heat extracted from the ground), and

$$\mathrm{COP}_H = \frac{H}{E} \tag{4.5}$$

We can, however, also relate the heat extracted from the groundwater to the flow rate Z, the temperature drop across the heat pump ($\Delta\theta \approx 5°\mathrm{C}$) and the specific heat capacity of water ($S_{\mathrm{VCwat}} = 4180\ \mathrm{J\ L^{-1}\ K^{-1}}$):

$$G = Z \cdot \Delta\theta \cdot S_{\mathrm{VCwat}} \tag{4.7}$$

Thus, if $Z = 1\ \mathrm{L\ s^{-1}}$, then

$$G = 1\ \mathrm{L s^{-1}} \times 5\ \mathrm{K} \times 4\,180\ \mathrm{J K^{-1} L^{-1}} = 21\,000\ \mathrm{J s^{-1}} = 21\ \mathrm{kW}$$

If we can obtain a heat pump with a COP_H of 4, this means that

$$G \approx H\left(1 - \frac{1}{\mathrm{COP}_H}\right) \tag{4.8}$$

$$H = 21\ \mathrm{kW} \times \frac{4}{3} = 28\ \mathrm{kW} \quad \text{and} \quad E = 7\ \mathrm{kW}$$

So, from a modest groundwater flow rate of 1 L s^{-1}, we can provide a heating effect of 28 kW, provided we have at least 7 kW electricity supply to power our heat pump.

We should note that we are also using an electrical pump to pump water from our well to the surface, consuming an electrical power E_{pump}. Clearly, if we wish to assess the efficiency of our entire heat pump system, we have to take this into account, together with any other power expenditure on circulation pumps, and so on. We can define a seasonal performance factor (SPF_H) for our heat pump, which is similar to our COP_H except that it is integrated over an entire heating season. We can also go one step further and define a system seasonal performance factor (SSPF_H) that takes into account all power expenditure in the system. Thus

$$\text{SSPF}_H = \frac{H}{(E + E_{\text{pump}} + \cdots)} \tag{4.9}$$

4.8 GSHPs for cooling

Small animals have a small volume (and a limited number of cells respiring and producing heat) but a surface area that is large relative to the volume. They are thus very susceptible to heat loss. They often have insulating fur coats, and in the winter they may curl into a ball (to reduce their surface-area-to-volume ratio) and have a long snooze in hibernation. Small buildings are much the same. In temperate Europe, most domestic houses and bungalows require some kind of heating in winter, but relatively few have any active air-conditioning during the summer.

Large animals, such as the elephant and hippo, have a low surface-area-to-volume ratio. They have loads of cells respiring and generating heat, but a limited surface area through which to get rid of that heat. Thus, large mammals have sparse coats, may spend a lot of time in the water and may even have large ears to act as radiator fins and assist in getting rid of waste heat. Large blocks of offices and apartments likewise have many 'cells' – rooms full of people and computer equipment, all generating heat. They have only a limited surface area, however, and may find it difficult to get rid of excess heat. Thus, large buildings tend to have net cooling rather than heating requirements, sometimes even in winter!

The good news is that many heat pumps can effectively be switched into reverse, so that heat from the inside of a building is pumped away to the outside. In fact, a standard air-conditioning unit is an air-to-air heat pump. Such air-conditioning units are fine, but they may operate relatively inefficiently because, on a hot summer's day, they are striving to reject heat to a 'sink' (the outside air) that is at a relatively high temperature (25–30°C). A heat pump would operate significantly more efficiently if it could reject heat to a cool reservoir, such as the ground at 11°C!

Indeed, we can switch GSHPs into reverse, so that they extract heat from the inside of a building and reject it to the ground or to a ground-coupled medium, such as groundwater. In the case shown in Figure 4.6, the groundwater acquires heat from the heat

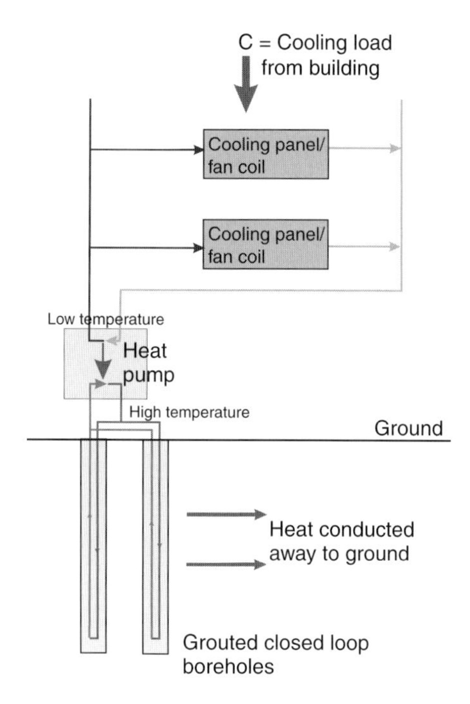

Figure 4.6 A GSHP in cooling mode. In this case, heat is transferred from the low-temperature building circuit to a high-temperature 'ground loop' or to groundwater. In this diagram, a 'closed-loop' (see Chapter 6) solution is shown where heat is conducted from the ground loop away into the ground via grouted boreholes (although an 'open-loop' solution would also be feasible, where heat is rejected to a groundwater flux).

pump and its temperature increases a few degrees (rather than decreases, as in heating mode, see Figure 4.5).

In cooling mode, we can also define a coefficient of performance (COP_C), which is sometimes also termed the energy efficiency ratio (EER):

$$COP_C = EER = \frac{C}{E} \tag{4.10}$$

where C is the total cooling effect, that is, the heat removed from the building (kW) and $E =$ electrical energy used to power the heat pump.

We can then state that the total heat rejected to the ground (G) is (Figure 4.6)

$$G \approx C + E = C\left(1 + \frac{1}{COP_C}\right) \tag{4.11}$$

If we compare this with Equation 4.8, for the heat pump in heating mode,

$$G \approx H - E = H\left(1 - \frac{1}{COP_H}\right) \tag{4.8}$$

we will see that the amount of heat energy rejected to the ground to deliver 1 kW cooling effect is significantly more the amount of heat energy extracted from the ground to provide 1 kW heating effect. This is because, in heating mode, the electrical energy powering the heat pump eventually turns up as useful heat, contributing to satisfying the heating load of the building. In cooling mode, the electrical energy simply turns up as yet more waste heat to be rejected.

We can thus question whether the use of GSHPs to achieve active cooling is 'environmentally friendly'. The answer must be 'not really'. The use of GSHPs for cooling is nowhere near as CO_2-efficient or cheap (in terms of running costs) as free/passive cooling (Figure 4.1), but may be 20–40% more efficient (Kelley, 2006) than conventional active cooling solutions (e.g. air sink heat pumps/air-conditioners). Nevertheless, with any form of active cooling, we are using up electricity to run heat to waste! We could, of course, redeem ourselves, if we could somehow utilise this waste heat elsewhere, or even store it until winter. We could, for example, dump waste heat to the ground during summer and then recover it again (via our GSHP in heating mode) during the winter months. This is termed underground thermal energy storage or UTES (more of this in Chapters 7 and 10).

4.9 Other environmental sources of heat

Thus far, we have considered air-sourced heat pumps and GSHPs. We have noted the potential drawbacks of air-sourced heat pumps (low efficiency due to fluctuating air temperature), but should probably mention their benefits in terms of lower capital cost and ease of installation. We should also note that there are other sources of reliable environmental heat than the ground. Efficient heat pump systems can be based on abstraction of heat from the sea, from fjords (below a certain depth, seawater has a relatively constant temperature) or from sewerage (Matte, 2002).

With a little imagination and an understanding of the heat flows within an industrial process, the possibilities are manifold. At dairies, there is typically a need to cool milk immediately following the early morning milking. Using heat pumps, we can chill the milk and transfer the heat to the Dairy Manager's office (a 'cow-sourced' heat pump!). At other farms, root vegetables may require cool storage after they are brought in from harvest – a 'potato-sourced' heat pump can convey heat from the cool store to the other farm buildings.

4.10 The benefits of GSHPs

We can list a host of reasons why GSHPs provide attractive sources of heating and cooling. First, they are visually unobtrusive: a typical water-to-water heat pump is a white box, not unlike a fridge. It can be tucked away in cellar or a plant room and nobody need know that it is there. Conventional air-conditioning and cooling systems may require large units to be bolted on to the side of buildings or mounted on the roof. Even other environmentally friendly energy sources can have a major visual impact: consider wind turbines, solar thermal panels or photovoltaic arrays mounted on the roofs of buildings. The low visibility of GSHPs can be particularly attractive to those requiring cheap 'green' energy in a national park or other area where planning regulations restrict visual impact of new developments. The low visibility of GSHPs has a down side as well: wind turbines and solar panels advertise themselves, saying 'Look at me, I'm a low carbon household'. GSHPs are not immediately obvious; possibly one reason for the initial resistance of the UK market to the technology.

Furthermore, a large office building that is cooled and heated by GSHPs will not need massive roof-mounted cooling arrays associated with conventional active cooling systems. This may have structural implications for the building: it will not have to bear the weight of the cooling arrays, possibly saving construction costs.

GSHPs present a minimal fire hazard and require minimal ventilation. They are also extremely low maintenance and have a long lifetime compared with many fossil fuel boilers. GSHPs produce relatively little noise, if properly mounted in an insulated cabinet, and can be placed in a household utility room or garage with minimal disturbance. (GSHPs, like fridges, do emit some noise, however. They should probably not be placed in a room that is regularly occupied.)

Probably, the most important advantages relating to heat pumps are (i) running cost and (ii) environmental impact in terms of CO_2 emissions. Before we proceed, it is important to realise that GSHPs provide a low-CO_2 source of heating, but not usually a zero-CO_2 source. GSHPs use electricity: a 6 kW unit may use 1.5–2 kW electricity to deliver its peak heating load. In the UK, electricity generation results in emissions of CO_2. In fact, heating your home by electricity can be very carbon-unfriendly, due to efficiency losses during generation and transmission through the national grid. Let us consider the UK's electricity supply (Figure 4.7). The bulk of the UK's electricity in

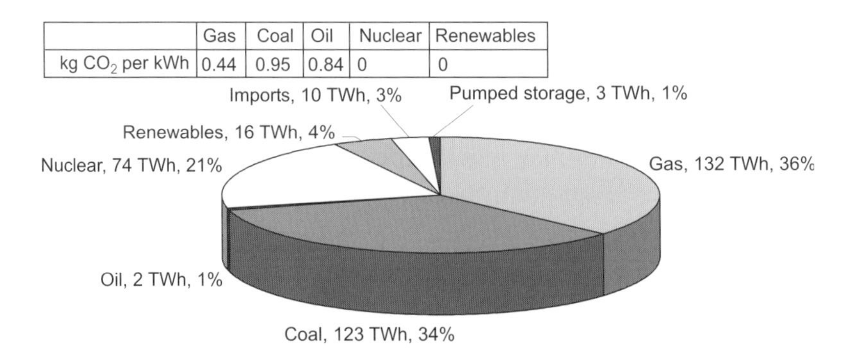

	Gas	Coal	Oil	Nuclear	Renewables
kg CO_2 per kWh	0.44	0.95	0.84	0	0

Figure 4.7 The percentage of electricity generated by various sources in the UK in 2005, according to DTI (2006). The pie diagram shows the total TWh of electricity contributed to the National Grid from the various main sources, and as a percentage of a total of ≈360 TWh. The table shows the kg of carbon dioxide emitted per kWh of electricity delivered from the Grid from various sources (Killip, 2005). Note 1 kg C = 3.66 kg CO_2.

the first decade of the twenty-first century is generated by combustion of natural gas and coal. For every kWh of electricity consumed that is generated by gas, 0.44 kg of CO_2 are released. Electricity generation by coal is even worse with 0.95 kg CO_2 emitted for every kWh consumed (Killip, 2005). Generation of electricity by nuclear or 'renewable' sources such as hydropower result in zero CO_2 emission at the point of generation (although we should remember all the CO_2 emitted during uranium mining and dam construction). The carbon 'footprint' of using electricity to heat your house thus depends on the 'mix' of fuels used to generate the electricity. In the UK, the average CO_2 emission for electricity consumed is around 0.42 kg per kWh.

Of course, elsewhere, the direct use of electricity for heating a house may result in much lower CO_2 emissions, depending on how the electricity is generated. Norway's electricity is dominated by hydroelectric power generation and electricity consumption is very CO_2 efficient. Sweden has long been dependent on a network of nuclear power stations: these may be undesirable for other political reasons, but are also CO_2 efficient.

Let us compare the costs and CO_2 emissions from using electricity to heat your house with other common domestic fuels (Table 4.2). In the UK, we can see that heating a house with electricity is both expensive and carbon inefficient. At the time of publication of SAP (2005), 1 kWh of electricity costs anywhere between 2.9 p and 7.8 p (depending on your specific tariff and when you use the electricity – the standard tariff was 7.1 p per kWh) and results in the emission of 0.422 kg CO_2. The most efficient conventional means of heating your house is with a modern condensing 'combi' gas boiler, which can be up to 90% efficient in converting the energy of combustion into useful space heating. Efficient usage of gas costs 1.63 p and emits only 0.194 kg CO_2 per kWh.

Why would anybody therefore choose to run an electrically powered GSHP rather than a combi gas boiler? Because, a GSHP is not using the electricity to heat your house directly: it is using electricity to transfer free, zero-CO_2 energy from the ground to your house. If we consider a GSHP with a seasonal performance factor of 4, for every

Table 4.2 The nominal costs per kWh of various heating fuels, together with the approximate mass of CO_2 emitted per kWh of energy delivered. Figures for gas, LPG, oil, coal and electricity are derived from the UK Government's standardised assessment procedure (SAP, 2005). Since publication of this document, gas and electricity costs have increased substantially: the table is thus for illustrative purposes only and should not be used for design. Note that modern condensing combi gas boilers can approach 90% efficiency. The figures for the GSHP are derived from those for electricity supply by assuming a seasonal performance factor of 4.

	Cost per kWh (British pence = £0.01)	**kg CO_2 per kWh**
Mains gas	1.63 p	0.194
Bottled liquid petroleum gas (LPG)	4.32 p	0.234
Heating oil	2.17 p	0.265
House coal	1.91 p	0.291
Electricity	7.12 p (standard) range 2.94–7.83 p	0.422
GSHP	1.78 p (standard tariff) and probably lower	0.106

1 kWh of heat provided to your house, the GSHP consumes only 0.25 kWh of electricity. This means that we can divide the environmental and economic costs of the electricity by 4, as shown in the bottom line of Table 4.2. Here, we see that the GSHP wins hands-down over mains gas on CO_2 emissions with a little over 0.1 kg CO_2 released per kWh heating effect. In terms of running cost, the GSHP compares very favourably with gas, too, especially if one can negotiate a cheap (e.g. nighttime) tariff for electricity supply.

Thus, we have demonstrated that, while GSHPs running off mains electricity are not a zero-carbon source of heat energy, they are significantly better than any conventional alternative. And if you live in a country such as Norway, where most electricity is 'green' hydroelectricity you can genuinely boast of having a zero-carbon heating system. Moreover, you can comfort yourself with the knowledge that you are using Norway's limited reserves of hydroelectric power three to four times more efficiently than if you had merely plugged an electric fan heater into the mains. (Norway's hydropower reserves are, of course, finite. In the late 1990s they effectively ran out, forcing Norway to import significant quantities of energy from its neighbours for the first time, sending energy costs spiralling during dry summers!)

Given that gas and electricity used in the home is responsible for 20–25% of the UK's carbon dioxide emissions (DEFRA, 2005), the environmental arguments for GSHPs are unassailable. They are also beginning to look increasingly attractive from an economic point of view. So why is everyone not rushing out to buy one?

4.11 Capital cost

The fact is that the initial capital cost of a GSHP system is rather high. A 6 kW water-to-water heat pump in the UK may cost around £3000 (at the time of writing in 2006 – see Figure 4.8). Including the cost of borehole-drilling or trench installation, pipework and manifolds, the total cost of a small domestic 6 kW GSHP installation may exceed £8000.

We have seen above that the running costs of a GSHP may well be lower than a conventional combi gas boiler. But are they low enough to justify that much larger capital cost (a combi gas boiler costs less than £1000)? A study was made of a group of around 60 similar bungalows in Nottingham, UK (Hill and Parker, 2005/06; Parker, 2007). Some of these were fitted with combi gas boilers and others with GSHPs. For the bungalows with gas boilers, and considering solely heating and hot water, the average gas bill (December 2005 tariffs) was £210–250 per year, with an annual CO_2 emission of around 2 tonnes of CO_2 per household. For the bungalows with GSHPs, the heat pump's electricity bill was £200 per year and just over 1 tonne of CO_2 was emitted. The GSHPs recorded an average overall seasonal performance factor (SPF) of 3.48 (with estimated values of 4.16 when delivering low-temperature waterborne space heating at 35–55°C and 2.47 when delivering domestic hot water at up to 65°C). As we have already concluded, the environmental benefits were overwhelming, in terms of CO_2, but the cost savings were modest (compared with gas combi boilers, but substantial

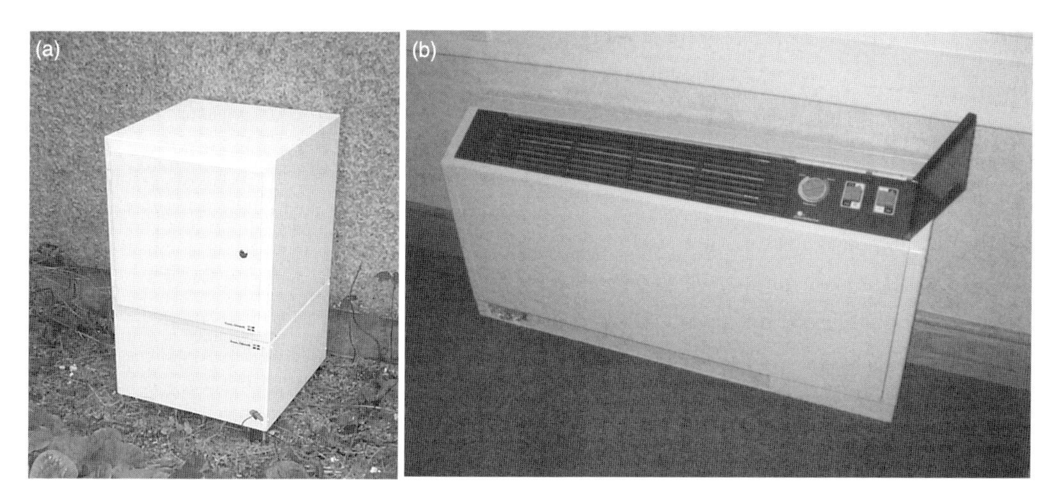

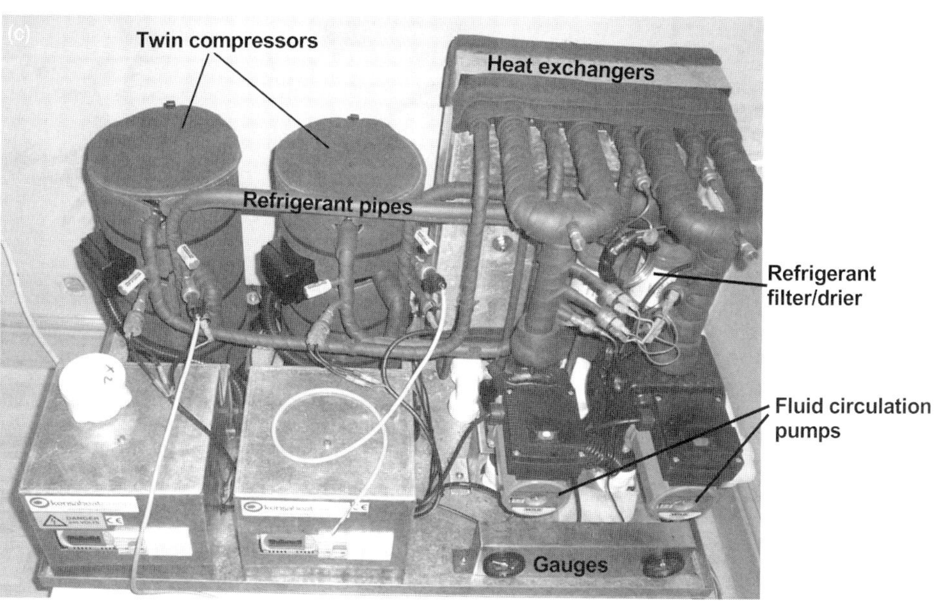

Figure 4.8 (a) A domestic 6 kW water-to-water heat pump that would supply a domestic water-borne central heating system or underfloor heating system, *reproduced by kind permission of Kensa Engineering Ltd.* (b) A console-type water-to-air heat pump: this extracts heat from a ground-coupled carrier fluid to a flow of warm air. It can be reversed in summer, to provide a current of chilled air, rejecting waste heat to the carrier fluid (*photo by David Banks*). (c) The innards of a twin-compressor water-to-water heat pump, with integrated fluid flow circulation pumps, *reproduced by kind permission of Kensa Engineering Ltd. and GeoWarmth Ltd.*

compared with other conventional fuels). With an annual saving in fuel bills of no more than £50, it would have taken decades to pay back the initial capital expenditure on the heat pump. We could, of course, argue that, over the 20–25 year lifetime of a heat pump, a gas combi boiler with a 12 year life may have had to be replaced once. Furthermore, we could argue that the GSHP owners would not have incurred the gas boilers' hefty annual maintenance bills, and also that the GSHP households may well have received a significant government subsidy towards the cost of the heat pump. Indeed, Hill and Parker (2005/2006) argue that, taken over the lifetime of the heat pump, the total costs (capital plus running costs) of a combi gas boiler and a GSHP may not be too dissimilar. Nevertheless, the economic arguments for domestic-scale GSHPs are not (currently) overwhelming and it begins to become clear why British citizens still tend to choose gas boilers over GSHPs! Indeed, we are forced to admit that, for domestic properties in Britain with access to mains gas, gas heating is still the cheapest form of space heating on economic grounds. In fact, those households and individuals who choose GSHPs tend to be motivated by moral, 'environmental' reasons, rather than financial ones. The exception to this observation is, of course, in rural areas of Britain where mains gas is unavailable and GSHPs are genuinely an economically attractive alternative to oil firing or bottled gas. (Of course, in other nations, with other energy pricing structures than Britain, economic analysis will provide wholly different results – in Norway, for example, most people probably do not have access to cheap mains gas!)

All is not lost, however, because there is an economy of scale that comes into play. The larger the heat pump you wish to buy, the cheaper it becomes per installed kW. Figure 4.9 shows the cost of a range of heat pumps supplied by one British heat pump manufacturer (2005 costs). It will be seen that while a small domestic 6 kW heat pump costs £3000 (£500 per kW), a 20 kW unit costs between £5000 and £6000 (i.e. £250–300 per kW). Thus, the bigger your building or project, the proportionately lower the capital cost of your GSHP system becomes and the more attractive the technology is from an economic point of view. Indeed, Banks *et al.* (2005) cited several medium-to-large projects, where a payback time of several years, rather than several decades, was calculated (see also Box 3.4, p. 41). A significant fall in payback times for commercial projects in recent years is also documented by Kelley (2006).

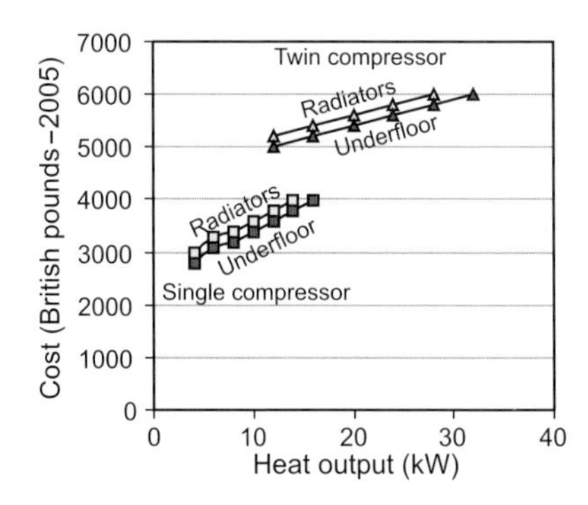

Figure 4.9 The cost of GSHP units of different heat output ratings, tailored for either underfloor or wall-mounted low-temperature radiators, supplied by a British manufacturer in 2005. Note that larger units (above 12 kW) are typically fitted with dual compressors. This reduces the peak electricity draw on start-up. There is relatively little price increase within the range of single-compressor GSHPs; the big increase in price occurs as one moves from single to dual compressor systems.

4.12 Other practical considerations

A householder considering the purchase of a GSHP system may be so overcome by considerations of cost and carbon emissions that he or she may forget some practical aspects. We would, in particular, encourage potential purchasers and installers to be mindful of the following:

- GSHPs are low-noise installations, but not no-noise. Make sure that they are installed in an insulated and vibration-proof cabinet in a separate room away from main living areas (cellar, utility room, cupboard, etc.)
- GSHPs use electricity very efficiently. A 10 kW_{th} unit may only consume 2.5 kW electrical energy – but that is still quite a lot of electricity: (11 A at 240 V). An electrician should ensure that the heat pump is powered by an appropriately rated circuit. Bear in mind that, on start-up, the compressor will draw significantly more power (may be as much as three times more) than its normal running load. The practical implication of this is that a 240 V domestic electricity supply may not tolerate more than a 10 kW_{th} heat pump unit. For heat pumps in the range 10–20 kW_{th} output, twin compressor constructions would be recommended (see Figure 4.9), such that both compressors do not start up simultaneously, limiting the current drawn at any one time. For heat pump installations in excess of 20 kW_{th}, a three-phase electricity supply may be necessary. The bottom line is 'check with your electricity supplier' – tolerable loads vary from location to location in the UK and between various utility companies.
- GSHPs operate most efficiently (highest COP_H) with a low-temperature output, running steadily for longer periods. Therefore, warm-air circulation or low-temperature underfloor warm-water heating systems (or even low temperature, high-surface-area wall-mounted radiator systems) ideally complement GSHPs. GSHPs can be retrofitted to older heating systems, but they may not provide a wholly comfortable level of heat.
- A simple, flexible solution for ground source cooling and heating in larger commercial or industrial complexes is to circulate a fluid, chilled or heated by the heat pump, around a so-called building loop within each heating/cooling block. Fan-coiled units, mounted on this loop, will distribute warm or cool air throughout the space.
- GSHPs can also be used to supply domestic hot water (DHW). However, the efficiency of most GSHP systems decreases at such high temperatures. For example, the COP_H of a heat pump delivering hot water at 55°C may not be more than 2 (K. Drage, *pers. comm. 2007*). Furthermore, domestic hot water should ideally be delivered at temperatures in excess of 60°C to avoid problems with *Legionella* (Box 6.3, p. 139), although there are other methods of controlling such bacteria. Thus, while heat pumps can deliver DHW directly, an alternative strategy might be to use the heat pump to raise water temperatures to around 45°C, and then a conventional resistive element immersion heater (on off-peak, low-tariff electricity) to boost the temperature above 60°C (see also Section 6.5, p. 139).

4.13 Summary

In this chapter we have seen that there is a tool called a heat pump, which can be used to transfer heat from one place to another. It can be used to pump heat up a temperature gradient: for example, from the cool ground to a warm living room. It can also be used in reverse mode to pump heat down the temperature gradient faster than would have occurred under passive conditions. For example, we can use a heat pump in summer to pump waste heat from a large building to the ground, to a lake or to the outside air.

This may seem an unfamiliar concept to us, but we are all happy to use heat pumps in practice – most of us have refrigerators. Larger-scale heat pumps, for heating and cooling buildings are available. They use a modest amount of electricity to transfer free, environmentally friendly energy from the outside environment into our homes. A heat pump that extracts heat from, or rejects heat to, the ground (or any medium thermally coupled to the ground) is termed a *ground source heat pump* (GSHP).

GSHPs offer cheap, reliable, heat energy with low visual impact. They also deliver heat with a low-CO_2 emission footprint: far lower, in fact, than all conventional space heating solutions. The main drawback is the high capital cost of GSHP systems, making them unattractive to many small-scale, domestic users. At larger scales, however, the capital cost of GSHP schemes decreases per installed kW, rendering them attractive to major commercial, governmental, community, leisure and industrial developments.

4.14 Challenges: the future

We may envisage that future heat pumps will be sleeker, quieter and use electrical energy more efficiently. Above all, we would hope that their capital cost will decline in line with increased demand and uptake.

We would also hope to see increased transfer of ground source heat technology to the rapidly developing, highly populated economies such as China, India, Brazil and Russia. The per-capita emission of CO_2 from these nations is still modest, but has the potential to increase dramatically in the forthcoming decades (Box 4.4). The outcome of the battle to retain some global control on atmospheric carbon emissions will likely be lost or won in those countries.

A less obvious, but intriguing, challenge is to investigate the potential that heat pumps can offer to deliver reliable, cheap and efficient space heating to poorer countries, maybe without a well-developed or reliable electricity grid. Electricity, which is so often regarded as being quintessentially 'modern', is actually a peculiarly energetically inefficient means of delivering space heating. Should we rather be looking at developing a new generation of non-electrical heat pumps whose compressors run on water power, biogas or diesel combustion? Or maybe even absorption heat pumps that utilise combustion of gas or wood fuel (Box 4.3)? These technologies could conceivably deliver

BOX 4.4 Energy consumption, CO_2 emission and standard of living.

It has been argued that access to energy resources should be a fundamental human right, but the fact remains that abundant energy is a privilege enjoyed by the few. Energy consumption can therefore be a good indicator of standard of living, and because most energy consumption involves emitting CO_2, per-capita emissions of CO_2 are also such an indicator. The UNDP (2004) have produced a fascinating chart linking energy consumption to a so-called Human Development Index, which can be considered a surrogate index for the standard of living (Figure 4.10).

We can see that inhabitants of nations such as Mozambique and Nigeria consume less than 1 tonne oil equivalents [1 metric tonne oil equivalent (mtoe) = 12.4 MWh] and emit less than ½ tonne CO_2 per capita per annum (Table 4.3). Russians and Italians consume around 4 toe per capita per year and emit 8–10 tonne CO_2 per capita per year. The USA sits high on the list, with a per-capita energy consumption of around 8 toe per annum and a CO_2 emission of around 20 tonnes.

Of course, there is some discrepancy between the energy consumption and CO_2 emission, depending on the source of energy. Rural Africans, depending heavily on biomass (wood/charcoal) for their energy needs, will have a different rate of CO_2 emission than British or Russians guzzling electricity from relatively inefficient coal- or oil-fired power stations. Icelanders enjoy abundant energy consumption of 12 toe per annum, but with a CO_2 emission rate considerably lower than the USA, owing to their reserves of high-temperature geothermal energy.

Two points are noteworthy. From the UNDP graph, a relatively high standard of living (HDI values of around 0.7) can be reached without energy consumption exceeding 2 toe per capita per annum. To achieve a HDI in excess of 0.8 requires enormously greater energy consumption – there is a marked 'kink' in the graph at this value.

Second, we should note that the vast populations of India, Brazil and China currently have a modest per-capita annual CO_2 emission (Table 4.3). These economies are expanding rapidly and their inhabitants understandably have expectations of commensurate increases in living standard. It is not unreasonable to suppose that the degree to which these mega-economies can achieve growth, while restricting their CO_2 emissions, will be decisive for the future of Planet Earth.

efficient ground source heat to mountain villages in Pakistan, Scotland or Bolivia, or to remote towns in Russia or China, without the need for massive capital investment in ultimately inefficient electrical infrastructure.

Such non-electric heat pumps are, of course, not a new concept: Heap (1979) notes, however, that a precondition for the efficient operation of fuel-engine-powered heat pumps is the ability to recover the exhaust heat from combustion and deliver this as useful space heating. Sumner (1976) proposed such oil-engine heat pumps, with recovery of heat from combustion. Both engine-driven (Worsøe Schmidt, 1982; d'Accadia *et al.*,

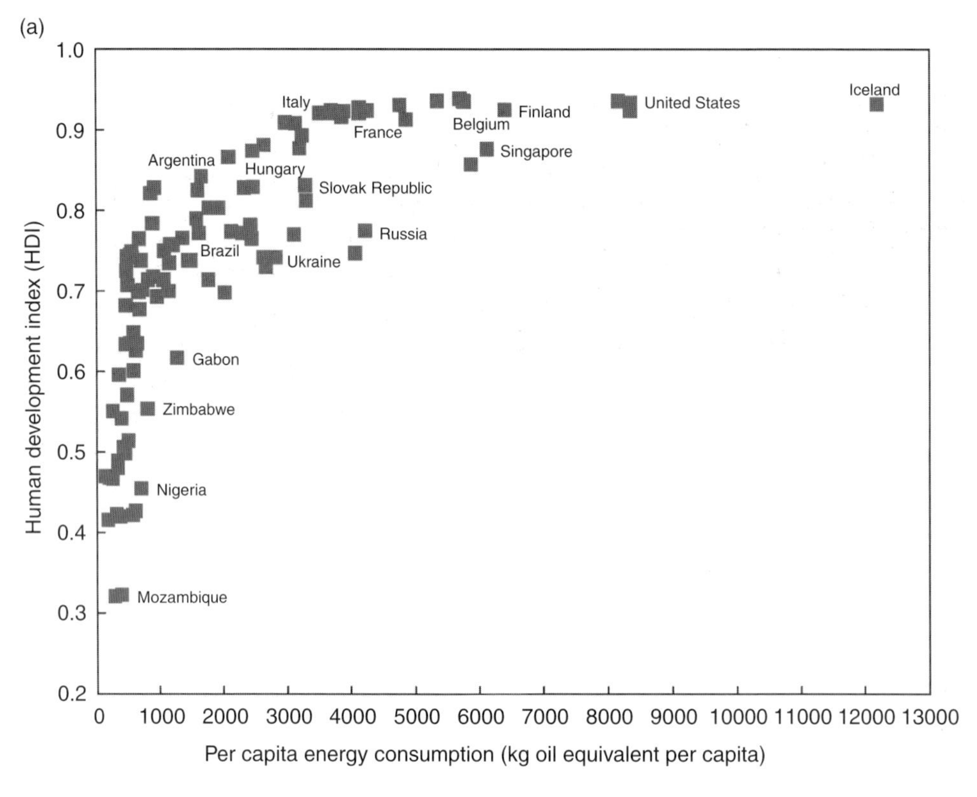

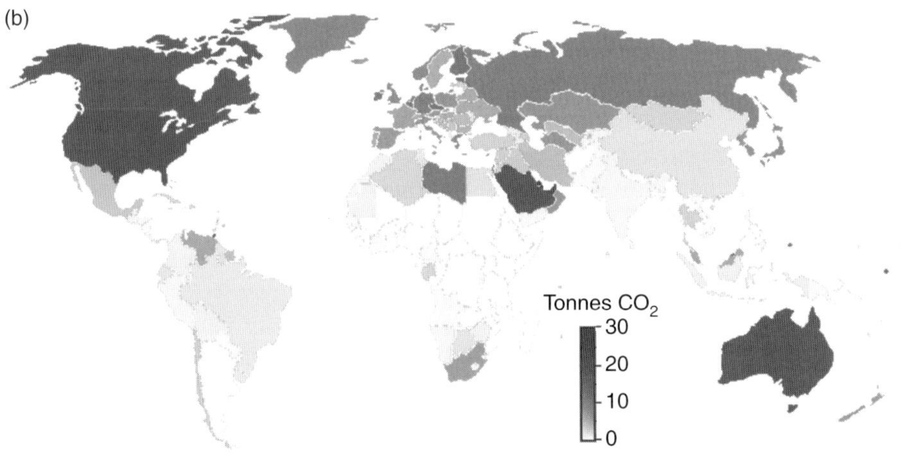

Figure 4.10 (a) The relation between per capita energy usage in kg oil equivalent and standard of living (HDI, human development index) in 1999–2000. (b) The per capita annual emissions of carbon dioxide, as metric tonnes CO_2, by country. *The top figure (a) is published by UNDP (2004) and reproduced by kind permission of the United Nations Development Programme. The map (b) appears on Wikipedia and is reproduced under the terms of the GNU Free Documentation License, Version 1.2 or any later version published by the Free Software Foundation.*

Table 4.3 Annual per capita carbon dioxide emissions by country for the year 2003, based on CDIAC (the US Department of Energy's Carbon Dioxide Information Analysis Center) data sourced from the UN Statistics Division website for Millennium Development Goals Indicators.

Country	Per capita annual emission of CO_2 in 2003 (tonnes CO_2)
Mozambique	0.08
Nigeria	0.42
India	1.2
Brazil	1.6
China	3.2
Sweden	5.9
Iceland	7.6
Italy	7.7
UK	9.4
Russian Federation	10.3
USA	19.8

1995; Lian *et al.*, 2005) and absorption cycle (Murphy and Phillips, 1984) heat pumps for space heating have been sporadically researched in recent decades, with the former even incorporating the technology of Stirling Engine cycles (Angelino and Invernizzi, 1996). In fact, engine-driven heat pumps are marketed today, largely as a result of the notorious recent 'brown-outs' in the power supply infrastructure along the US western coastline (Chris Underwood, *pers. comm. 2007*). To date, however, there has been no significant emphasis in the utilisation of non-electric GSHPs as 'intermediate technology'.

5

Heat Pumps and Thermogeology: A Brief History and International Perspective

> The iceman cometh ...
>
> Eugene O'Neill (1939)

> As the world supplies of fossil fuels diminish, the value and the monetary saving by the use of the heat pump will increase proportionately. Over a century ago Lord Kelvin foresaw and warned us about a world fuel shortage and devised a machine to alleviate that shortage.
>
> John A. Sumner (1976)

The history of heat pumping is intimately connected with the history of artificial refrigeration. The development of both was stimulated first by the growth of a market for ice as a luxury item consumed by the upper classes in Europe, America and 'the colonies' and second by the needs of the marine transport industry, which was searching for means of keeping meat and food produce cool during long ocean voyages. Richard Trevithick, James Harrison and Lord Kelvin all had interests in the maritime industries, and even today, one of the UK's major ground source heat pump (GSHP) manufacturers evolved from the production of marine heat pumps (heat pumps designed for boats and ships to extract heat from or dump heat to the sea).

5.1 Refrigeration before the heat pump

Historically, of course, refrigeration did not entail the use of heat pumps. In fact, throughout temperate climes, the source of refrigeration was ice, collected in winter

Figure 5.1 The ice house at Syon Park, Brentford, south-west London. This was constructed in the 1820s by order of the third Duke of Northumberland to replace earlier seventeenth/eighteenth century structures. The ice was tipped into the storage chamber via the roof and packed in straw. The store had a sump draining to waste via a sloping floor. Access to the ice house was via an 'airlock' of two doors. The ice house was not dug vertically downward to any great extent but was subsequently covered by soil to form a landscaped mound. The ice, being collected from lakes and the River Thames, was impure and not consumed directly. It was, however, used by the aristocrats to cool drinks and produce sorbets and ice-creams (by adding salt to form a freezing mixture). *Photo by David Banks.*

and stored until summer. Many manor houses and country estates had so-called icehouses (Buxbaum, 2002) – covered sunken chambers in the earth that were used to store ice collected from frozen lakes during winter (Figure 5.1). Such structures made use of thermogeology: they recognised that the subsurface is significantly cooler than the air during spring and summer and could be used to slow down the rate at which the ice melted. In fact, it has subsequently been shown (Weightman, 2003) that the subsurface nature of the icehouse was not the main determining factor for their satisfactory function. Other factors were more important: good insulation and, above all, the ice had to be kept dry and well drained. Most icehouses allowed meltwater to drain away from the ice blocks to a sump, where it could be extracted. Furthermore, the icehouse may have been constructed using double courses of bricks in the subsurface (a 'cavity wall') to minimise conductive heat loss to the ground (Buxbaum, 2002). The ice may also have been packed in an insulating material such as straw or sawdust. Additionally, it was recognised that air circulation conditions in the headspace above the ice 'pile' would have been of importance.

BOX 5.1 Applied thermogeology in Siberia.

The Siberians have a tradition of using the subsurface as a natural coolbox. Many country cottages would have a **Ледник** (*lednik*) or ice well. Ice would be harvested in the harsh winter and packed away in the ice well, maybe together with fish and meat, to act as a deep freeze, from which frozen goods could be recovered in summer.

The cottage may also have had a **Погреб** (*pogreb*) or subsurface larder (basically a hole, several metres deep), which would simply have been used to store potatoes, jams and mushrooms. Alternatively, these may have been placed in the underfloor area of the cottage (**Подпол** – *podpol*). Even today, in Russian cities, it is not uncommon to find Soviet-era blocks of apartments with a central subsurface 'bunker' in the courtyard, where residents can store their vegetables and pickles in cool conditions.

Thanks to Igor Serëdkin of Tomsk for this information.

Such solutions were not unique to the UK, of course. Similar structures are traditional in Russia, for example (Box 5.1), while the Mediterranean nations are believed to have harvested ice from the Alps during winter for transport to lower elevations, to be consumed during summer.

5.2 The overseas ice trade

With the development of world trade, some northern lands realised that they had a huge resource locked up in their winter ice. Could it be exported overseas? Weightman (2003) tells the extraordinary story of Frederic Tudor, a Boston American, who dared to believe that it could. He invested, lost and eventually partially recouped astronomical sums of money in setting up a system for harvesting New England lake ice, packing it with sawdust into insulated ships' holds and transporting it to the Caribbean Isles (the first cargo was sent to Martinique in early 1806) and the southern states of the USA. Tudor quickly realised that transport was not necessarily the major problem – an equally important issue was storing the cargo efficiently on arrival. He thus set out to design and construct a network of icehouses in the ports he supplied, although he seems to have been hindered by employing a string of icehouse keepers of dubious reliability. Tudor found that it was unnecessary to construct subterranean icehouses: with his Havana icehouse he demonstrated that a timber structure, built above ground, with cavity walls filled with sawdust or peat, and with excellent drainage and appropriate ventilation, was equally good (or better) than conventional subterranean structures.

In 1833, he pulled off the unthinkable – he transported over 100 tons of ice from Massachusetts to Calcutta, India, on board the *Tuscany*. The ship departed in May and arrived in September, and the quantity of ice lost to melting during the voyage was believed to be no more than one-third of the original total.

Once Tudor had demonstrated the ice trade to be both viable and profitable, others were keen to join in and the American ice trade expanded hugely, supplying a domestic market from New York to New Orleans and even exporting ice to England and India. The size of ice blocks harvested from the lake varied depending on the distance they had to be transported (heat loss is related to surface-area-to-volume ratio and is thus proportionately less for larger chunks): New York cubes measured 22 in. on each side while those exported to India were larger. By 1879, the American Ice Industry was estimated at 8 million tons harvested per year, with 5 million tons reaching the consumer (Weightman, 2003). In warm winters, onshore 'ice famines' would force the ice harvesters to chase freshwater icebergs in the Northern Atlantic and Greenland Sea.

The British market for American ice was opened up by Charles Lander in the 1840s and, subsequently, this ice was marketed as Wenham Lake Ice (a brand name originally derived from Lake Wenham, near Salem). It achieved a reputation for high purity, such that it could be consumed directly in drinks such as mint juleps and sherry cobblers. The Norwegians observed this development with interest and believed that they could supply cheaper ice to London than the Wenham Lake Ice Company – all they needed was a catchy brand name. They solved this problem in an uncharacteristically sneaky way – by renaming (at least, according to anecdote) their own unpronounceable Lake Oppegård as 'Wenham Lake' and using the same brand name to befuddle the British public! At any rate, the Norwegians became increasingly dominant in the British ice trade from around 1850, and continued exporting ice to London until well into the twentieth century. At the peak of the trade (around 1900), the UK's annual import from Norway was around 500 000 tons (Weightman, 2003).

The Norwegian ice was harvested (as in New England), by sweeping the surface of the lake free of debris and then using Nathaniel Wyeth's horse-drawn 'ice plough' to carve deep grooves into the lake surface. Blocks were then levered from these grooves and usually temporarily stored in an icehouse near the lake. In Norway, blocks of ice may have been slid down wooden 'ice railways' from the mountain lake to the harbour for export (Figure 5.2). On arrival in London docks, ice was typically loaded on barges and transported into the heart of London via the Regent's Canal. Ice wharves were located along the canal for the unloading (Figure 5.2) and storage of the ice in 'ice wells'. These phenomenal structures were huge shafts in the London Clay, several metres in diameter and several tens of metres deep. The ice importer, Carlo Gatti (1817–1878), dug two ice wells, around 1857 and 1863, at Battlebridge Wharf (now the London Canal Museum in Kings Cross). They were around 12.5 m deep, 10 m in diameter and had a capacity of 750 tons ice per well. As early as 1856, Gatti had signed a contract with the entrepreneur Johan Martin Dahll of Kragerø in Norway for import of ice harvested from lakes on the Toke watercourse and possibly also from glacier falls. Gatti's more established competitor, William Leftwich, had constructed similar ice wells near the Canal in Camden. His largest, at 34, Jameston Road, was built in the late 1830s and was reportedly London's deepest at 30.5 m, with a 13 m diameter and a capacity of 4000 tons. It initially received local ice from Camden ponds, but was later supplied from Norway.

Figure 5.2 (a) Ice ploughing on a Norwegian Lake; (b) an ice railway conveying ice from an inland Norwegian lake to the harbour; (c) and (d) reminders of the now defunct ice trade is found in the place names along the Regent's Canal of London. [*Photos (a) and (b) kindly provided by the London Canal Museum. Photos (c) and (d) by David Banks*].

As in the large American cities, the London 'Ice Kings' such as Carlo Gatti would have employed 'icemen' to operate a fleet of carts to deliver ice supplies to richer households, hotels and so on. Subscribing households would often have placed the ice in an insulated 'ice box' – a simple refrigerator that could be used to keep fresh foodstuffs cool. The iceman appears to have achieved a degree of notoriety (similar to Benny Hill's naughty milkman in a later decade) as a seducer of housewives, if we are to believe Eugene O'Neill's play *The Iceman Cometh*, set in 1912.

5.3 Artificial refrigeration: who invented the heat pump?

Even before the heat pump was invented, other techniques were employed to produce an artificial refrigerating effect. For example, it had long been known that a volatile liquid, such as ether or alcohol, produces a cooling effect when it evaporates. Think of how spilled vodka cools the skin of your hand. This is because the process of evaporation consumes a significant quantity of heat energy to effect the transition from liquid to gas (for water, this latent heat of evaporation is 2272 kJ kg^{-1}). Traditional methods of effecting cooling and producing ice slush in the Middle East and India involved evaporation of water from shallow ponds or porous earthenware pots during low-humidity nights (Weightman, 2003). Thus, the earliest forms of artificial refrigeration did not use the compression–expansion cycle at all. As early as 1748, the Scotsman William Cullen had demonstrated refrigeration at Glasgow University, simply by allowing ethyl ether to boil away into a vacuum.

In the early nineteenth century, it was also recognised that expansion of a pressurised gas could result in a drop in temperature (see Box 3.3, p. 39). This principle, together with that of the latent heat of evaporation, forms the basis of the heat pump cycle. By the early nineteenth century, the stage was set for its gradual development.

There seems to be a general consensus that the 'grandfather of the fridge' was the American Oliver Evans (1755–1819). He was an engineer who designed many improvements to the textile industry and also an improved high-pressure steam engine. He produced a design for a refrigerator using a compression–expansion cycle in 1805, although the contraption was never built.

Patriotic Cornwalesians claim that Richard Trevithick invented refrigeration, despite all evidence to the contrary! Trevithick was born in Illogan, Cornwall, in the year 1771. He worked at the Cornish tin mines (Wheal Treasury and Ding Dong mines), developing engines for raising ore. He is also famous for having constructed and run what many believe to have been the world's first steam locomotives. Another, less well-documented interest was in the marine transport industry. His claims on the heat pump are rather tenuous, but there is evidence that he became interested in the production of 'artificial cold' during the last 5 years of his life, having been struck by the magnitude of the ice-harvesting economy. In a single letter, written in June 1828, he outlined a means of producing active refrigeration by means of a steam engine and a compression–expansion sequence (in effect, a steam-powered heat pump). It is perhaps significant that the Frenchman, Sadi Carnot, had pointed out only few years prior to

Trevithick's letter that a heat engine (see Section 4.1, p. 58) could, in theory, be reversed to provide a heat pump. Trevithick died, near destitute, at Dartford, Kent, in 1833.

In 1834, an American-born resident of London, the prolific inventor Jacob Perkins (1766–1849), filed a patent entitled *Improvement in the Apparatus and Means for Producing Ice, and in Cooling Fluid* for a refrigerator employing ether in a compression–expansion cycle, apparently using a design similar to Evans'. A prototype was built in London by John Hague (Biblioteca ETSEIB, 2003). Reliable accounts of Perkins' apparatus are sparse, but Weightman (2003) goes as far as claiming that the power behind the compression was derived from the River Thames' tidal fluctuations.

It was left to the Scottish-born Australian, James Harrison, and the American, Alexander Twinning, to run with Perkins' ether vapour-compression concept and construct commercially functioning refrigerators during the 1850s. Harrison constructed such ether-based vapour-compression ice-making machines in Australia in the early 1950s and was granted a patent in 1855 (Richardson, 2003). His machine utilised a 5 m flywheel and produced some 3000 kg ice per day. Harrison applied his machines to the meat and brewing industries and in 1860 one was installed at the 'Glasgow and Thunder' brewery in Bendigo, Australia. By 1861, around 12 of Harrison's machines were in action, including one in a petroleum refinery at Bathgate, Scotland (Biblioteca ETSEIB, 2003). By the late 1870s refrigeration heat pumps were being installed on ships carrying meat between Australia, the USA and Britain (Heap, 1979).

In the 1840s, the American medical doctor, John Gorrie (1803–1855), had also been experimenting with the compression and expansion of air to produce ice. The ice was then used to chill the air circulating in wards for malaria and yellow fever patients. His contraption involved compressing air, resulting in a temperature increase. The compressed air was then cooled by circulating water in pipes, which served to draw heat out of the compressed air. The air was then allowed to expand, resulting in the formation of ice. He was awarded British and American patents in 1850 and 1851, respectively, and is regarded by many as the 'father of air-conditioning'.

Meanwhile, in 1857, Peter Ritter von Rittinger had applied the principles of the heat pump cycle to industrial processes at the salt works of Ebensee in Austria. Green (2007) reports that waterpower was used to compress steam, raising its temperature for utilisation in evaporating brine for salt extraction.

The final key players in the development of refrigeration were the French Carré brothers. As early as 1850, Edmond had figured out the principles of absorption refrigeration (Box 4.3, p. 66) using water and sulphuric acid. His brother, Ferdinand, perfected an ammonia–water absorption refrigerator in 1858–1859. This system was at that time particularly suited to the large-scale production of ice, although the toxicity of ammonia rendered it dangerous to use.

At around the same time, we note that William Thomson, who later became Lord Kelvin, was developing the theoretical foundations for the heat pump. In 1852, he suggested in his article '*On the economy of heating or cooling of buildings by means of currents of air*' that an open air-cycle heat pump concept could be used to absorb heat from the atmosphere and deliver it at a higher temperature to a building, resulting in a supply of heat greater than the mechanical energy input to the system. He noted that a vapour compression–expansion cycle could also do the same job (Thomson, 1852). At the time, however, there was little practical interest in this idea. The only flicker of interest came from British colonists in India, who were intrigued by the possibilities offered by heat pumps for *cooling* their public buildings (Thomson, 1852; Sumner, 1976; Curtis, 2001).

5.4 The history of the GSHP

The idea of using Kelvin's concept of the heat pump to extract environmental heat lay dormant for some time. The first patent for a GSHP was granted in Switzerland to an engineer named Heinrich Zoelly in 1912 (Ball *et al.*, 1983; Spitler, 2005; Kelley, 2006). By the late 1920s and 1930s, groundwater was being abstracted from wells on Brooklyn and Long Island and being used for air-conditioning, although Kazmann and Whitehead (1980) do not clarify whether this was by the use of heat pumps or simply passive cooling.

During the mid-to-late 1920s, T.G.N. Haldane constructed an experimental heat pump installation for his home in Scotland (Heap, 1979), utilising both outdoor air and mains water as a heat source (may be something that should be considered by water utilities companies!). His system appears to have been rather sophisticated with an electrically powered compressor (the electricity being hydroelectrically generated) and ammonia refrigerant. It supplied low-temperature waterborne central heating and domestic hot water. Bhatti (1999) notes that Haldane constructed another heat pump system for his office in London. Sumner (1948) also cites Haldane's pioneering work in Britain and draws attention to another small heat pump installation by one S.B. Jackson in 1938.

During the late 1930s several environmental heat pump systems were installed for heating buildings in Zurich, Switzerland (Heap, 1979; Schaefer, 2000). These included an R-12 refrigerant heat pump system for the City Hall, based on water from the River Limmat, providing a nominal heating capacity of 70 kW and a cooling capacity of 55 kW (IEA, 2001a).

The date of the first truly GSHP installation is somewhat hazy. Heap (1979) claims that 15 commercial environmental heat pump installations were in operation in the USA in 1940 – some of them using well water as the heat source. Furthermore, Arnold (2000) tells us of a building in Salem, Massachusetts, heated and cooled by well water in 1935.

We do know, however, that, in around 1945, Robert C. Webber of Indianapolis, USA, was recovering the heat rejected from his cellar deep-freezer and using it to heat his house. He was so enchanted with this idea that he constructed what may be one of the earliest direct-circulation closed-loop GSHP systems to replace his coal-fired furnace (Kelley, 2006). He built a dedicated heat pump, using circulating Freon refrigerant to extract heat from buried copper pipes in trenches (Sanner, 2001; IGSHPA, 2007).

At around the same time, Britain's great visionary proponent of heat pumps, John Sumner, assisted by (the probably grossly underrecognised) Miss M. Griffiths, flew the flag for the technology. In 1945, Sumner (Box 5.2) constructed a full-scale

BOX 5.2 John Sumner's Norwich heat pump.

John Sumner was the City Electrical Engineer for Norwich, UK. He is also known as a prophet for the technology of environmental heat pumps in the UK. In 1940, the Norwich City Council Electrical Department built new premises in Duke Street, on the banks of the River Wensum. It was originally intended to use a heat pump to heat the building, but in the wartime climate there was a lack of appetite for such an 'innovative' venture and coal boilers were installed. It was only after the Second World War, in 1945, that a decision was made to replace the coal boilers with a heat pump based on abstraction of river water from the Wensum. The original central heating system was designed for hot water at 70–80°C, but with minor modifications was rejigged to run at a temperature of 50–55°C. The heat pump itself was largely (by Sumner's own admission) cobbled together out of salvaged parts in the post-war 'austerity' economy, and used an SO_2 refrigerant. Even this non-optimal design, running on cold Wensum water (which could dip as low as 1°C during the winter) achieved a seasonal performance factor (SPF) of 3.45. Sumner was confident that, with a well-tailored design, this could easily be increased to >4. The pump ran at an average thermal delivery of 5 therms hr^{-1} (147 kW) and a peak capacity of 8 therms hr^{-1} (234 kW). Sumner compared the economy of the heat pump with the conventional coal-fired system and considered that the heat pump represented the clearly cheaper option, even taking into account the larger capital cost of the heat pump. He also considered the non-monetary benefits and recognised the lower pollution from the heat pump and its attractiveness in view of the finite resource of coal. He recognised clearly, however, that the success (or otherwise) of the heat pump technology had a political dimension, inextricably linked with the national coal economy of the time. Sumner even performed a calculation akin to that presented in Table 4.2 (p. 75) of this book, and concluded that, to compete efficiently with conventional heating by coal, an electrically powered heat pump needed to have a COP_H of at least 3.3. Sumner also recognised the desirability of some form of thermal storage in the system, such that off-peak electricity could be used to run the heat pump at night, with the heat being stored for use during the daytime.

Sumner's findings are presented in a hugely entertaining paper read to the Institution of Mechanical Engineers (Sumner, 1948). Even more interesting is the discussion following the paper's presentation, which could be a replica of the debates still taking place today. Sumner's audience divided itself clearly into 'believers' and 'sceptics'. The 'believers' immediately understood the potential of the heat pumps and proposed innovative solutions based on the UK's canal network, and on the groundwater supplied from the wells of London's Metropolitan Water Board (this suggestion from Messrs Lupton and Westbrook). The 'sceptics' queried the economy of the heat pump compared with coal: they asked awkward (but cogent) questions such as 'Would the heat pump have been more expensive if it had been properly constructed rather than cobbled together from spare parts?' and 'Did Mr Sumner include the costs of his time in the capital cost of the heat pump?'

Figure 5.3 John Sumner's Norwich heat pump, from 1946 (see Box 5.2). Not so much a unit as a whole room! *After Sumner (1976).*

ground-coupled heat pump system in Norwich, based on water from the River Wensum (Figure 5.3). In 1948, the British philanthropist millionaire, Lord Nuffield, employed Sumner as a consultant to build 12 heat pump systems using 'ground coils' for a 2-year period of testing in selected houses. These achieved a heat output of 9 kW_{th} each and a COP_H of around 3 (Sumner, 1976; Curtis, 2001). Sumner was also involved in a project in 1951 to heat London's Royal Festival Hall using a two-stage gas-powered heat pump, based on compressors modified from Merlin aircraft engines, using the River Thames as the heat source. The 2.5 MW project is sometimes described as a bit of a failure, although it is not always made clear that the problem was that it actually delivered too much heat (Neal, 1978; Heap, 1979; Tovey, *pers. comm.*), with a COP_H of 2.5–3, and supplying hot water at up to 82°C!

In the early-to-mid 1950s, Sumner also installed a closed-loop ground source heat system at his own house (Sumner, 1976; Neal, 1978). The ground loop was initially constructed using copper pipe buried at around 1 m depth and filled with circulating antifreeze (later supplemented with an additional lead piping coil). Sumner later experimented with black plastic pipe laid on the ground surface, to absorb solar energy. Sumner's heat pumps may have been technical triumphs, but they proved not to be a marketing breakthrough. Britons showed next to no interest in the prototypes; at that time, abundant coal seemed a cheap and limitless source of energy. In particular, Sumner (1976) was not quite able to contain his bitterness towards the Electricity Boards, whom he suspected of not being enthusiastic enough about a technology designed to cut their sales by 60%!

Interest in GSHPs in Britain may have been low, but elsewhere the story was different: a large groundwater-based open-loop heating and cooling scheme was installed at an office block in Portland, Oregon in 1948 (ASME, 1980; Arnold, 2000; Box 5.3) as a direct result of the British and Swiss experiences. Nevertheless, it took the OPEC

BOX 5.3 The Equitable Building Heat Pump System of Oregon, USA.

The Equitable Building ground source heating and cooling scheme (ASME, 1980; Arnold, 2000) was constructed between 1946 and 1948 in Portland, Oregon. Not only it was the first major installation of its kind in a commercial building in the USA, but it was also a genuinely ground-sourced scheme, being based on groundwater abstraction from wells (an open-loop scheme). The building in question had 14 storeys and had no opening windows. It therefore had not only a heating demand in winter, but also an air-conditioning (cooling) demand.

The Chairman of the Board of the Equitable Savings and Loan Company was the sweetly named Ralph H. Cake. He had heard of the practical application of heat pumps in the UK in around 1946 (surely this must have been via contact with John Sumner?). Cake insisted on a ground source heat pump solution at Equitable and the air-conditioning and heating scheme was eventually designed by J. Donald Kroeker.

The four water-to-water heat pump units (using Freon 11 and Freon 113 as refrigerants) utilised groundwater pumped at 17–18°C from two wells of depth ~46 m and yields of 195–450 gpm (12–28 L s^{-1}). The used water was re-injected to a single 155 m deep well, at a nominal temperature of ~14°C.

The building had separate heating and cooling circuits. The pumped groundwater passed over the evaporator of the heat pump array and was chilled. The chilled groundwater effected cooling via fan coil units before being bled off to the re-injection well. The heating circuit was a closed circuit, passing through the condenser of the heat pump array and utilising the heat extracted from the groundwater. Condenser units could be coupled in and out, in response to heating demand. Moreover, the system was designed to recover heat from waste ventilation exhaust air, and to precondition incoming air. Thus, the heat to the building could largely be considered a 'by-product' (ASME, 1980) of the cooling load (heat recovered from cooling). No auxiliary heating was required for the building – indeed the heat pump capacity required to provide the necessary cooling load was far in excess of the heating requirement.

As of 1980, the scheme was still running and was satisfying energy efficiency standards in force at that time. One change that had taken place since its original construction was that the direct circulation of well water into the heat pumps had been replaced by the use of prophylactic heat exchangers (see Section 6.3.2, p. 115), separating the groundwater circuit from a building loop. This was found to be necessary due to scaling and corrosion damage attributable to the groundwater.

oil crisis of the 1970s to really focus people's attention on non-fossil energy sources. The Swedes (and, later, the Swiss, Austrians and Germans), in particular, took up the GSHP torch. Switzerland's native natural energy resources were limited to hydroelectric power and there was thus considerable motivation to use that as efficiently as possible (Sumner, 1948). Likewise, Sweden, which had very few sources of native power, was actively developing new atomic reactors and was looking for efficient means of heating homes by electricity rather than paraffin. The post-1970 period was characterised by the development especially of closed-loop systems employing polyethene, rather than metal, pipe. The first borehole-based vertical closed-loop systems were reported from Germany and Switzerland in around 1980 (Sanner, 2001, 2006).

5.4.1 Sweden and Switzerland

By the late 1970s, around 1000 GSHPs had been installed in Sweden (Rawlings and Sykulski, 1999). Between 1980 and 1986, 50 000 GSHPs were installed in Sweden, with a total installed capacity of some 500 MW. In the period from the mid-1980s to mid-1990s, development slowed down due to relatively low energy prices. From 1995 onwards, the market has grown steadily, thanks in large part to a system of subsidies initiated by the Swedish government, until by the early 2000s, over 90% of all new Swedish dwellings were reportedly being installed with heat pumps, of which at least three-fourths are ground-sourced (Bouma, 2002; Lund *et al.*, 2004; Nowak, 2006). This ground source heat is believed to supply some 15% of Sweden's total 100 TWh year^{-1} heating and cooling demand (Hellström, 2006). By 2004, almost 200 000 GSHP units were believed to be operational, with around 30 000 new systems being installed each year. The majority of such systems are small (average 10 kW), and 80% are believed to be closed-loop systems installed in boreholes to an average of 125 m depth, typically using 40 mm diameter 6.3 bar polyethene pipe (Lund *et al.*, 2004; Hellström, 2006).

In Switzerland, in 2000, one-third of all newly built single-family homes utilised a heat pump, and of these, around 40% were ground-sourced (Rybach and Sanner, 2000). Vertical, borehole-based closed-loop systems have gained in popularity hugely in recent years and, as in Sweden, account for the majority (65–75%) of the GSHP installed capacity, whereas horizontal closed loops account for less than 5% and groundwater-based open-loop systems 30% (Lund *et al.*, 2004). In fact, there is, on average, one GSHP installation per 2 km^2, reportedly the highest density anywhere in the world (Rybach and Sanner, 2000). One major factor in the popularity of GSHPs in Switzerland is believed to be the willingness of utility companies to develop GSHP schemes and sell space heating energy to subscribers at a competitive rate per kWh; in other words, the utility company bears the risk and absorbs the initial capital outlay of the scheme (Lund *et al.*, 2004).

5.4.2 North America

The heat pump market in the USA took off for different reasons: Sumner (1948) suggests that they were actively promoted by power utilities seeking new markets for consumption of excess power generation capacity. As Sumner was writing his book *Domestic Heat Pumps* in 1976, he was able to relate that there were over 1 million domestic heat pumps used for heating or cooling in the USA (admittedly mostly air-sourced or air-conditioning units), and that he had recently stayed in a hotel where 400 heat pump units could provide guests with heating or cooling at the flick of a switch.

The GSHP also expanded rapidly in the USA, with some 28 000 GSHPs being installed annually in 1994, rising to 50 000 in 1999 (Bouma, 2002). By 2004, Lund *et al.* (2004) estimated 80 000 units being installed per year, of which 46% are vertical closed loop, around 38% horizontal closed loop and 15% open loop. A total of up to 600 000 units are believed to be in operation, with strong growth in the governmental and public (e.g. schools) sectors (Bouma, 2002). Admittedly (except in the northern states), such GSHPs are often dominated by a cooling demand rather than a heating demand and are thus typically rejecting heat to the subsurface. A typical residential cooling capacity is around 10.5 kW. In order to demonstrate that GSHPs are not solely for woolly-jumpered environmentalists, Lund *et al.* (2004) note that even George W. Bush has installed a GSHP at his Texan ranch!

North of the border, in Canada, around 30 000 GSHPs were installed annually in the 1990s, with 20% of these being in the commercial/industrial/schools sectors, although Bouma (2002) reports that the market slowed significantly in the latter part of the decade.

5.4.3 Meanwhile ... back in blighty

What was going on in Britain, which had played such a crucial role in the theoretical and practical development of heat pumps, while tens of thousands of installations were being completed elsewhere in Europe and North America? A survey in 1999 (Rawlins and Sykulski, 1999) could document a total of 10 GSHP systems in the UK! In the early 2000s, the market for GSHPs has expanded tremendously in the UK: Britain is still near the bottom of European GSHP league table (Table 5.1), but it possesses one of the most rapidly growing markets: in 2001, for example, some 150 new GSHP systems were commissioned and the market was growing at over 100% per annum (Figure 5.4).

We have seen how the development of the GSHP market differs hugely from country to country, even within Europe. In Britain, GSHP technology has low penetration but high growth. In Sweden, the market is tending towards saturation and one can speculate what proportion of the Swedish sales in Figure 5.5 reflect sales to replace first-generation units.

The critical factors in uptake of GSHP technology from country to country are likely to be

- access to and cost of conventional fossil fuels;
- cost of electricity;
- level of subsidy provided by government for installation of GSHPs;
- marketing and packaged solutions offered by power utilities;
- level of awareness of GSHP technology;
- prioritisation of environmental (especially CO_2 emission) concerns at both political and individual levels.

Table 5.1 European GSHP market status in 2001 (after Bouma, 2002) and in 2005 (after EREC, 2006). It is clear that there are some potential inconsistencies in the data: they do, however, give some impression of the diversity in GSHP uptake in Europe.

Country	Annual GSHP sales (units/year) 2001	Market growth rate for GSHPs (%) 2001	New installations 2005	Installations in operation 2005	Installations in operation MW 2005
UK	150	>100	750	–	10
Czech Republic	350	25	4 000	–	200
Poland	500	5	1 465	8 000	104
Norway	650	10	–	–	–
Switzerland	2 800	6	–	–	–
Italy	–	–	13 000	–	120
Germany	3 600	20	25 486	48 662	633
Sweden	27 000	6	61 350	185 531	1 700
Europe (total)	41 000	–	–	–	–
EU-25	–	–	140 144	368 843	4 377

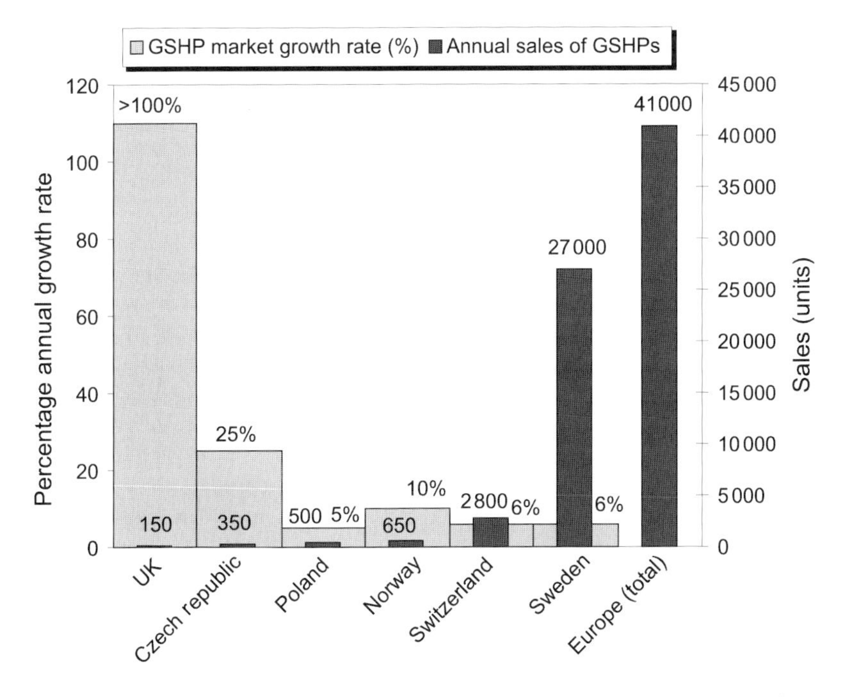

Figure 5.4 The European market status for GSHPs in selected European countries for 2001. *Based on data provided by Bouma (2002).*

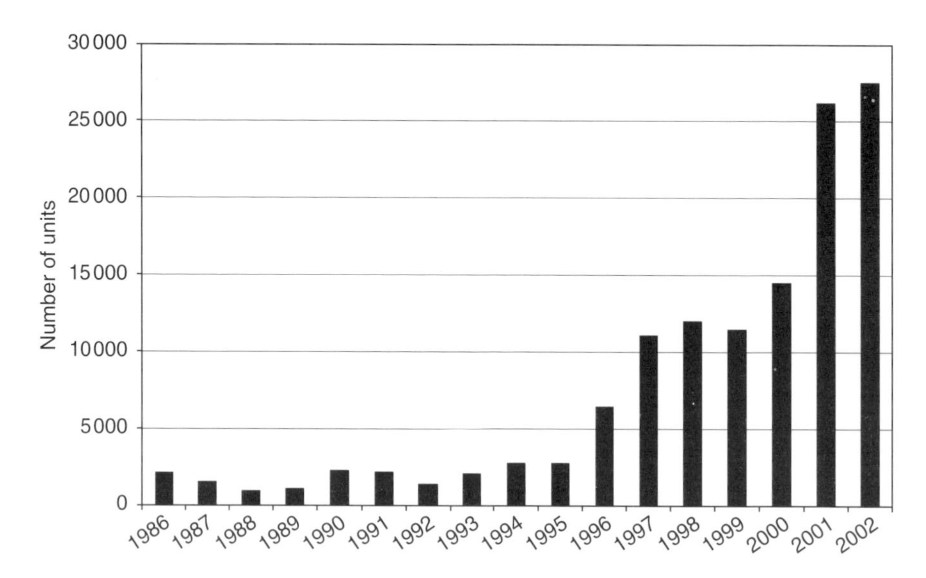

Figure 5.5 Number of annual GSHP sales in Sweden. *After Lund et al. (2004) and reproduced by kind permission of Dr John W. Lund and the GeoHeat Center, Klamath Falls, Oregon.*

5.5 The global energy budget: how significant are GSHPs?

Statistics for global use of energy are notoriously slippery things. In addition to the obvious difficulties in collating statistics from nations as diverse as China, Nigeria and Luxembourg, it can be difficult to distinguish between figures for actual energy delivery and figures for installed potential capacity. For GSHPs and other small-scale alternative energy sources, the problems become particularly acute as, in many countries, nobody is really keeping track of GSHP installations (in the UK, closed-loop systems currently need no license from the Environment Agency and may not require specific planning permission). Therefore, take the following statistics with a pinch of salt and recognize that there may be significant error margins on the figures cited.

It is claimed (Bouma, 2002) that there may be as many as 100 million heat pumps (not just GSHPs) installed worldwide, with a total annual output of 1300 TWh. (Here we face our first query – to what extent does this figure include standard air-conditioning devices?). It is further claimed that this results in a global saving in annual CO_2 emissions of 0.13 Gt, in the context of a global total CO_2 emission of 22 Gt (Bouma, 2002).[1] (Here again, we must ask: if heat pumps are being used for active cooling, in what sense are they saving energy?)

In 1999, Rawlings and Sykulski estimated the total number of GSHP systems installed worldwide as no less than 400 000, while 5 years later Lund *et al.* (2004) place the figure at 1 100 000 (600 000 of which are in the USA and 230 000 in Sweden).

[1] Henson (2006) cites a figure of 26 Gt CO_2 for 2002.

Table 5.2 Global energy status today: energy delivery as installed capacity and average usage, by source. The year for which the data apply is provided in parentheses. Data compiled from [1]Bouma (2002), [2]UNDP (2004), [3]BP (2005) and [4]Lund *et al.* (2004).

	Installed capacity (GW)	Average usage (GW)
GSHPs	6.7 GW_{th} (2002)[1]	0.7 GW_{th} (2002)[1]
	(4.8 GW_{th} in USA)[1]	
	12 GW_{th} (2004)[4]	2.3 GW_{th} (2004)[4]
	(6.3 GW_{th} in USA)[4]	
'Geothermal'	11 GW_{th} (2001)[2]	6.3 GW_{th} (2001)[2]
	8 GW_e^2	6.0 GW_e^2
Solar PV (photovoltaic)	1.1 (2001)[2]	0.1 (2001)[2]
	1.8 (2003)[3]	
Solar thermal	57 (2001)[2]	6.5 (2001)[2]
Wind	23 (2001)[2]	4.9 (2001)[2]
	48 (2004)[3]	
'Modern' biomass	250 (2001)[2]	103 (2001)[2]
Hydroelectric	715 (2001)[2]	308 (2001)[2]
		844 (2004)[3]
Nuclear		831 (2004)[3]
Natural gas		3221 (2004)3
Coal		3698 (2004)[3]
Oil		5014 (2004)[3]

Bouma (2002) estimates the worldwide total GSHP installed capacity to be ≈6.7 GW_{th}, of which the majority (4.8 GW_{th} or 500 000 units) is in the USA (these figures in GW represent the installed capacity as thermal output). Lund *et al.* (2004) cite a figure of 12 GW_{th} installed capacity worldwide, with 6.3 GW_{th} in the USA and 2.3 GW_{th} in Sweden. Estimates of the actual thermal energy provided by GSHPs range from 23 270 TJ $year^{-1}$ (0.7 GW_{th} average; Bouma, 2002) to 72 000 TJ $year^{-1}$ (2.3 GW_{th} average; Lund *et al.*, 2004).

Let us look at more mainstream statistics (Table 5.2), culled mainly from two sources, the UNDP (2004) World Energy Assessment and British Petroleum's Statistical Review of World Energy (BP, 2005). UNDP claims that there are 11 GW_{th} of geothermal heat energy delivery capacity installed worldwide. This term includes (presumably) heat pumps, and also thermal energy delivered directly by high temperature, truly geothermal energy schemes, such as those in Iceland. In the context of this, Bouma's 6.7 GW_{th} GSHP capacity seems reasonable. Note the difference between the delivery of heat (thermal energy = GW_{th}) from geothermal sources and delivery of electrical energy (GW_e) from turbines in power stations powered by high-temperature geothermal sources. UNDP (2004) claims a further 8 GW_e geothermal electrical generating capacity installed worldwide.

Table 5.2 compares these statistics with installed capacities and actual usages of other alternative and conventional energy sources. We will see that, in term of installed capacity, GSHPs and geothermal energy are considerably more significant than solar photovoltaic (PV) power generation, but appear to lag a short way behind solar thermal and wind. We should note, however, that the actual usage of wind and solar thermal energy (21% and 11% of installed capacity) illustrates the dependence of both these on climatic and sidereal considerations. The actual usage of ground source heat (19% of installed capacity, according to Lund *et al.*, 2004) does not reflect external limitations (the earth's temperature is stable and predictable), but rather reflects annual and daily patterns of demand.

If we consider non-fossil-fuel energy supply, however, we see that wind, ground source heat and solar energy account for only less than 1% of the energy generated by the mainstream low-carbon sources of energy (nuclear and hydroelectric, weighing in at several hundred GW generated on average). These in turn account for a small proportion of the energy generated by the three fossil fuels: gas, oil and coal, which provide energy at a combined rate of over 10 TW on average throughout a year.

5.6 Ground source heat: a competitor in energy markets?

One view of ground source heat (which Table 5.2 appears to promulgate) is that it is an alternative, renewable source of heat energy, competing in the energy market with other 'green' energy sources such as wind power and solar thermal systems. In some respects, this point of view is correct: a householder faced with a choice of 'green' energy solutions may be forced on economic grounds to make a concrete choice between solar thermal cells mounted on the roof and a ground source heat borehole drilled in the garden.

In other respects, this point of view is an oversimplification. After all, GSHPs do not work by themselves. They require an electricity supply and must be seen as a 'complementary technology' to electricity, enabling electric power to be used more efficiently. That electric power may, in itself, be generated by low-carbon or 'green' technologies, such as hydroelectric or nuclear. In a sense, one can thus regard GSHPs as an integral part of a national strategy for electricity generation by nuclear power or hydroelectric turbines, as has been the case in Sweden and Norway, respectively.

One could even conceptually envisage a GSHP being powered by an array of wind turbines or photoelectric cells although, in practical situations, the unreliability and limited capacity of these sources means that the electricity generated by them would not be used directly. It would typically be sold to a utility company, and repurchased to power the heat pump. Furthermore, we will see in the case studies of Chapter 6, that some ground source heat schemes operate in parallel with other technologies, such as solar thermal cells and conventional heating, to provide total solutions to domestic hot water and space heating. I would encourage readers not to just regard ground source heat as an alternative, 'green' technology competing with other conventional and renewable energy sources: rather, it is a technology that complements them and that can even be attractive to the most die-hard petrol-head!

6

Options and Applications for Ground Source Heat Pumps

> A statement made by a university engineering lecturer that the description [of a heat pump] given by Lord Kelvin is impossible to accept, or by another lecturer that [it] offends thermodynamic law, is a less pleasant feature [of life].
>
> John A. Sumner (1976)

This chapter considers the various ways in which we can use ground source heat pumps (GSHPs), and includes a short discussion of design parameters and cost. After examining briefly how we can assess what size of heat pump scheme we will require, we will continue by looking at the two fundamental ground source heat options: the *open-loop* and *closed-loop* systems. We will then go on to consider whether, in addition to space heating, GSHPs can also provide domestic hot water (DHW). Finally, we will examine the ways in which we can deliver flexible solutions to buildings that may have simultaneous heating and cooling needs.

6.1 How much heat do I need?

A building during winter can be considered as a warm box placed in the middle of a cold environment. If the temperature outside is colder than the temperature inside, then the building will lose heat. The greater the temperature difference ($\Delta\theta$), the faster the building will lose heat. The better the standard of insulation of the building, the lower its thermal conductance (U) and the lower the rate of heat loss. As a very coarse simplification, we can say that the rate of heat loss from the building (Q in W_{th}) is given by

$$Q \approx \Delta\theta \times U \tag{6.1}$$

6.1.1 Degree days

If we consider a period of time t (say, a whole month), we can estimate the total conductive heat loss (which will give us a first estimate of the heat demand required to maintain a comfortable indoor temperature) as follows:

$$\text{Conductive heat loss} \approx U \int_{0}^{t} \Delta\theta \, dt \tag{6.2}$$

Of course, temperature varies with time, so to calculate this value we need to consider the entire period of time that the outside temperature falls beneath a critical 'baseline' value (i.e. the comfortable indoor living temperature). We do this (Figure 6.1) by calculating the total area between the real temperature curve and the baseline value for the period in question. The value is an expression of both the severity and duration of cold weather and is expressed in so-called degree days (Carbon Trust, 2006b).

In Britain, for example, the 'baseline' temperature is often taken as 15.5°C, although this will depend on the use of the building. This is because many buildings generate enough internal heat (from electronic equipment and respiring human bodies) to approximately balance the small heat loss when the outdoor temperature is 15.5°C and still maintain a comfortable living environment. If we experienced 48 continuous hours of an outside temperature of +1°C, this would be equivalent to 2 days × 14.5°C = 29 degree days. On average during the past 20 years, for example, the Thames Valley area has experienced 114 degree days in the month of October. In the harsher climate of north-east Scotland, 195 degree days is the norm in October (Carbon Trust, 2006a).

6.1.2 Thermal resistance

If we continue to accept our rather simplified relationship that

$$\text{Conductive heat loss} \approx U \int_{0}^{t} \Delta\theta \, dt = U \times (\text{Degree days}) \tag{6.3}$$

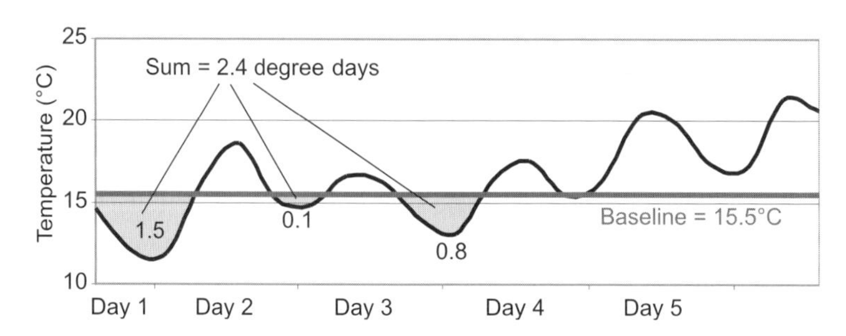

Figure 6.1 Calculation of degree days for a 5-day period. The area of the shaded region gives the total number of degree days of heating demand in the 5 days.

then we now need to approximate the thermal conductance of our house. This is normally done by considering our house as consisting of a number of thermal conductances – walls, roof, doors, windows, floor – coupled in parallel. For each of these elements, a 'U-value' can be calculated (Box 6.1).

BOX 6.1 Thermal conductance, thermal resistance and U-values.

Thermal conductivity (λ) is an intrinsic property of a material (albeit one that varies a little with temperature). Its inverse is thermal resistivity $= 1/\lambda$ in m K W^{-1}.

We can also talk about an extrinsic property called thermal conductance. It is extrinsic as it depends on the dimensions (specifically, the thickness) of the body in question. Let us consider an expanded polystyrene (EPS) sheet of thickness 10 cm. The thermal conductivity of EPS is around 0.035 W m^{-1} K^{-1}. The thermal conductance (Λ) of the sheet is given by

$$\Lambda = \frac{\lambda}{L} = \frac{0.035\ \mathrm{W\,m^{-1}\,K^{-1}}}{0.1\ \mathrm{m}} = 0.35\ \mathrm{W\,m^{-2}\,K^{-1}}$$

In other words, our sheet of EPS will conduct 0.35 Watts of heat per m^2 area for every degree Kelvin temperature difference across it.

$$\text{Thermal resistance } (R) = \frac{1}{\Lambda} = \frac{L}{\lambda} = 2.86\ \mathrm{m^2\,K\,W^{-1}} \text{ for our EPS sheet.}$$

Crucially, thermal resistance is an additive property. Thus, if we have a wall comprising a course of 125 mm bricks, an EPS-filled cavity of 75 mm and another brick course 125 mm thick, we can calculate the total thermal resistance (the R-value) of the wall (assuming the thermal conductivity of our bricks is 0.4 W m^{-1} K^{-1}):

$$R_{\mathrm{total}} = R_{\mathrm{brick}} + R_{\mathrm{EPS}} + R_{\mathrm{brick}} = 2 \times \frac{0.125}{0.4} + \frac{0.075}{0.035} = 2.8\ \mathrm{m^2\,K\,W^{-1}}$$

Thus, the total thermal conductance for our composite wall is 0.36 W m^{-2} K^{-1}. If it is 20°C indoors and –5°C outside, we can thus estimate that our house is losing 9 W of heat through every square metre of wall (neglecting convection effects at the skin of the wall).

The U-values cited in building regulations for various constructions of walls, windows and roofs are based on exactly this concept of thermal conductance as the inverse of the sum of the thermal resistances (R) of the various layers in a composite structure. U-values are sometimes referred to as *composite thermal conductance* or thermal transmittance. They may also factor in components of heat loss due to convection (skin effects) and radiation, as well as simply conduction, and are referred to as *overall heat transfer coefficients*. The British Standard Assessment Procedure (SAP, 2005) currently sets reference U-values for dwelling walls at 0.35 W m^{-2} K^{-1} and for roofs at 0.16 W m^{-2} K^{-1}.

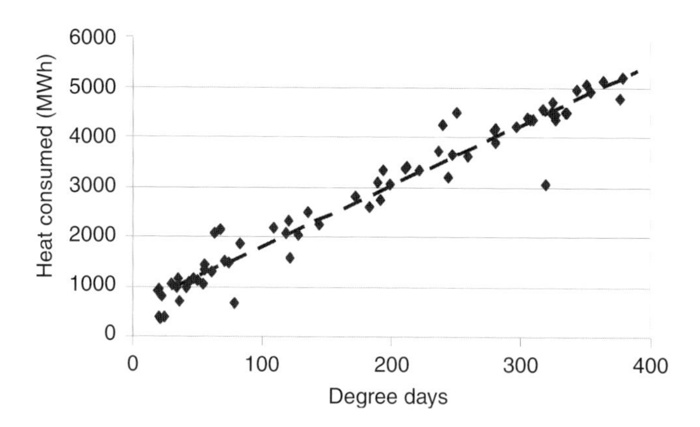

Figure 6.2 Thermal response plot for relatively modern buildings in Norwich, England. The low scatter of points around a linear trend line indicates a well-managed building – that is, heat is released in response to demand (*modified after Tovey, 2006*). *Reproduced by kind permission of Keith Tovey and the University of East Anglia.*

This, if we know the outdoor temperature pattern for a given period (in degree days) and the thermal performance of our house, we can calculate how much heat we will require to keep our house warm. (Actually, reality is a bit more complex: our house loses heat not only by conduction, but also by convection – i.e. draughts – through openings such as chimneys, door jambs and vents, so we need to quantify these losses as well!). We can plot the actual amount of heat used against the theoretical heat demand to produce a thermal response plot, as in Figure 6.2. If the points are widely scattered around a trend, it means that our building is not being responsively managed (too much or too little heat is often supplied). If the points fit well to a linear trend (as in Figure 6.2), the building is being well managed. Ideally, the plot should pass through the origin: however, if the intercept on the 'heat supplied' axis is positive (as in Figure 6.2), it means we require heat even when the outside air temperature is at the baseline condition. This may mean that we have overestimated the building's internal heat generation (respiration, electrical equipment) and we should maybe select another baseline temperature. If the intercept on the 'heat supplied' axis is negative, it likely means we have underestimated the building's internal heat generation (Carbon Trust, 2006b).

6.1.3 *What does all this mean for us?*

For a given climatic zone and a given standard of building insulation, we would thus expect that the amount of heating required by a house would be roughly related to the external surface area (walls, windows and roof) of the building. We would thus also intuitively expect some relationship with the size of the building. Such a relationship exists but it is not as simple as one might expect because, broadly speaking, heat losses increase with the surface area of the building, while internal heat generation (by respiration and electrical equipment) roughly increases with the volume of the building. Thus, the bigger the building, the lower the surface area-to-volume ratio and the smaller the need for additional sources of heating. Indeed, in very large buildings there may even be a net cooling need for much of the year – that is, a need to get rid of waste heat, rather than to provide heat.

However, for modest-sized buildings we can roughly relate peak heating demand to the size of building. Sumner (1976b) states that, for a reasonably well-insulated house, we can base our estimate on a calculation of the total living volume of the house (floor area × height for each room). For a comfortable living temperature of 20°C and an outdoor temperature of 0°C (which is not unreasonable for Britain), he estimated a demand of 36 W m^{-3}. Sumner reckoned that this would be adequate to cover the heat demand for 95% of the heating season. He furthermore estimated that the average demand throughout the heating season would be 65% of this figure. If we assume that an average room has a height of 2.4 m, we can reduce Sumner's (1976b) relationship to an effective peak heat demand of 87 W m^{-2}. It is perhaps an indication of how far the technology of house construction and insulation has progressed since Sumner was writing, that engineers in Britain today tend to target a figure of around 50 W m^{-2} floor space for new buildings. All of this means that a typical new small-to-medium sized domestic house will usually have an effective peak heating demand of 6–10 kW.

6.1.4 *The heat demand of a real house*

Figure 6.3 shows how we can learn something about our heating requirements from a utilities bill. The diagram shows the gas used by a small (≈90 m^2 floor area, three storey) semi-detached Victorian property in the East Midlands of the UK. A non-condensing combi gas boiler provides central heating via wall-mounted radiators in the heating season and DHW all year round.

During the peak quarter of the heating season, we can see that between 6000 and 7000 kWh of gas are consumed for space heating, equating to an average of around 3 kW. If we bear in mind Sumner's 'rule of thumb' that average heating demand in the heating season may be around 65% of peak demand, we can estimate that peak demand could be as high as 5 kW. For floor area of 90 m^2, this equates to an estimated peak demand of 56 W m^{-2}. Not bad for a poorly insulated Victorian house – but we should remember that patterns of heat usage also come into play. In fact, the building is often unoccupied for much of the day and the occupants typically actively utilise the heating for around 6–8 hr day^{-1}. This may lead us to revise our estimate of the peak heat demand to somewhere nearer 6–9 kW.

If we average the gas demand throughout the year, we find an average consumption of 13 000 kWh $year^{-1}$ or 1.5 kW. Thus, the average heat demand over the course of a year in such a building may be less than 25–30% of the peak demand.

Of course, the estimates above are rather simplified – we have not taken into account the efficiency of the gas boiler in converting gas to heat, and we have rashly assumed that gas consumed for DHW will be the same throughout the year and can be estimated

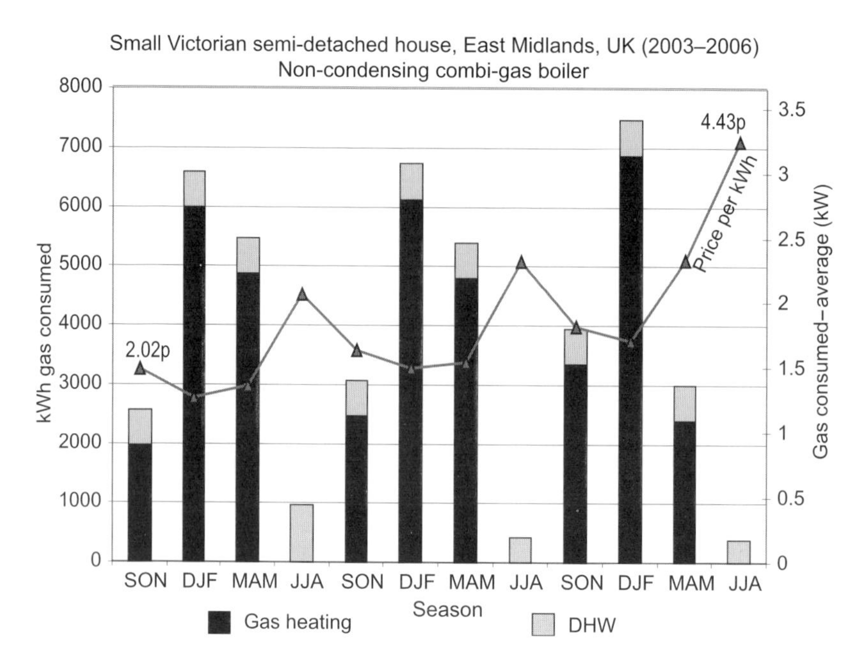

Figure 6.3 The heating supplied by a gas combi boiler to a poorly insulated three storey, semi-detached house in the East Midlands of England, over the period 2003–2006. The line shows the trend in gas prices in pence per kWh during that period (1 p = £0.01).

from summertime gas usage (it will, in fact, be larger during winter) – but they do hopefully render the heat budget of your home a little less obscure.

Finally, note the price of gas throughout the 3-year period of Figure 6.3. It is most expensive when purchased in small quantities (summer), at least in the UK! Furthermore, the peak yearly price has risen from 2.8 p kWh^{-1} in summer 2004 to 4.4 p kWh^{-1} in summer 2006, reflecting a drastic trend in energy prices throughout the UK. Given this, you might think that the house's occupants would be considering buying a GSHP to cut the costs of their energy bills. In fact, this may be a case where a GSHP is not an ideal solution. Technically, of course, it could be done, but

i. It may mean replacing the old high-temperature wall-mounted central heating system with a low-temperature one, or even underfloor heating. This is another significant capital investment on top of a borehole and heat pump (the garden is not big enough for a trenched system).
ii. The building likely has a low thermal mass – it heats up relatively rapidly and cools down rapidly.
iii. The building is rather poorly insulated (and capital may be better spent on improved insulation, rather than on a GSHP).
iv. The building is only occupied for limited periods during the day. The more a heat pump is used, the quicker savings accumulate and the sooner the capital expenditure is down-paid.

A GSHP runs most efficiently at low temperatures for long periods. It is thus an excellent solution in houses with a suitable low-temperature heating system and good standard of insulation, with a high thermal mass that can retain, for example, heat released at night. In the case of Figure 6.3, with the current pattern of occupancy, a combi gas boiler, rather than a GSHP, may be the most sensible solution.

6.2 Sizing a GSHP

We have seen in Section 6.1 how we can examine climate statistics and estimate our peak heating requirement on the coldest days of a typical year. If our peak heat demand is 8 kW, should we simply rush out and buy an 8 kW heat pump?

If we consider the distribution of winter temperature (Figure 6.4), we typically find that the coldest temperatures (and the peak heating demand) occur on a very few days each winter. For example, in southern UK, the temperature in January may plunge to –5°C on a couple of days, but otherwise have a daily minimum somewhere close to 0°C. Given that the capital cost of a GSHP system is related to its capacity, it may make sense to choose a heat pump that is not dimensioned to supply the maximum heat demand (on the very few nights when the temperature is –5°C), but one capable of satisfying the demand on the majority of January days when the temperature does not fall much below 0°C. On the very coldest days, we could have a supplementary source of heat – a coal or wood fire or paraffin stove – to top up the heat provided by the 'undersized' GSHP.

In fact, Rosén *et al.* (2001) demonstrate that a GSHP with a rated output of around 60% of the calculated peak heat demand of a Swedish house can supply over 90% of the heat energy required by the house over the course of a heating season (Figure 6.5), assuming that the heat demand of the house is directly related to the outdoor temperature. We can express this relationship by defining two parameters:

$$\text{Effect Coverage (\%)} = \frac{\text{Rated output of heat pump (kW)}}{\text{Peak heat demand of building (kW)}} \times 100\% \qquad (6.4)$$

$$\text{Energy Coverage (\%)} = \frac{\text{Total heat supplied by GSHP (kWh)}}{\text{Total heat required by building over heating season (kWh)}} \times 100\% \qquad (6.5)$$

Skarphagen (2006) notes that it has formerly been common practice in Scandinavia to select GSHPs rated at around 60% of peak demand, partly because this supplies the majority of property's energy needs and partly because it means that the heat pump is usually running against a demand (i.e. protracted running time).

A heat pump that is oversized or dimensioned at 100% of the peak demand, and which has little 'thermal mass' or thermal storage to act as a buffer, runs the risk of cutting repeatedly in and out, resulting in excessive wear on the compressor. Furthermore, a GSHP operating against a high thermal mass, while energetically efficient, is not especially responsive to short-term temperature fluctuations. The combination

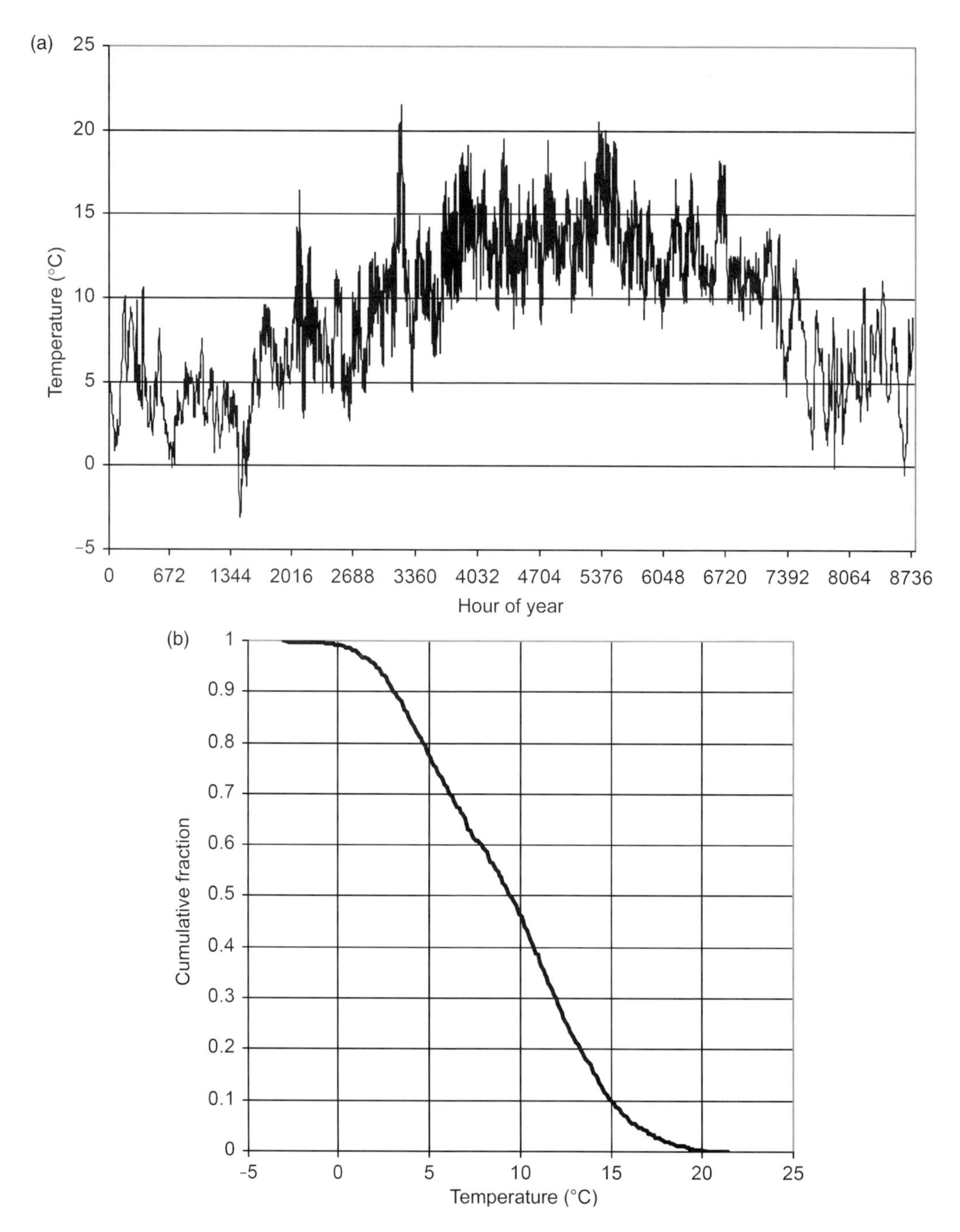

Figure 6.4 (a) Air temperature variations throughout a year (6-hr average). It will be seen that the peak temperatures (and hence peak heating and cooling demands) occur for a tiny proportion of the time. (b) The cumulative frequency diagram shows that the 6-hr average air temperature is $<0°C$ for around 1% of the year but $>3°C$ for 90% of the year. *Figure derived from generic temperature data for mid-Britain, sourced from NCEP Reanalysis data provided by the NOAA/OAR/ESRL PSD, Boulder, Colorado, USA, from their Web site at http://www.cru.uea.ac.uk/cru/data/ncep/.*

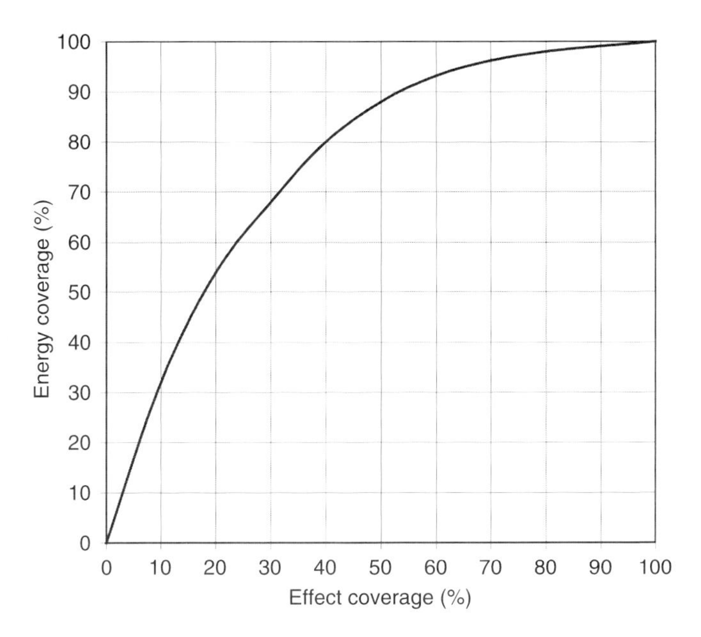

Figure 6.5 The relationship between Energy Coverage and Effect Coverage (see text) for a GSHP supplying a typical Swedish residence (prepared by G. Hellström for the publication by Rosén *et al.*, 2001). *Reproduced by kind permission of Göran Hellström.*

of a base load provided by a GSHP and heat demand peaks being satisfied by a more conventional source does improve the degree of system responsiveness that can be achieved.

The typical running time (i.e. equivalent load hours) of a GSHP will also depend on the duration of the heating season and thus on the climate. In Sweden, running times of 3200–4000 hr year^{-1} are common for domestic heat pumps, dropping to 2200 hr year^{-1} in Switzerland and 1800 hr year^{-1} in Central Europe (Rosén *et al.*, 2001).

Skarphagen (2006) notes, however, that Effect Coverage has been creeping up in recent years and that, now, it is common practice in Norway to select a heat pump rated at 80% of the peak demand. The reasons for this are as follows:

- There is relatively little difference in capital cost (e.g. of an 8 kW heat pump relative to a 6 kW heat pump) – see Figure 4.9 (p. 78).
- Control of heat pump output has become increasingly sophisticated in recent years. A heat pump does not now necessarily need to run at a single 'speed', but the output can very in response to demand.
- A higher Effect Coverage is increasingly economically attractive due to the increasingly sophisticated methods that utilities companies use to monitor (and charge) electricity consumption.

For larger buildings in Norway, the Effect Coverage selected may be as low as 40%: this results in a large running time (effective load hours) for the heat pump and an Energy Coverage as high as 80% (H. Skarphagen, *pers. comm.*).

Note that much of the above refers to dimensioning for domestic houses in Scandinavia. Although the same general principles may apply further south in Europe, the exact figures may not. Furthermore, different countries have different cultures for providing heat: most Scandinavian homes will have more than one heat source and it will be very common for a house to have either a paraffin stove or wood fire to provide supplementary heat on the coldest days. This may not be the case in Britain (and indeed, some UK heat pump manufacturers are recommending selecting a heat pump on the basis of peak heating demand). Other dimensioning practices may be applicable to larger commercial or public buildings depending on their specific needs.

At the other extreme, it is quite common in much of the central and southern USA to use GSHPs to provide domestic air-conditioning (i.e. active chilling) as well as a small amount of winter heating (domestic air-conditioning is uncommon in most of central and northern Europe). In the USA, therefore, it is common practice to select a heat pump on the basis of summertime cooling load rather than heat demand.

6.2.1 Thermal storage

It is usually desirable to have some form of thermal storage within a GSHP system, for exactly the same reasons that one incorporates water storage (e.g. a header tank or a pressure tank) into a groundwater supply scheme. The thermal storage provides a buffer in the system, such that a sudden, short-term demand for heat can be supplied by the thermal store. The heat pump can then operate for a protracted period to replenish the thermal store when its temperature falls below a certain level. This avoids the heat pump's compressor repeatedly cutting in and out at short intervals, leading to wear on the compressor. (In the same way, in a groundwater supply scheme, the well's pump will be activated by the water level or pressure in the storage tank falling below a certain level.)

The thermal storage may simply be a large tank of warm water on the central heating circuit (sometimes called an 'accumulator tank'). Alternatively, it may be the thermal mass of the building itself, which will retain heat and release it gradually, damping out externally imposed fluctuations in demand. The thermal mass or storage of the building may be enhanced by, for example, laying the underfloor heating pipes in a thick concrete bed. This acts like a storage heater, enabling the GSHP to be run on a cheap electricity tariff at night to heat up the floor slab, which will then release heat during the working day, as at the Hebburn Eco-Centre (Box 4.2, p. 63).

A similar principle can be used in cooling schemes, where we may wish to use the heat pump on a night-time tariff and store up 'coolth' that can be used for air conditioning during the following day (DoE, 2000). In this case, we can store 'coolth' simply by freezing water to form chunks of ice or an ice 'slush' (as in the Southampton scheme – Box 2.3, p. 19). This is especially effective as we are using the latent heat released during the phase change from water to ice to store 'coolth'. More sophisticated 'phase-change' storage media are now under development – alternative materials include salt hydrides and organic compounds such as esters, fatty acids or hydrocarbons.

We have now gained some overview of how we might tailor a heat pump to the needs of our building. Let us now introduce the two main means by which we can convey heat from the ground to our heat pump: open-loop and closed-loop systems. We will examine these in considerably more detail in Chapters 7–10.

6.3 Types of ground source heat system: open-loop systems

Open-loop systems are those where we physically abstract water from a source: this can be a river, the sea or a lake. In the context of thermogeology, however, we are primarily concerned with groundwater abstracted (usually pumped) from springs, dug wells, drilled boreholes or flooded mines. Heat is extracted from this flux of pumped water or, in cooling mode, dumped into it. In cooling mode, we do not necessarily need to use a heat pump: cool groundwater at 11°C can be circulated through a network of heat exchange elements in a building to provide 'free' or 'passive' cooling (Figure 4.1, p. 58). Often, however, we will use a heat pump to provide space heating or active cooling. The amount of heat (G) that we can extract from a flow of water is given by Equations 4.7 and 6.6:

$$G = Z \cdot \Delta\theta \cdot S_{\mathrm{VCwat}} \tag{6.6}$$

where Z = flux of water in L s^{-1}; $\Delta\theta$ = the temperature drop (or rise, in cooling mode) of the water flux, in K; S_{VCwat} = specific heat capacity of water 4180 J L^{-1} K^{-1}.

If we use a heat pump with a coefficient of performance COP_H to extract heat from this water, the total heat load (H) delivered for space heating can be derived by combining Equations 4.7 and 4.8:

$$H \approx G + \frac{H}{\mathrm{COP}_H} = \frac{Z \cdot \Delta\theta \cdot S_{\mathrm{VCwat}}}{1 - (1/\mathrm{COP}_H)} \tag{6.7}$$

If we are dumping heat from passive cooling (Figure 4.1, p. 58) to our water flow, then the cooling load (C) that can be effected is simply given by

$$C = Z \cdot \Delta\theta \cdot S_{\mathrm{VCwat}} \tag{6.8}$$

If we are using a heat pump, with efficiency COP_C, to perform active cooling, the cooling effect is given by

$$C \approx \frac{Z \cdot \Delta\theta \cdot S_{\mathrm{VCwat}}}{1 + (1/\mathrm{COP}_C)} \tag{6.9}$$

Let us consider a GSHP that is able to achieve a $\mathrm{COP}_H = 4$ for heating and let us assume that it transfers heat, resulting in a temperature decrease $\Delta\theta = 5$°C in the groundwater, which is a typical value for many heat pump systems (although Sumner's Norwich heat pump seems to have operated with a $\Delta\theta$ as low as 1°C, based on the River Wensum's near freezing temperature in winter – Box 5.2, p. 92). If we pump groundwater at 11°C from a drilled well at a constant rate of 1 L s^{-1} through our heat pump, it could effect

$$\text{a heating effect } H \approx \frac{1 \times 5 \times 4180}{3/4}\ \text{J s}^{-1} = 28\ \text{kW} \tag{6.10}$$

of which 21 kW comes from the groundwater and 7 kW from the heat pump's compressor. In cooling mode, let us assume a $COP_C = 3$. If the heat pump is ideally reversible, it will extract a cooling load of 21 kW from the building, add 7 kW of heat from the compressor and reject 28 kW to groundwater. This will result in a temperature increase in the groundwater of 6.7°C. Equation 6.9 applies

$$\text{Cooling effect } C \approx \frac{1 \times 6.7 \times 4180}{4/3}\ \text{J s}^{-1} = 21\ \text{kW} \tag{6.11}$$

Here we see that, for a given flow of water and for a given change in temperature, the potential active cooling effect via a heat pump is typically lower than the potential heating effect, all other things being equal. This is because we are rejecting the electrical energy powering the heat pump's compressor in cooling mode, whereas we are using it in heating mode. In practice, however, in cooling mode the $\Delta\theta$ may well be higher than in heating mode. Via the use of a relatively high-temperature building loop and the judicious use of heat exchangers, we may in fact be able to achieve a $\Delta\theta$ significantly higher than 6–7°C and hence a somewhat greater cooling effect than Equation 6.11 might indicate. Nevertheless, it is fair to say that active cooling using a heat pump is less efficient than heating, as the compressor's energy turns up as waste heat in the former mode and usable heat in the latter. Note that the above calculations neglect the fact that COP refers to the instantaneous performance of the heat pump under given temperature conditions – in reality, it will vary somewhat. The equations also neglect any energy expenditure in submersible pumps in our groundwater well, or circulation pumps in the building. Strictly speaking, if we wished to include these factors and to average over an entire heating or cooling season, we should use a system seasonal performance factor rather than COP (Equation 4.9).

6.3.1 The abstraction well

We cannot construct a groundwater-based open-loop heating or cooling system wherever we want. The fundamental criterion is that we need an aquifer – a permeable body of rock or sediment in the subsurface that has adequate transmissivity and storage properties to yield a usable and constant flux of groundwater. We thus need specialist knowledge in order to ascertain the existence and properties of an aquifer in the subsurface. We will touch on this further in Chapter 7, but it is not the purpose of this book to delve too deeply into the science of hydrogeology. To obtain good advice on this topic, you can usually contact your national Geological Survey, Ministry of Water or Environment Agency. You are also likely to require advice, however, from a professional hydrogeological consultant.

Having ascertained the existence of an aquifer and assessed its potential to yield water, you will need ask your tame hydrogeologist or groundwater engineer to assist in designing abstraction (and, if necessary, re-injection) wells. The fundamental types of information required to design a well are as follows:

- *The design depth of well.* This will depend on the depth of the aquifer stratum, the groundwater level in the aquifer and, to some extent, the hydraulic conductivity of the aquifer. Wells may be a few metres deep in shallow alluvial deposits, or may extend to a depth of several hundred metres.
- *The design diameter of the well.* This will ultimately depend on the yield of the well, which will in turn affect the diameter of the required pump (which must fit comfortably within the well).
- *The design yield of the well (see above).* This will be constrained by the hydraulic properties of the aquifer and by the desired heating/cooling load.
- *Aquifer lithology.* This will govern the type of well required and hence its cost.

It is, of course, possible to dig a well by hand, but nowadays, the job is usually done by a drilling rig to produce a narrow diameter 'drilled well' or 'borehole'. In hard, lithified aquifers, such as granite, limestone or chalk, it may be adequate to drill an 'open hole' well. Here (Figure 6.6), the well is installed with a relatively short length of plain casing (usually, a steel tube) at the top of the well to support any loose or crumbly strata near the surface and to shut out any poor-quality surficial water and prevent it from entering and contaminating the aquifer. The casing should normally be grouted in with a bentonite–cement mix to provide a good seal in the annulus (again to prevent surface water entering the aquifer). The remainder of the well is drilled unlined or 'open hole'. Groundwater enters via natural cracks, fractures and fissures in the borehole wall (see Figure 7.1, p. 150).

In loose or poorly cemented lithologies, such as sands, gravels and many sandstones, we cannot leave our well unlined – the sediments would simply crumble into it. We need a means of supporting the wall of the well, shutting out the sediments, but allowing the groundwater to seep in. The solution here is to use a well screen (Figure 6.6). In its most primitive form, this may simply be a length of steel casing with slots cut into it. In its most developed form, it will comprise stainless steel wire wrapped around a framework of struts (Figure 6.7). This wire-wrapped screen can have an open area as high as 30–50%. Clearly, the size of the slots in our screen will be related to the size of the grains in our sediment or rock: the larger the grains, the bigger our slots can be. In fine grained or very uniform sediments, we may choose to install a gravel pack (or filter pack) outside the well screen. This acts as a filter medium, keeping fine particles out of the well. It allows us to have a larger slot size in our well screen than would otherwise have been the case, and thus to improve the hydraulic efficiency of the well.

Well design and drilling is a specialist business. Obtain advice from a consultant hydrogeologist or groundwater engineer and use a driller with experience of drilling

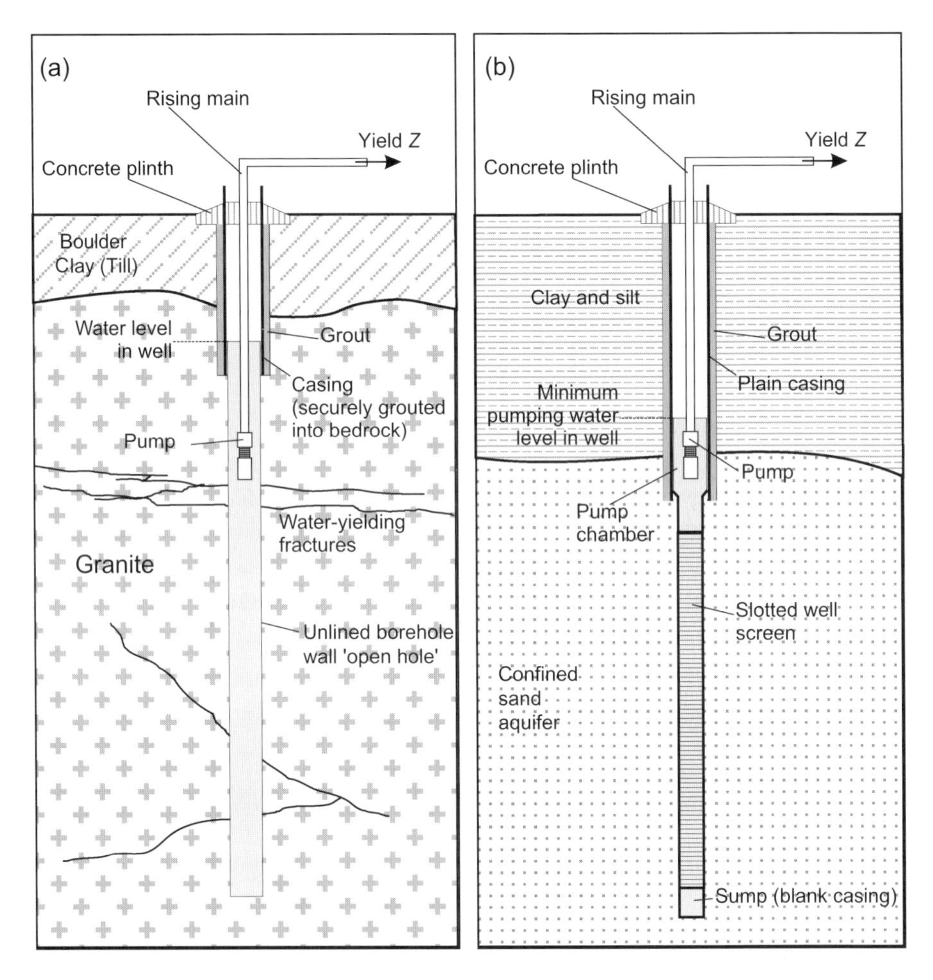

Figure 6.6 A schematic diagram of (a) a well that has been drilled 'open hole' in a hard lithology such as limestone or granite and (b) a well that has been cased and screened in a poorly competent rock or sediment.

wells for water supply. It is possible to make serious and expensive mistakes (of which Figure 13.1, p. 276, shows a small variety), which can incur the wrath of regulatory or environmental agencies and which will be discussed further in Chapter 13.

In most countries, the construction of a well for groundwater supply will normally require a permit or license from the relevant regulatory authority. This authority may specify a number of conditions for granting the license:

- an environmental impact assessment or water features survey;
- a pumping test to demonstrate that the required volumes of water can be obtained without unacceptable environmental impact and without detriment to other aquifer users;
- evidence that the abstracted water will be utilised appropriately or efficiently.

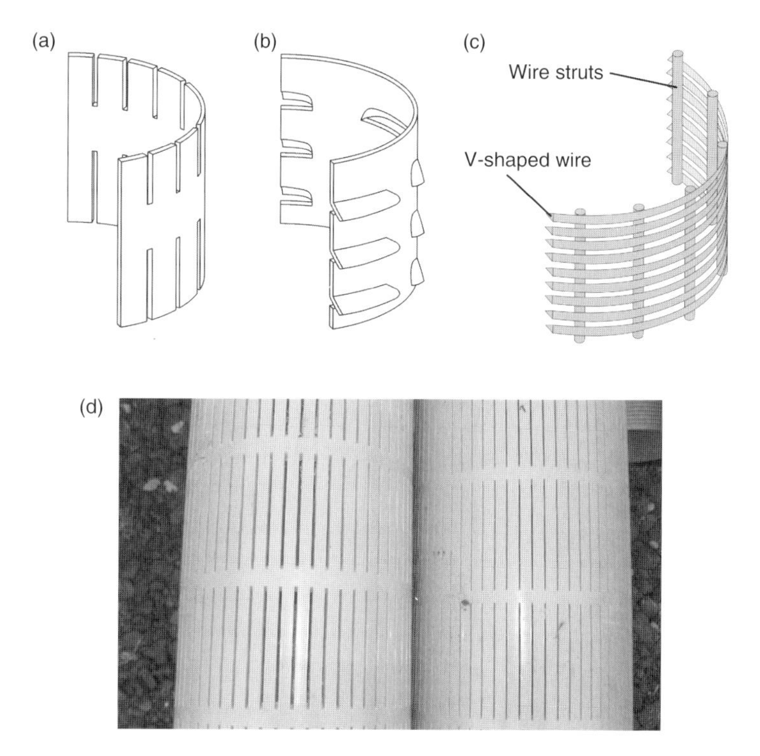

Figure 6.7 Types of well screen. (a) and (d) Vertically slotted screen; (b) louvred screen; (c) continuous wire-wrapped screen. *(a), (b) and (d) After Misstear et al. (2006) and reproduced by kind permission of John Wiley and Sons Ltd.*

The regulatory authority will usually make a charge for administering the license and may levy an annual charge for the water abstracted.

6.3.2 Hydrochemical compatibility and prophylactic heat exchangers

The subsurface geological environment is chemically very different from the atmosphere. It is basic and reducing, whereas our atmosphere is acidic and oxidising. As groundwater is brought to the surface it may experience

- degassing of CO_2, which can increase the water's pH and lead to precipitation of carbonate minerals such as calcite ($CaCO_3$);
- exposure to, and dissolution of oxygen. This may lead to oxidation of dissolved metals such as manganese (Mn^{II}) and ferrous iron (Fe^{II}), resulting in the formation of poorly soluble precipitates of manganese and ferric oxides or oxyhydroxides.

If we circulate our pumped groundwater directly through a heat pump, we are taking a number of risks:

- that any particles in the water may clog or abrade the heat pump's pipework;
- that minerals such as calcite or iron oxyhydroxide may precipitate within the evaporator of the heat pump;

- that, if the groundwater is saline enough, reducing enough or contains sufficient dissolved gases (CO_2, H_2S), it may promote corrosion. Note that the heat pump at the Hebburn Eco-Centre (Box 4.2, p. 63) has a marine-grade corrosion-resistant evaporator. The Centre has, however, experienced corrosion of associated pipework due to the saline nature of the pumped groundwater;
- that the groundwater circulation may promote the formation of biofilms: slimes of non-pathogenic bacteria, such as *Gallionella*, that are commonly found in the geological environment. These biofilms can clog up well screens, pipes or heat exchange elements in the heat pump.

In order to avoid our groundwater causing such problems in our new, expensive heat pump array or in heat exchange elements within the building, we may choose not to allow the groundwater to enter the heat pump at all. We may choose to place a 'prophylactic' heat exchanger (see Box 6.2 and Figure 6.8) between the groundwater flow and a separate loop of circulating carrier fluid. This carrier fluid absorbs heat from the groundwater via the heat exchanger and carries it to the heat pump (Figure 6.9). A modern heat exchanger can be highly efficient, such that minimal heat loss occurs to the system as a result of using it. Of course, there is still a risk that the prophylactic heat exchanger may become fouled or corroded. It is usually cheaper, however, to temporarily decommission (or replace) such a heat exchanger than a heat pump. The risk of particulate clogging can be reduced by installing removable filters prior to the heat exchanger. The risk of chemical or biological fouling of the heat exchanger can be reduced by

i. Maintaining a high pressure within the groundwater circuit to prevent degassing of CO_2 within the exchanger.
ii. Preventing contact between the groundwater and oxygen in the atmosphere (i.e. closed systems).
iii. Addition of small amounts of biocidal chemicals or reducing chemicals (e.g. sodium thiosulphate; Dudeney *et al.*, 2002) to prevent the formation of biofilms and the oxidation of ferrous iron, respectively.
iv. Regular maintenance. This might involve flushing of the exchanger with acid or proprietary detergents or reagents to remove build up of calcite or manganese/iron oxyhydroxide deposits. If a system is likely to be high maintenance, it may also be wise to select a heat exchanger that can be taken apart for cleaning.

Let us consider the active cooling system depicted on the right-hand side of Figure 6.9. From Box 6.2, we can see that the heat transfer rate (Q) across the prophylactic heat exchanger is given by

$$Q = (\theta_{r2} - \theta_{s2}) \cdot S_{VCcar} \cdot F = (\theta_{ginj} - \theta_{gout}) \cdot S_{VCwat} \cdot Z \quad (6.12)$$

BOX 6.2 Heat exchangers.

Heat exchangers are devices that efficiently transfer heat between two fluids. A car radiator, an elephant's ear and the grid on the back of a refrigerator are all heat exchangers. The most common forms of heat exchanger have the two fluids circulating on either side of a dividing wall. In an efficient heat exchanger, the dividing wall will have as large a contact area for heat transfer as possible, and as low a thermal resistance as possible. Arguably the simplest form of heat exchanger is the single pass counter-flow exchanger (Figure 6.8a). As the two fluids flow past each other, a heat flow rate Q passes from the warm to the cool flux and the temperatures of the fluids change as shown in Figure 6.8a. The temperature difference $\Delta\theta_a$ is known as the *approach temperature*.

If the heat exchanger has no external losses, the heat gained by the cool stream (fluid 2) should equal the heat lost by the warm stream (fluid 1):

$$Q = (\theta_{1i} - \theta_{1o}) \cdot S_{C1} \cdot \dot{m}_1 = (\theta_{2o} - \theta_{2i}) \cdot S_{C2} \cdot \dot{m}_2$$

Where θ = temperature, S_C = specific heat capacity (J kg^{-1} K^{-1}), $\dot{m}$ = mass flux of fluid (kg s^{-1}). The subscripts 1 and 2 refer to the warm and cool fluids, while the subscripts *o* and *i* refer to outflow and inflow temperatures.

We can define an overall heat transfer coefficient U (W m^{-2} K^{-1}) for the heat exchanger, such that

$$Q = U \cdot A \cdot \Delta\theta_{mean}$$

where A is the exchange area and $\Delta\theta_{mean}$ is some form of measure of the mean difference in temperature between the two fluids. For a simple parallel flow or counter-flow heat exchanger of the type considered above, it can be shown that this is best expressed as the logarithmic mean temperature difference (LMTD) – see Figure 6.8a:

$$\Delta\theta_{mean} = \text{LMTD} = \frac{\Delta\theta_a - \Delta\theta_b}{\ln(\Delta\theta_a/\Delta\theta_b)}$$

where S_{VCcar} = specific volumetric heat capacity of the carrier fluid (J L^{-1} K^{-1}), S_{VCwat} = specific volumetric heat capacity of groundwater (J L^{-1} K^{-1}), F = flow rate of carrier fluid in the intermediate loop (L s^{-1}), Z = groundwater flow rate (L s^{-1}), and θ_{gout} and θ_{ginj} are the abstraction and waste (injection) temperatures of groundwater (K).

We can thus see that we have the opportunity, with a well-designed choice of heat exchanger, to trade off groundwater rejection temperature against groundwater flow rate. If the carrier fluid is water-based, Equation 6.12 reduces to

$$(\theta_{r2} - \theta_{s2}) \cdot F = (\theta_{ginj} - \theta_{gout}) \cdot Z \qquad (6.13)$$

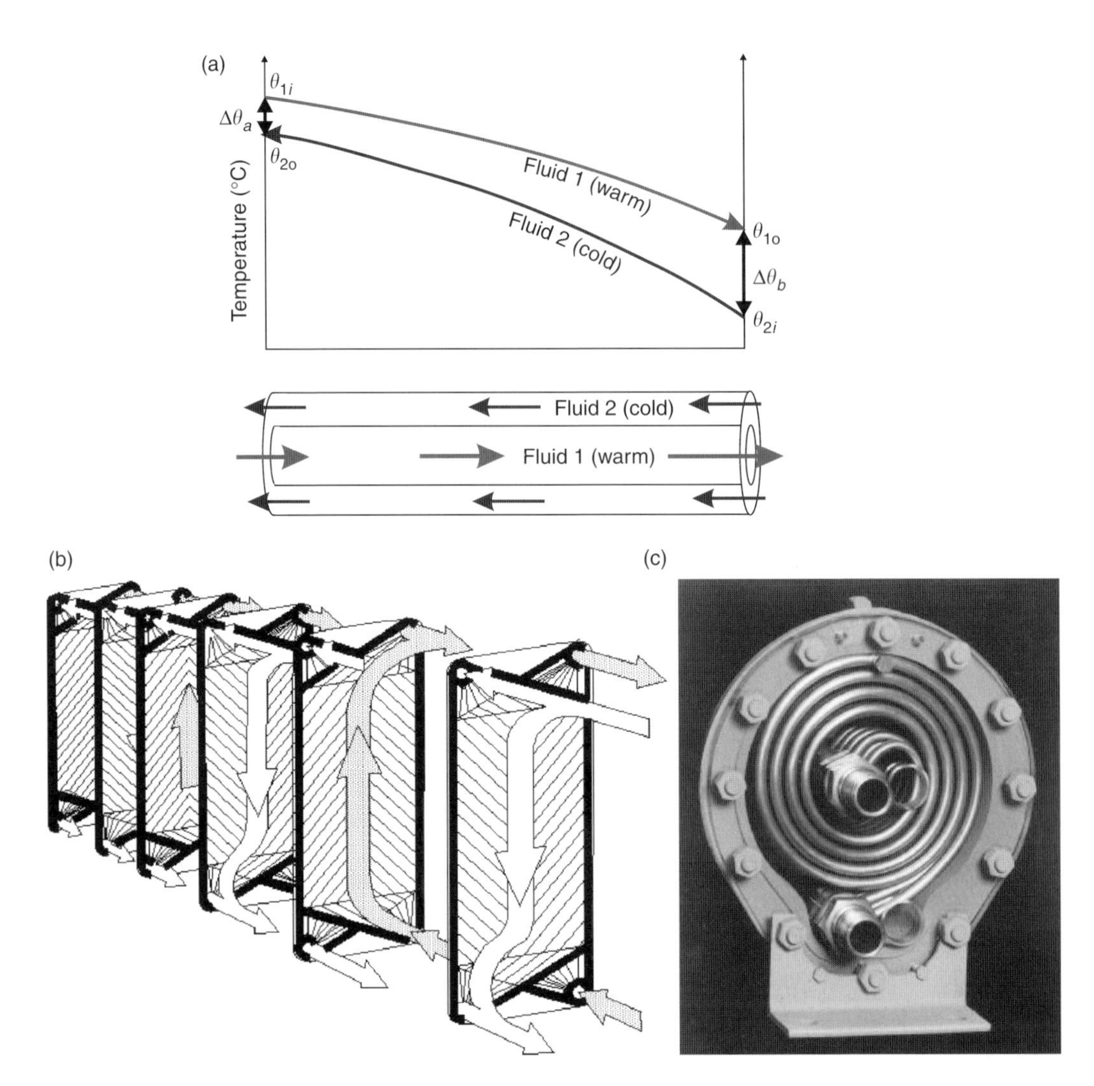

Figure 6.8 (a) Schematic diagram of a single pass, coaxial, counter-flow heat exchanger (see Box 6.2); (b) a parallel plate heat exchanger, *after Rafferty and Culver (1998) and reproduced by kind permission of the GeoHeat Center, Klamath Falls, Oregon*; (c) helical heat exchanger (Copyright 1999 from *Graham Corporation – Evolution of a Heat Transfer Company by R.E. Athey. Reproduced by permission of Taylor & Francis Group, LLC., http://www.taylorandfrancis.com*).

Thus, if our well yield or aquifer resources are limited, we can reduce Z. The price we pay is that θ_{ginj} increases, and a groundwater rejection temperature that is too high may be regarded as unacceptable by many environmental authorities. If a regulator imposes a stringent upper limit on θ_{ginj}, we can increase the pumped yield from our abstraction well. Note, however, that θ_{ginj} can never be higher than θ_{r2}, and will usually be at least 1–2°C below it (in a cooling mode).

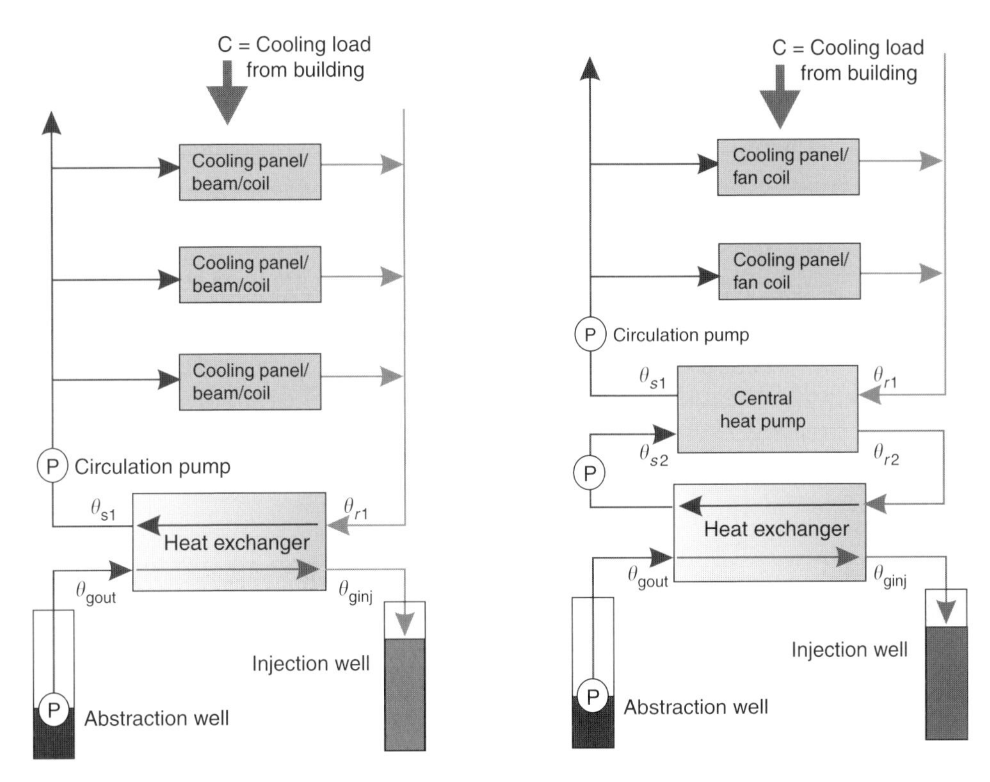

Figure 6.9 Two schematic diagrams showing how a 'prophylactic' heat exchanger could function to provide free cooling (compare with Figure 4.1, p. 58) and active heating. Note that the use of such a heat exchanger results in the need for additional circulation pumps. Note also, that in a real situation there would typically be some form of hydraulic storage (a tank) on the groundwater circuit, prior to the heat exchanger, to 'buffer' variations in flow.

Let us assume that our intermediate carrier fluid operates at a temperature in the mid-thirties, in °C. Heat is transferred from the building's chilled water loop to the carrier loop, raising the carrier loop's temperature from, say, a flow temperature $\theta_{s2} = 31°C$ to a return temperature $\theta_{r2} = 38°C$. Let us further assume that our θ_{gout} is 11°C. If our groundwater flow is the same as the flow in the intermediate carrier loop, then θ_{ginj} will be approximately 18°C. If, however, our groundwater flow is doubled, the θ_{ginj} decreases to 14.5°C.

A heat exchanger should, of course, be tailored to the temperature and flow regimes that are specified for the heating/cooling system. Broadly speaking, as the efficiency of the heat exchanger increases (the U-value increases) and the approach temperature drops, the more the exchanger will cost. Kavanaugh and Rafferty (1997) contend that it is perfectly feasible to design modern plate heat exchangers with approach temperatures as low as 1.5–2°C.

6.3.3 Disposal of wastewater

Having passed our groundwater through the heat pump or heat exchanger, we are left with the problem of how to dispose of it! If the groundwater had an original water temperature of 11°C and we have used it for heating, we may now have a water whose temperature is as low as 6°C. If we have used it for cooling, the wastewater will be warm. Note that in some countries (such as England and Wales), two sets of permits – abstraction licenses and discharge consents – may be required for an operation involving abstraction and disposal of groundwater. We will usually have the following options for water disposal:

Disposal to sewer. Obviously, this depends on having a sewer or storm drain to hand. It must have the excess capacity to accept our waste flow. We will require the permission of the utility that owns the sewer and will usually have to pay a significant charge.

Disposal to a surface water body (e.g. a river). In order to do this, we will usually require the permission of a national or regional authority with responsibility for regulating the water environment. We may have to pay a discharge fee. We will usually have to perform some form of risk assessment, which will address questions such as:

- How will the temperature (warm or cold) of the discharged water affect the ecology of the surface water body, or its utility value to other users? Remember that waste heat (or cold) may be regarded as a potential pollutant by many environmental authorities.
- Is the discharged groundwater geochemically compatible with the water in the recipient? Groundwater may, for example, be poor in dissolved oxygen, or rich in dissolved iron, or somewhat saline. All these factors could significantly impact the life in a watercourse.
- Will the additional discharge of groundwater to the surface water increase the flooding risk in the watercourse?

If the discharge is to the sea or to an estuary, the discharge of groundwater from a heat pump will in some cases be less tightly regulated.

Re-injection to the abstracted aquifer. A further option is to re-inject our waste groundwater back to the same aquifer that it was abstracted from. This is attractive because it has few or no water resources implications (there is no net abstraction of the water from the aquifer) and will often minimise any risk of ground settlement that can occur in some soils due to prolonged net abstraction of groundwater. This solution may thus be attractive to environmental regulatory agencies charged with ensuring that water resources are protected. In some countries, such an abstraction/re-injection operation may not require licensing. In other countries, such a license may be less complex than for purely consumptive water use (i.e. abstraction only), and may attract lower charges.

The re-injection of limited quantities of water may be feasible via a soakaway structure. Larger quantities will usually require the use of re-injection wells. Their construction and operation is a specialist task; injection wells may need special well screens. The injected water may need to be sterilised to hinder bacterial clogging of the well screen or of the aquifer; it will need to have a very low particle content and the water pressure and its gas content will need to be controlled such that exsolving gas bubbles do not clog the aquifer and decrease its permeability.

Finally, we need to take care that we do not re-inject our groundwater so close to our abstraction well that we start to get 'short circuiting' – our cold (or warm, if it is a cooling scheme) re-injected groundwater breaking through into our abstraction well. If this occurs, the temperature of our abstraction well water will start to drift downwards (or upwards) with time, compromising the efficiency of the system. This will be dealt with further in Chapter 7.

Disposal to another aquifer. If there is more than one aquifer below the site, we can abstract groundwater from one aquifer stratum and dispose of it (after passage through the heat pump system) by re-injection or re-infiltration to another. If we do this, we are consuming water from one body of water and re-injecting it to another hydraulically distinct body. A regulatory authority will regard this in much the same light as disposal to surface water and will need to be satisfied that abstraction from the first aquifer does not unacceptably compromise its resources or impact negatively on other users or on the environmental features supported by it. Likewise, the regulators will need to be satisfied that

- the injection of water will not cause an unacceptable rise in groundwater levels in the second aquifer;
- the injected water is geochemically compatible with the natural water of the second aquifer;
- any heat plume (i.e. body of warm or cold injected groundwater that is migrating with natural groundwater flow) does not constitute unacceptable heat pollution and will not adversely impact any users.

Disposal of wastewater to the abstraction well. In some smaller schemes, a proportion of the waste groundwater, having served our heating or cooling scheme, may be re-injected back into the upper part of our abstraction well. The thinking here is that the cool re-injected water (if it is a heating scheme) takes a finite time to flow down the well to the pump, and that its temperature will re-equilibrate on the way down by conductive transfer of heat from the borehole walls. Furthermore, in the well it will be mixed with a percentage of 'new' groundwater entering the well from the aquifer such that, eventually, an equilibrium situation will be established. Such an arrangement is called a standing column well (see Chapter 11 for further information).

6.3.4 *Disadvantages of open-loop heat pump systems*

Disadvantages of groundwater-based open-loop systems can be listed as follows. Such systems

- are geology-dependent. They require the site to be underlain by an aquifer, capable of providing an adequate yield;
- require a significant degree of design input from a hydrogeologist or groundwater engineer;
- need one or several properly constructed, durable (i.e. expensive) water wells, with pump installations, monitoring and control mechanisms;
- incur pumping costs associated with abstracting the groundwater from the well (as a rule of thumb, power consumption by water/circulation pumps and other auxiliary gear should be $<10\%$ of the total electrical energy budget of a GSHP scheme);
- generate a used water flow that must be legally disposed of;
- will usually require formal consent from a regulatory authority to abstract groundwater and to discharge the used water to a recipient. A fee may also be levied;
- may need to be monitored for water chemistry and turbidity and will have a maintenance requirement to prevent clogging, fouling or corrosion of heat pump, heat exchangers or wells.

6.3.5 *Advantages of open-loop heat pump systems*

On the other hand, open-loop groundwater-based heating and cooling systems have a number of persuasive advantages.

- They utilise a natural medium (groundwater) that occurs at a constant temperature in the subsurface and has a huge specific heat capacity ($4180\ \text{J L}^{-1}\ \text{K}^{-1}$).
- They extract heat by forced convection of groundwater rather than by subsurface conduction. They thus tend to extract more heat per borehole/well than closed-loop systems do (Figure 6.10).
- The abstraction well may provide a resource of potable water, as well as a heat resource. There is no reason why we cannot utilise good quality groundwater, having passed it through at heat pump scheme, instead of running it to waste. Even if the groundwater is of poorer quality, it may still be used as 'grey' water (i.e. for flushing and washing purposes) in a building.
- For use in heating and cooling schemes, the quality of the abstracted groundwater is not necessarily an important issue. Thus, open-loop heating and cooling schemes can be based on groundwater from flooded mines, from dewatered mines/excavations or even from contaminated sites. If groundwater is being pumped in connection with the remediation of a contaminated site ('pump and treat' schemes), there is no reason why useful heat could not be gained from it. Open-loop systems can also be based on natural saline groundwaters – for example, in coastal areas (see Box 4.2, p. 63).

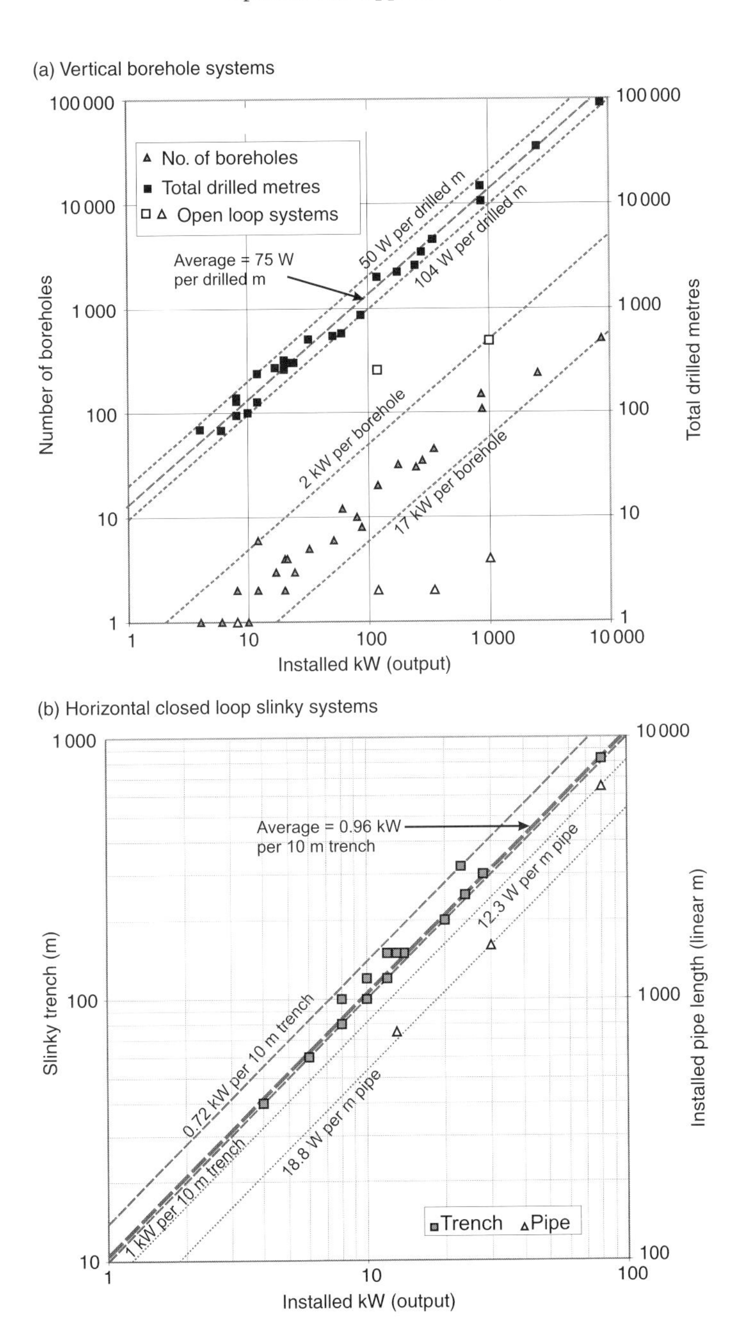

Figure 6.10 (a) The number of boreholes drilled and the total drilled metres for a variety of vertical closed-loop GSHP schemes (and a small number of open-loop schemes), related to installed heat pump kW delivery. (b) The installed meterage of slinky pipe (as linear metres) *or* the installed metres of trench for a number of slinky-based horizontal closed-loop schemes. For both (a) and (b) the majority of schemes are heat-dominated, though some of the larger (especially >60 kW) schemes are cooling-dominated or provide both cooling and heating. Most of the schemes are in the UK and are actually constructed (a few only reached the planning stage). They are sourced from case studies documented by GeoWarmth Ltd., Earth Energy Ltd., Kensa Engineering Ltd. and Drage (2006).

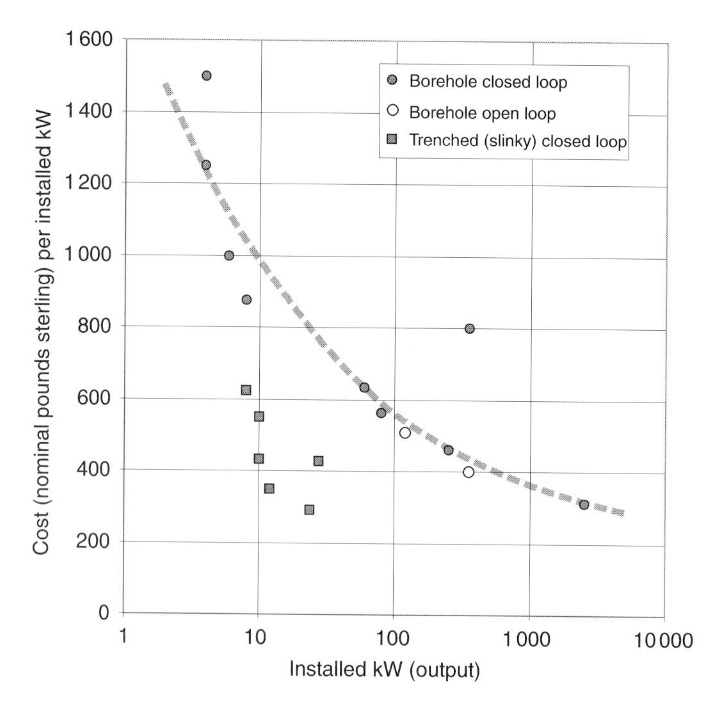

Figure 6.11 The cost, per installed kW, of the schemes shown in Figure 6.10. The cost is shown in 'nominal' pounds sterling and many of the schemes are 3–7 years old. The prices likely reflect only the capital cost of the trenches, boreholes, heat pump and hardware. They are unlikely to reflect consultants' fees, taxes and other additional costs. The prices are thus likely to underestimate the total cost of commissioning a GSHP installation by around 50%.

From Figure 6.10, we see that very often the lesser number of boreholes/wells required for open-loop schemes tends to balance the greater cost of their design, construction and licensing, such that the capital cost per installed kW turns out to be comparable to closed-loop schemes (Figure 6.11).

6.4 Closed-loop systems

Hitherto, we have considered heating and cooling schemes that are based on extraction of water from a well (or spring, flooded mine, river or lake). But aquifers and surface water bodies are not ubiquitous. Fortunately, there is another way to extract heat from the ground that does not require any water to be abstracted or re-injected at all. Such schemes are called closed-loop schemes, and they can be constructed practically anywhere: in granites, clays, waste tips, permafrost or abandoned mines. Closed-loop schemes are of two types: direct circulation and indirect circulation. Direct circulation (DX) schemes were more common in the early years of GSHP systems, although they are still being installed by some companies today. However, indirect circulation

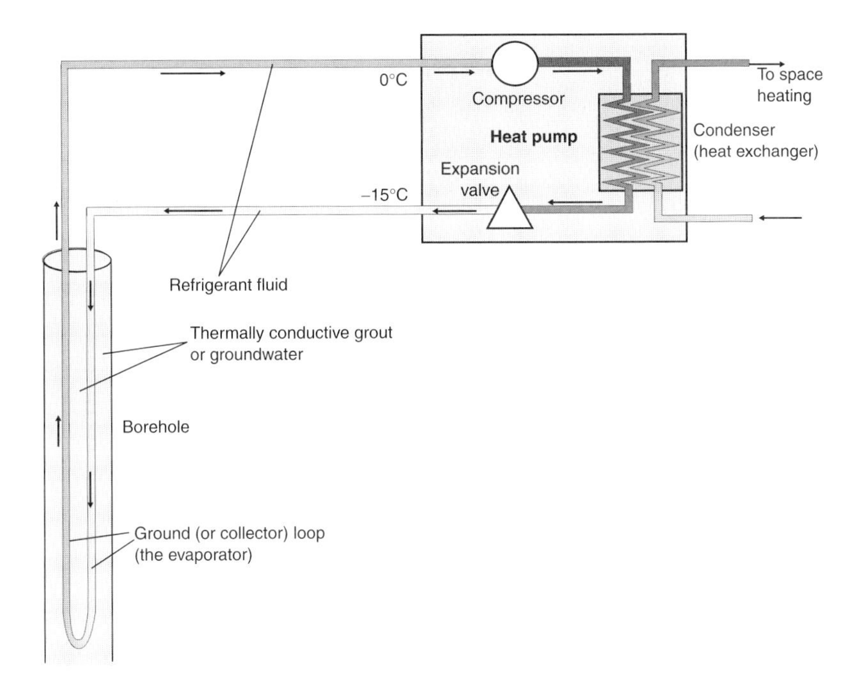

Figure 6.12 A schematic diagram of a direct circulation closed-loop scheme, installed in a borehole.

schemes have become far more widespread than DX schemes in today's European GSHP market.

6.4.1 *Direct circulation systems*

In DX schemes (Figure 6.12), the heat pump's refrigerant is circulated into the subsurface in a closed loop of copper tube. The tube may be buried in the subsurface in a shallow trench or installed in a vertical borehole. In effect, the subsurface ground loop acts as the evaporator of the heat pump. The chilled refrigerant from the heat pump enters the subsurface in the loop, absorbs heat from the relatively warm earth and returns to the compressor of the heat pump, for the temperature of this heat to be boosted.

The major advantage of DX systems is that the ground loop can operate at a rather low temperature (refrigerant entry can be as cold as −15°C). This large temperature differential between the loop and the ambient ground, and the direct transfer of heat from ground to refrigerant (and, to a lesser extent, the high thermal conductivity of the copper tube) means that heat can be absorbed rather efficiently from the ground (Halozan and Riberer, 2003). Such DX heat pumps have fallen somewhat out of favour in some countries, however. This is not because such systems do not work well [although there have been practical issues to resolve regarding the mechanical operation of the compressor unit – Sumner (1976a) drew attention to the potential for the refrigerant flow to become blocked in long ground loops by accumulations of oil from the compressor]. It is rather that many installers have become nervous about circulating refrigerant substances into the ground in a copper tube, which may be susceptible to mechanical damage and which, under some geochemical conditions, may be liable to corrosion. Refrigerants, such as the commonly used R407c, are fluorocarbons: environmental regulators usually regard such as halogenated hydrocarbons as potential groundwater contaminants. Indeed, the Environment Agency of England and Wales classifies organohalogens as 'List 1' substances, prohibited from entering groundwater. It is feared by some that any future refrigerant leakages may thus be regarded as a pollution incident by regulators (regardless of the fact that the practical consequences of such a small leakage could be argued to be minimal).

In other countries, there is a more upbeat attitude towards DX systems. Halozan and Rieberer (2003) describe the healthy market for them in Austria and also document new developments using CO_2 as a refrigerant fluid, thus negating regulatory concerns regarding halogenated organics circulating in the subsurface.

6.4.2 *Indirect circulation systems*

We can also avoid circulating the refrigerant fluid directly into the ground by using an 'indirect circulation' closed-loop scheme. Here, a carrier fluid circulates in a closed loop of pipe, which passes into the subsurface, either via a borehole drilled vertically into the earth (Figure 6.13), or via a horizontal coil or trench. In a heating scheme, the chilled

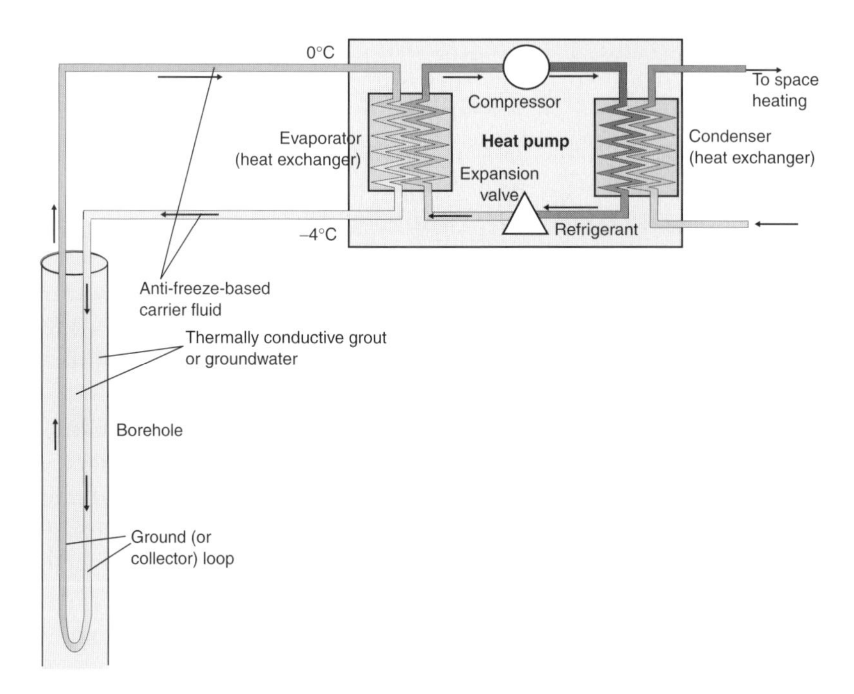

Figure 6.13 A schematic diagram of an indirect circulation closed-loop scheme, installed in a borehole.

carrier fluid exiting the heat pump absorbs heat (by conduction) from the subsurface and conveys it back to the heat pump, where the heat is extracted. The carrier fluid is thus chilled again and ready to start its next circuit through the earth. In a cooling scheme, the opposite applies: the warm carrier fluid from the reversed heat pump (or from a free cooling scheme in a building) descends into the subsurface and conducts a portion of its heat load to the relatively cooler earth. It emerges cooled to re-enter the heat pump or the building's cooling network. The closed loop thus functions as a 'subsurface heat exchanger'. Some practitioners refer to such solutions as 'geoexchange' of heat.

Sumner (1976a) initially used copper pipe in his pioneering British closed-loop installations, but concluded that the thermal resistance of the pipe was probably of little significance to the system performance as a whole. Thus, nowadays, we typically use polyethene (PE) pipe, which has a lower thermal conductivity than copper but is much cheaper, more durable and corrosion-resistant. The pipe is typically of 26–40 mm OD (outer diameter), although pipe diameters as high as 50 mm have been utilised in Scandinavia (Skarphagen, 2006). The pipe material is usually rated to withstand fluid pressures of 10–16 bar, although, during operation, the carrier fluid is pressurised to around 2–3 bar (C. Aitken, *pers. comm. 2007*). Occasionally, where large pressure loads are not expected, pipes with a pressure rating as low as 6 bar are sometimes used.

The carrier fluid is usually water-based, and is typically an antifreeze solution, allowing the fluid to be chilled to temperatures below 0°C, if necessary. The antifreeze may be a solution of ethylene glycol, or ethanol, or of salt (amongst others – see Section 8.3, p. 189). Carrier fluid freezing points of between –10°C and –20°C are typical; monoethylene glycol has a freezing point of –14°C at 25% strength and –21°C at 33%. Regulatory authorities tend to be somewhat less concerned with the possibility of leakages of substances such as these: the alcohol-based antifreezes are not especially environmentally toxic and will biodegrade rapidly. The fluid flow rate and pipe diameter are selected such that (a) turbulent flow conditions are achieved in the subsurface closed loop (turbulent conditions facilitate heat transfer from the ground to the fluid) and (b) the fluid can convey the required amounts of heat. Flow rates of 3–3.5 L min^{-1} per kW of heat transfer are typical. Under typical operating conditions in heating mode, one might aim to achieve an average carrier fluid temperature of –1.5°C to –2°C. For example, the downhole flow temperature might be –3°C to –4°C and the return temperature from the borehole might be around 0°C. Under peak loading conditions, however, the system will usually be designed to cope with even colder carrier fluid temperatures. Note that the fluid viscosity (and thus the threshold for turbulent flow) will depend both on antifreeze type and on temperature: this may be an important consideration in designing carrier fluid antifreezes and flow rates (Box 8.1 and Sections 8.3 and 8.4, p. 185 and 189–192).

Note that there is a trade-off between loop operating temperature and heat pump efficiency. In heating mode, low loop temperatures increase conductive heat transfer from the ground, but will result in a lower heat pump COP_H.

6.4.3 Horizontal closed loops

One of the cheapest forms of indirect closed-loop scheme is the horizontal closed loop, installed in a trench. The optimal depth for such a trench is regarded as ≈1.2–2 m (although some will argue that a depth as shallow as 1 m is adequate). This depth is

- one that can be practically excavated using a mechanical excavator;
- deep enough to provide a sufficient (though modest) thermal storage to support a heating scheme during a winter, a reasonable soil moisture content and to isolate the loop from the worst winter frosts;
- shallow enough to allow solar and atmospheric heat to penetrate and replenish the thermal storage around the loop during the summer months.

Horizontal loops thus function largely as subsurface solar collectors.

As a rule of thumb, a single, straight polyethene (PE) pipe installed in a trench is judged (Rosén *et al.*, 2001) capable of supporting a installed heat pump capacity of 15–30 W m^{-1} of trench in a heating scheme (i.e. 33–67 m per installed kW). The drier the soil, the lower the output. It is worth noting, however, that Sumner's (1976a) colleague, Miss M. Griffiths, obtained impressively high steady-state heat absorption rates of 29–58 W per linear metre to pipes buried in London Clay.

To increase this output, it is common practice to bury not simply a single PE pipe in our trench, but overlapping coils of pipe. This arrangement goes by the name of a 'slinky' (Figure 6.14). Such slinkies are formed from pre-coiled PE pipe and are typically installed in relatively wide trenches, with a coil diameter of ≈0.6–1 m. Different sources claim slightly different outputs for slinkies, although they typically cite figures of around 10 m of slinky-filled trench to support each 1 kW of installed heat pump. Figure 6.10b suggests that UK practice results in between 10 and 14 m trench being excavated per installed kW output, with an average of 10.5 m. These figures are dominantly based on heating only schemes. Where the horizontal loop is used for cooling rather than heating, a few more metres of trench (depending on the heat pump COP and the operational temperature) may be required per kW of installed cooling load, as discussed in Section 6.3. Where more than one trench is installed, parallel trenches should be at least 3 m, and preferably 5 m, apart. Chapter 8 contains more detailed consideration of the reasons why such 'rules of thumb' appear to exist and the different configurations of horizontal closed loop that are practicable.

6.4.4 Pond and lake loops

Coils of PE pipe can also be installed in deep ponds or lakes. For this to be an appropriate solution, the lake should ideally be at least 3 m deep, to ensure that natural temperature variations at its base are low. The lake should also be large enough, so that the heat extracted (or dumped) by the heat pump does not change the temperature of the lake water by an unacceptable amount. Remember that some aquatic species

Figure 6.14 The installation of a slinky-based horizontal closed-loop system in a trench. *Photo reproduced by kind permission of GeoWarmth Ltd.*

may be sensitive to large temperature variations for several reasons: (i) they may be directly sensitive to temperature, (ii) increasing temperature will reduce the solubility of gases such as O_2 and (iii) temperature changes may affect the productivity of the water (e.g. reproduction of algae). Pond loops will be considered in greater detail in Chapter 9.

6.4.5 Vertical closed-loop arrays

If we have a large amount of available space at our development site, horizontal trenched installations may be the cheapest means of installing a ground loop. At many sites, however, ground area is at a premium. A far more space-efficient means of installing a ground-loop array is via vertically drilled boreholes.

Such boreholes are not nearly as sophisticated as the water wells discussed in Section 6.3. Although a short length of permanent casing should be installed and grouted in the uppermost section of the borehole, to prevent surface contamination entering the subsurface, such boreholes are typically drilled either 'open-hole', or using only temporary casing in loose rocks or sediments.

Drilling techniques

Several different drilling techniques are available for drilling closed-loop boreholes. The drilling method of choice will depend on lithology. In Sweden and Norway, where

Figure 6.15 A down-the-hole hammer (DTH) drilling rig, operating in Flatanger, Norway. Note that many such rigs can drill at a substantial angle from the vertical. The cuttings and groundwater are blown back up the hole by the compressed air and can be seen gushing out to the left of the borehole (*photo by David Banks*).

very hard, crystalline rock is overlain by a relatively thin veneer of loose Quaternary sediments, a drilling method known as 'down-the-hole hammer' (DTH) can be used in almost all circumstances. This is essentially a drilling bit, embedded with tungsten carbide buttons, mounted on a slowly rotating compressed air hammer. The hammer is powered by a stream of compressed air pumped down the drilling stem. This airflow returns up the annulus of the borehole, removing drilling cuttings (Figure 6.15). The method is good for many different rock types and is capable of drilling a 100 m borehole in granite in less than 1–2 days. The method is not especially good in heavy clay strata or strata containing large boulders or fragments. Because the geology in FennoScandia is more uniform (in terms of geotechnical properties) than the UK, many Nordic firms operate a single (DTH) drilling method to a highly standardised procedure, leading to relatively cheap drilling rates.

In more geologically varied countries, such as the UK, the driller will usually need to maintain a greater range of drilling rigs. The greater prevalence of poorly consolidated rock means that temporary casing may be needed more often. The cost of drilling is thus generally higher, and more variable, than in Scandinavia. While DTH techniques may commonly be applicable, other techniques may be preferred by drillers. These include 'conventional' rotary drilling (using mud, foam or air as drilling fluids) and even percussion (cable tool) drilling. In this latter technique, a large chisel-shaped bit is repeatedly dropped on the end of a cable: it sounds primitive but can achieve reasonable results in certain lithologies, such as the British Chalk. It is probably worth noting that the drilling industry should be tending in the following directions if it intends to service the ground source heat market:

- Track- or trailer-mounted drilling rigs with a small footprint that can fit into domestic properties (e.g. small DTH units with a separate compressor that can be parked at some distance from the well-head).
- High turnover of specialised GSHP drilling by dedicated crews.
- Flexible drilling rigs which can offer a range of techniques. Such trailer-mounted rigs are now available offering DTH compressed air drilling, in combination with conventional rotary drilling using air, foam or mud-flush.
- Pre-agreed pricing and guaranteed installation of U-tubes. Drilling firms may have to include geological risk and drilling failure as a component in their pricing strategy, rather than simply off-loading all risk onto the customer.
- Punctual delivery of services, to fit in with a rigorous building completion schedule.

Number and depth of boreholes

We have seen earlier in this chapter that, for open-loop systems, there is a proportionality between rate of groundwater flow and heating load serviced (Equations 6.7–6.9). We have also seen that there is a relationship between the length of trench required and the heating load in horizontal closed-loop systems (Section 6.4.3). We would intuitively expect the number of drilled borehole metres in vertical closed-loop schemes to also be in rough proportion with the heating load delivered. Figure 6.10a indeed demonstrates that there is a relationship between the number of closed-loop boreholes per scheme and the installed heat pump capacity in kW. Indeed, the installed capacity per borehole ranges from 2 kW to 17 kW. The smallest installed capacities are related to the shallowest boreholes (40 m) and the highest to the deepest boreholes (180 m). If we divide the installed heat pump capacity of the scheme by the total number of drilled borehole metres, we obtain a much more linear relationship, with specific installed thermal outputs of between 50 and 104 W per drilled metre (average of 75 W per drilled metre). If we assume that a heat pump scheme has a typical coefficient of performance of 3.4, then these figures equate to specific peak heat absorption rates (from the ground) of 35–73 W m^{-1} (average 53 W m^{-1}). Note that the systems depicted in Figures 6.10 and 6.11 are mostly heating-dominated or heating-only schemes, although a few of the larger ones are cooling-dominated.

But surely the nature and thermal conductivity of the ground is important as well? It is, but if we look at Table 3.3 (p. 53), the thermal conductivity of most British rocks ranges over a factor of less than 3, which approximately reflects the variability in our calculated specific heat absorption rates! Rules of thumb, such as '50–100 W m^{-1} of installed GSHP capacity in heating mode', are only possible because most rocks have rather consistent thermal properties and their behaviour is not strongly lithology-dependent.

Thus, there is no hard and fast rule about how deep a ground source heat borehole should be. In other words, two boreholes to 50 m depth (provided they are situated sufficiently far apart) should provide an approximately equivalent heat output to one borehole to 100 m. However, it costs money to move a drilling rig from one site to another and to complete every new borehole (surface casing, grouting, manifolds, couplings, etc.). Moreover, two boreholes have a greater areal space requirement at a site than a single borehole. These factors may argue for fewer, but deeper, boreholes. On the other hand, drilling penetration rates become slower with depth and drilling becomes more expensive. Thus, there will come a cut-off point where it becomes cheaper to commence a new borehole than to continue drilling ever deeper in a single borehole. In practice, unless there are overriding considerations of space, ground source heat boreholes tend to be drilled to between 70 and 120 m depth (although, in Scandinavia, greater depths are more frequent). Is this because the economic cut-off point occurs at around 100 m? Or could it be that many drillers only carry 120 m of drill string?

Finally, we should be wary of 'rules of thumb' relating drilled metres to thermal output. First, ground thermal properties will have some influence. Second, heat transfer rates per metre of borehole will depend on our design loop operating temperature. A warmer loop (in heating mode) means less conductive heat transfer from the ground (but probably a higher heat pump COP_H). Third, patterns of heat usage will be important. A greater number of borehole metres will be necessary to support a 12 kW heat pump with 3200 operational hours per year than one with 500 hours. Fourth, we should be aware that in large, complex buildings, the total installed heat pump capacity may not be identical to peak heating load: it may be greater, because the building may be divided into different zones, whose peak heating demands occur at different times of day (Kavanaugh and Rafferty, 1997). In summary, 'rules of thumb' are a good starting point for design, especially for smaller, simple systems, but system design involves more complex considerations. More of this will be discussed in Chapter 10.

Emplacement of U-tubes

Once the drilling of a borehole for the closed-loop scheme has been completed, a 'U-shaped' closed loop (or 'U-tube') is usually emplaced down the length of the borehole (although, other configurations of exchange tube are possible – Chapter 10). U-tubes are usually made of high-density polyethene (HDPE) tubing of diameter 32–40 mm (although medium-density polyethene – MDPE – is sometimes used). A shank spacing (distance between the centres of the uphole and downhole tubes) of around 50–60 mm is typical. Thus, the U-tube installation will have a width of 90–100 mm (Figure 6.16). The diameter of boreholes drilled for closed-loop ground source heat schemes is typically around 125–130 mm (5 in.). To prevent thermal short-circuiting of heat between the upflow and downflow tubes, shank-spacers or clips should be placed down the length of the U-tube to maintain an acceptable shank spacing. The 'U' on the base of the tube is typically a pre-fabricated element and will usually be weighted to render emplacement of the tube into a water- or mud-filled borehole easier. In a water-filled borehole, there will also be buoyancy effects to overcome when emplacing the tube: the weighted 'U' may assist here, but it will usually also be necessary to fill the U-tube with water during emplacement (Figure 6.17).

Figure 6.16 A newly installed polyethene U-tube in a grouted borehole. The two shanks of the tube can be seen, as can the top of a length of steel casing (which was employed in this rather unstable, mined ground). Subsequently, the borehole would typically be completed below ground level. *Reproduced by kind permission of Pablo Fernández Alonso.*

Figure 6.17 Persuading a U-tube to descend a fluid-filled borehole can be trickier than it sounds, and it can also be messy. *Photo reproduced by kind permission of GeoWarmth Ltd.*

Once the U-tube is in place, it is common practice (though not the only way of completing a borehole) to grout the space between the U-tube and the borehole wall with some form of grout. The grout is usually pumped down to the base of the borehole using a 'tremie pipe' – the tremie pipe is gradually withdrawn as the borehole fills up with grout (Figure 6.18). When grouting the U-tube in place, be aware that grout is

Figure 6.18 Pumping grout into a closed-loop borehole. The reel of tremie pipe and the tank containing the grout are mounted on the lorry. *Photo reproduced by kind permission of GeoWarmth Ltd.*

denser than water and liquid grout may thus exert a huge pressure on the U-tube at the base of a deep hole. In some circumstances, it may thus be necessary to pressurise the U-tube during grouting. Other methods of backfilling are discussed below.

Alternatives to U-tubes

Other geometries of closed loop may be installed in vertical boreholes, with the objective of increasing heat exchange area and improving the overall heat transfer (and thus reducing the required drilled length for a given thermal output). The 'Double U' tube has two upflow and two downflow tubes, for example. It is somewhat more efficient than a 'single U' (but very difficult to emplace in a borehole). Similarly, Rosén *et al.* (2001) report the 'vertical slinky' or 'Svec' coil – a helical coil of tubing designed for emplacement in somewhat larger-diameter drilled boreholes.

Completion of boreholes

Several options exist for borehole completion (Figure 6.19):

Open, water-filled hole. The U-tube may be suspended in a groundwater-filled borehole (usually only employed in competent lithologies where the borehole walls are self-supporting). The water provides a thermal contact between the rock and the U-tube. Although water only has a modest thermal conductivity, the efficiency of heat transfer can be dramatically improved by (a) the formation of convection cells in the

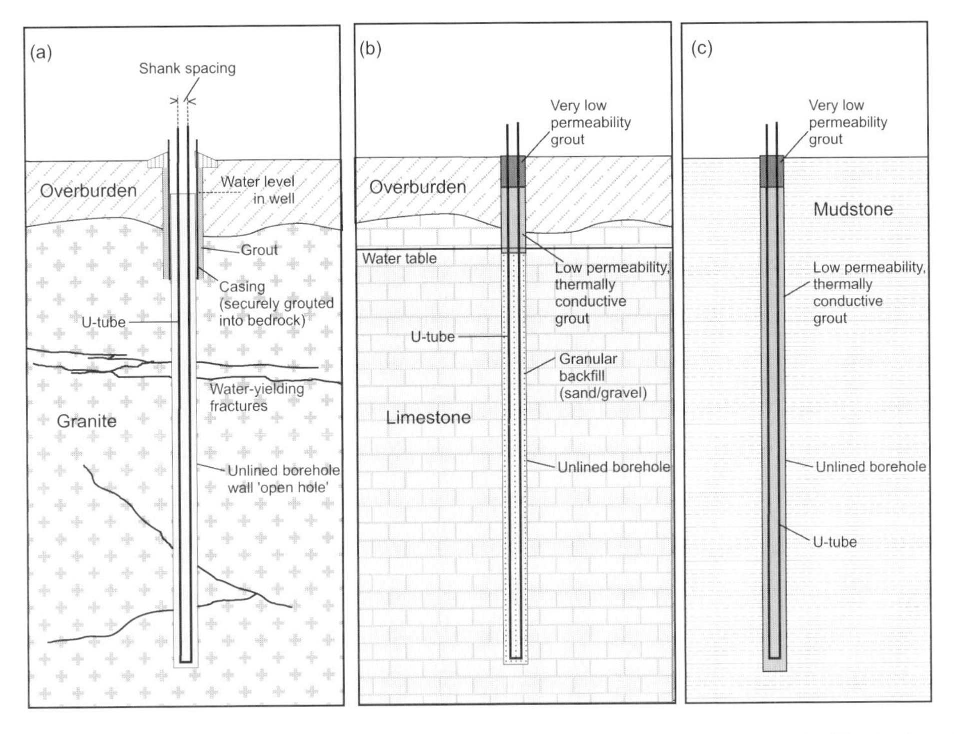

Figure 6.19 Schematic diagrams illustrating possible options for installation of a U-tube in a borehole: (a) suspended in a groundwater-filled well, (b) backfilled with sand/gravel, (c) backfilled with a thermally efficient grout. For (a) and (b) to function well, the groundwater level must be high.

groundwater column of the borehole, (b) the formation of high-conductivity ice around the loop and (c) heat replenishment by bulk groundwater flow through the borehole array. This solution has been favoured in Sweden and Norway.

Porous backfill. The borehole might be backfilled with quartz-rich gravel or sand and the upper portion sealed with a cement-based grout (to prevent ingress of any surficial pollution to borehole). This allows any temporary casing to be extracted, if the borehole was drilled in poorly lithified strata. Both the quartz grain matrix (remember quartz's high thermal conductivity) and the mobile groundwater filling the pore spaces provide efficient heat transfer between the borehole wall and the U-tube.

Grout backfill. The U-tube can be grouted into place with a grout. Ideally, the grout should have a high thermal conductivity (to facilitate the transfer of heat) and a low hydraulic conductivity (to prevent contaminant migration within the borehole or down the borehole). The former property can be provided by a high quartz content (Table 3.1, p. 35), while the low hydraulic conductivity can be provided by a clay matrix, such as bentonite. In fact, a so-called thermal grout, comprising a mixture of fine quartz silt/sand and bentonite, is a commonly used material. Swedish researchers (Rosén *et al.*, 2001) have cast some doubt on the use of bentonite in grouts, suggesting that water molecules, trapped within the bentonite structure, might form discrete pockets of ice if the grout freezes, compromising the grout integrity and imposing stresses on the U-tube. It is claimed that more conventional cement/sand/bentonite-based grouts provide an alternative option, with quite a reasonable thermal conductivity. Again, this is discussed further in Chapter 10.

The *open hole* and *porous backfill* options require that the borehole be largely filled with groundwater in order to obtain good thermal contact. *Grout backfill* is the only realistic option of providing a good thermal contact for a 'dry' borehole or a borehole above the water cable. The grout backfill option may often be preferred by environmental authorities (and may even be obligatory), for several reasons: (i) the grout prevents uncontrolled movement of groundwater from one aquifer horizon to another; (ii) the grout hinders pollution from the surface entering the geological environment; and (iii) the grout provides an extra barrier against antifreeze contamination in the event of a rupture of the U-tube.

Fluid flow rates and manifolds

The rate of flow of a carrier fluid (e.g. antifreeze) should satisfy two criteria:

- It should be high enough to induce turbulent flow in the downhole U-tube at the lowest design temperature (highest carrier fluid viscosity). While this leads to some additional energy expenditure on overcoming hydraulic resistance to flow, it greatly enhances the downhole transfer of heat from borehole to carrier fluid. Remember that viscosity of fluid increases with decreasing temperature.
- It should be adequate to convey the quantity of heat required by the heat pump. If we assume that a typical GSHP effects a temperature drop of some $\Delta\theta = 4\text{–}5°\text{C}$ in the carrier fluid, the necessary fluid flow rate can be estimated by

$$\frac{(1-(1/\text{COP}_H))\times 1000\ \text{W}}{S_{\text{VCcar}}\times\Delta\theta} = \frac{0.75\times 1000\ \text{W}}{(4000\ \text{J}\,\text{K}^{-1}\,\text{L}^{-1}\times 4\ \text{K})}$$
$$= 0.047\ \text{L}\,\text{s}^{-1}$$
$$= \approx 3\ \text{L}\,\text{min}^{-1}\ \text{for every 1 kW of peak thermal output.} \quad (6.14)$$

For example, for a domestic 6 kW scheme, we might estimate that a carrier fluid flow of 18 L min^{-1} will be required. We should always check that this is adequate to result in turbulent flow within our U-tube – this depends on the value of a calculated parameter named the Reynolds Number, which in turn depends on the diameter of the pipe and the density and viscosity of the carrier fluid. Carrier fluid flow would typically be achieved using a small electric pump that may be built into the heat pump (for small schemes), or be an independent feature (Figure 6.20).

Where we have a ground array comprising several boreholes, these are usually plumbed in parallel, via a manifold, to flow and return header pipes (Figure 6.20). When

Figure 6.20 (a) Carrier fluid circulation pumps at the St Mary's Health Centre closed-loop GSHP scheme, Isles of Scilly, UK (*photo by David Banks*); (b) Carrier fluid manifolds in a subsurface chamber. *Photo reproduced by kind permission of GeoWarmth Ltd.*

designing the network of header pipes and their connections to ground loops (U-tubes), we will typically be interested in

a. Achieving turbulent flow in the ground loop (to optimise heat transfer), but laminar flow in the header pipes (to minimise heat loss and hydraulic resistance). Thus, header pipes will usually be of substantially larger diameter than ground loops and may be insulated.
b. Being able to isolate any given ground loop in the event of a problem (e.g. a leakage) developing. Thus, the various boreholes will be fitted with individual isolation valves at the manifold so that any one borehole can be taken out of service without affecting the remainder of the system.
c. Obtaining balanced flows (i.e. similar flow rates in each borehole of a closed-loop array, assuming all are of similar depth). This can be performed by adjusting the valves at the manifold (this can be done automatically in response to loop temperature), or by adjusting each borehole circuit to a similar length (and thus a similar hydraulic resistance) by adding small additional coils of ground-loop pipe at the head of each borehole.

Skarphagen (2006) has argued that, for very large arrays of boreholes, an 'octopus' configuration of boreholes and ground loops around a manifold is a particularly effective means of minimising pipe length and ensuring relatively balanced flows (Figure 6.21).

Capital cost

Figure 6.11 shows the reported capital cost for a number of real or projected GSHP schemes, mostly in the UK. We can note that capital costs per installed kW ranged from around £400 to £1500. Further, we should note that the capital cost per installed kW

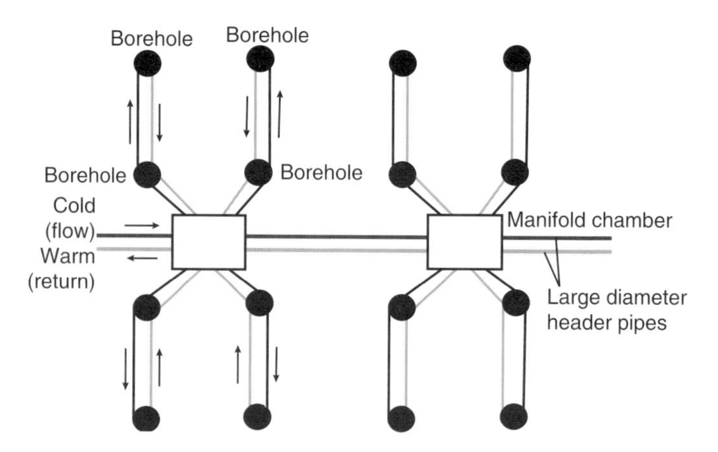

Figure 6.21 An 'octopus' configuration of closed-loop boreholes around a manifold.

is scale-dependent, declining with increasing size of scheme. These data support our assertion that ground source heating and cooling becomes increasingly economically attractive the larger the scheme (Chapter 4, Section 4.13, p. 80). This is partly due to the decreasing cost of the heat pump itself (per installed kW – Figure 4.9, p. 78), but is also due to the mobilisation costs of a drilling rig being spread over many holes on a site in a large scheme. The monetary values cited in Figure 6.11 should be taken as indicative only, for several reasons: (i) they are out of date (some of the cited schemes were developed over 5 years ago) and (ii) they are largely reported by consultants or installers who may not have included all of the real costs. For example, I suspect that the cited costs tend to reflect mainly the cost of the ground loop, boreholes and heat pump and do not necessarily reflect the consultant's time costs or profit margin, or any necessary ancillary works. I estimate that the cost of commissioning works from a typical British installer may currently be at least 50% greater than the values indicated by Figure 6.11. According to one British firm, the breakdown of a typical large closed-loop contract worth £150 000 might comprise 60% drilling, 20% heat pump supply, 10% labour, 7.5% manifolds, headers and antifreeze, and 2.5% design consultancy.

6.4.6 *Energy piles*

Some engineers have noticed that, in many geotechnical situations, large buildings require foundations composed of drilled reinforced concrete pile structures. If it is necessary to drill such holes for purely structural reasons, why not 'kill two birds with one stone' and install a closed-loop system within the pile structure? Such 'energy piles' or 'geo-piles' will often be 15–40 m deep and may be over 1 m in diameter. In principle, they function in exactly the same manner as purpose-drilled vertical closed-loop borehole schemes. However, they may perform less efficiently because they will usually have a larger diameter than a typical purpose-drilled closed-loop borehole (usually 125–150 mm) and will be filled with concrete, which has a low thermal conductivity

(Table 3.1, p. 35). The pile structure itself will thus have a high thermal resistance. To partially overcome this, we would install multiple U-tubes within a single pile and ensure a large shank spacing between the individual pipes. Furthermore, energy piles may not be able to operate at as low a temperature as purpose-drilled closed-loop boreholes. Being situated beneath a building, it will usually be important to prevent ground freezing for geotechnical reasons.

6.5 Domestic hot water by ground source heat pumps?

We have seen in Chapter 4 that GSHPs are particularly good at producing warm fluids (water or air) at temperatures of up to 50°C, which can be used for space heating. We have also learned that they can produce temperatures higher than this, but that the efficiency of the heat pump declines significantly with increasing temperature of delivery.

Because of the risk of legionellosis, caused by the proliferation of *Legionella* bacteria in warm (20–45°C) water systems, many countries will insist that DHW, at least for certain industrial or public service sectors, is stored at a temperature in excess of 60°C, above which the bacterium is killed (Box 6.3). If we have already chosen to use a GSHP to provide space heating, several strategies can be used to provide DHW.

BOX 6.3 *Legionella.*

Legionellosis (including the forms known as Legionnaire's Disease, Pontiac and Lochgoilhead Fever) is a lung infection caused by the genus of bacterium *Legionella.* The bacterium is common in nature and exposure to it need not necessarily cause any symptoms. However, it can proliferate in water whose temperature is elevated (20–45°C) and people may be at particular risk where high concentrations are transmitted in aerosol form, sometimes over large distances, and subsequently inhaled. Thus, the type of installations that can lead to a risk of legionellosis includes hot water systems and some components of large-scale waterborne cooling systems, such as wet cooling towers or evaporative condensers. The disease can usually be controlled by antibiotics.

A combination of several strategies is often recommended for controlling the risk from legionellosis, including system maintenance and regular cleansing or replacement of any filters or strainers. As *Legionella* cannot survive in water at temperatures in excess of 55°C, one element of a risk-control programme is to ensure that any store of domestic hot water is maintained at a temperature in excess of this value. Indeed, at temperatures greater than 60°C, the bacterium is killed in less than 30 min. In some countries and for some industrial or public service sectors there may even be legal requirements to enforce this policy.

Legionella is named after a fatal outbreak of the disease at a convention of the American Legion in Philadelphia in 1976.

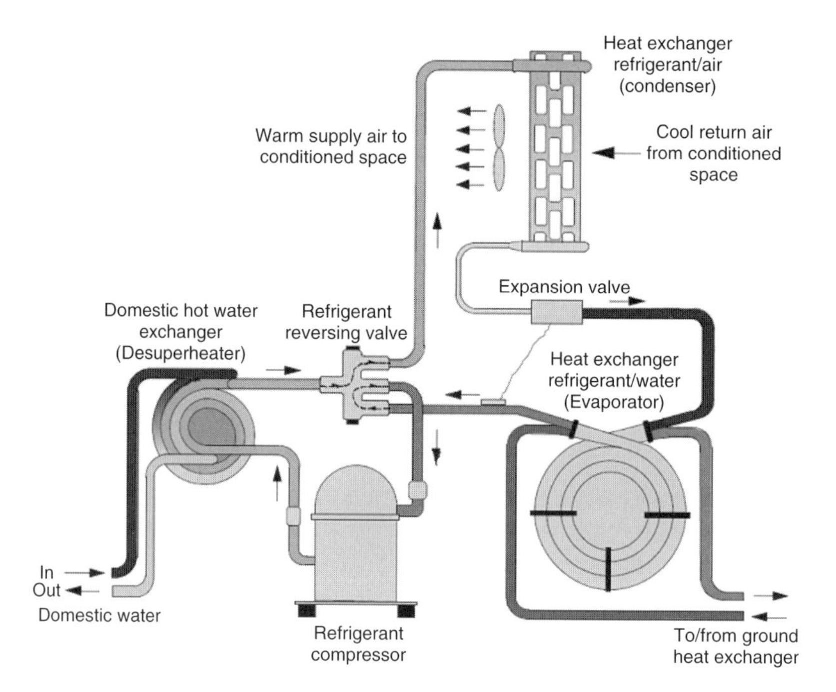

Figure 6.22 A heat pump with a 'desuperheater'. Note that the heat pump is reversible and can be switched from heating to cooling mode via the 'refrigerant reversing valve'. *After IGSHPA (1988) and reproduced by kind permission of © the International Ground Source Heat Pump Association/Oklahoma State University.*

6.5.1 Heat pumps that deliver DHW – the desuperheater

The amount of heat that we typically use for DHW (Figure 6.3) is often substantially less than that used for space heating. Thus, it is possible to buy a heat pump with a gadget called a 'desuperheater' (Figure 6.22). This is a small heat exchanger, located between the compressor and the condenser, where a small amount of high-temperature heat is skimmed off from the hot refrigerant gas and transferred to the DHW system. This provides householders with a total solution to space heating and DHW, although it introduces another layer of complexity to the system and, inevitably, the coefficient of performance of the heat pump system suffers.

6.5.2 A two-stage approach

Imagine that we use a conventional GSHP to transfer heat to a building loop of fluid at, say, 45°C that delivers space heating via an underfloor heating circuit. We could envisage a spur of the building loop that supplies warm fluid to a second heat pump, which extracts a small amount of heat from the building loop at 45°C and transfers it to a DHW circuit at >60°C. Because the temperature 'step' is relatively modest, this secondary heat pump performs relatively efficiently (Figure 6.23a). However, again, the

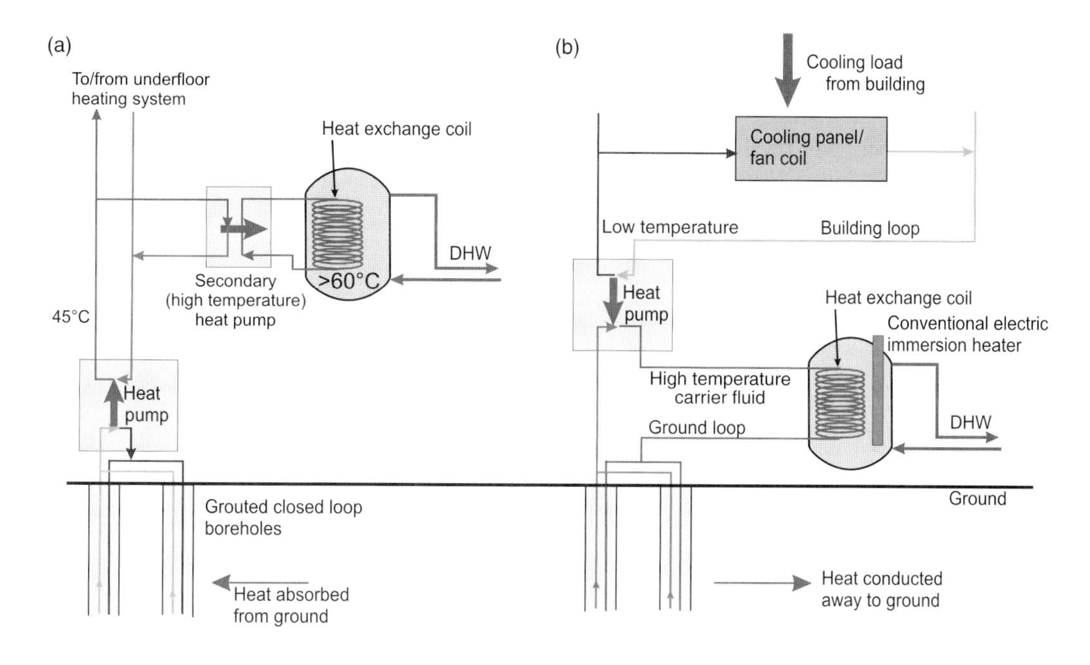

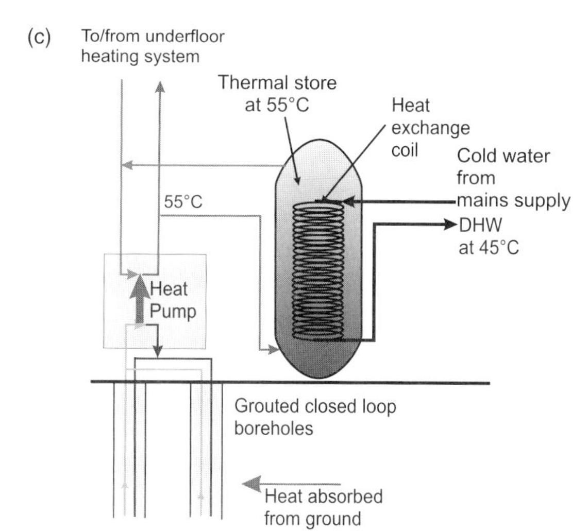

Figure 6.23 Three plausible means of providing DHW by means of GSHPs, in highly schematic form: (a) using a two-stage heat pump system (shown in a system providing space heating); (b) using the heat pump to pre-heat DHW (shown in a system in cooling mode); (c) using the heat pump to maintain a large sealed thermal store at 55°C, to heat cold mains water to 45°C on demand.

total system performance will still be lower than for a system solely providing space heating at 45°C. We should also note that the refrigerants that performed best in vapour compression heat pumps with high evaporator temperatures have now been restricted as being ozone-unfriendly. Most modern environmentally benign refrigerants prefer evaporator temperatures up to 20°C. However, Rakhesh *et al.* (2003) have demonstrated that, by careful choice of refrigerant, efficient systems can be constructed to operate at high evaporator temperatures.

6.5.3 *Using a heat pump to pre-heat DHW*

Alternatively, we could bow to the inevitable and recognise that, if we use a GSHP to supply hot water at 60°C, its performance will suffer. We could, thus, simply use the GSHP to elevate the temperature of a DHW cylinder to 45°C (say) and use an alternative source of energy, such as a conventional electrical resistance element, a gas boiler or a solar thermal panel, to raise the temperature on a regular basis to >60°C. In this case, a separate spur of the building loop would circulate warm carrier fluid through a heat exchange element (such as a copper coil) within the DHW tank (Figure 6.23b).

A third solution (Figure 6.23c) involves using the heat pump to warm a large insulated thermal store of water to, say, 55°C. As the store is a sealed system (and biocidal agents could also be added, if necessary), exposure to *Legionella* is minimised. When DHW is demanded, it is drawn from the cold mains water supply through a heat exchange element in the tank. The exchange surface is sufficiently large to warm the water to maybe 45°C for immediate use. As this DHW is not stored, but produced 'on demand', *Legionella* concerns are again minimised.

Note that, in discussions of the above solutions, we have considered production of DHW from heat pumps that also deliver space heating. If we are considering heat pump systems that deliver space cooling, the whole exercise of generating DHW begins to look increasingly attractive. Here we would not be merely 'stealing' heat that would otherwise have been used for space heating: we would have been recovering rejected heat that would otherwise have been regarded as waste (Figure 6.23b).

Finally, of course, after deep consideration, we may decide that we should not try to get ground source heat to do everything! We may decide that GSHPs are good at providing space heating. We may then decide to select a complementary renewable energy technology, such as solar thermal roof panels, to provide our DHW.

6.6 Heating and cooling delivery in complex systems

6.6.1 *Ground source heat pumps to provide cooling – a summary*

We have already established (Chapter 4, Section 4.8, p. 71) that GSHPs can be reversed to provide for active cooling. The larger the building, the smaller the surface-area-to-volume ratio (typically) and the more likely the building is to have a cooling requirement for prolonged periods of the year. We have seen that promoting active ground-sourced cooling as an 'environmentally friendly means of cooling' is a slightly dubious practice: true, using GSHPs for cooling may be 20–40% more efficient (Kelley, 2006) than conventional cooling (disposing waste heat to air) and have a lower visual impact. Never- theless, we are still essentially using energy to throw heat away! Cooling solutions by means of 'passive' or 'free' cooling, which utilise the natural temperature gradient between the building and the earth are preferable in terms of environmental impact but may have a greater capital cost.

The situation where active cooling (using a heat pump to dispose of heat to the ground) begins to look attractive is where we have approximately balanced heating and cooling loads to the ground. In this case, the subsurface can be used to store the summer's waste heat, such that it can be re-extracted during the winter. Such underground thermal energy storage (UTES) schemes will be discussed further in Chapters 7 and 10. If (as is often the case for large buildings in temperate or southern Europe) the waste heat from cooling is larger than heat extracted for space heating (i.e. a net cooling scheme), we can bring the heat fluxes to and from the ground back into balance by dumping a proportion of the waste heat to air via conventional cooling towers. Such hybrid schemes increase the likelihood of obtaining the subsurface thermal balance that is important for the long-term sustainability of UTES concepts (Spitler, 2005).

6.6.2 Centralised systems

A 'centralised' GSHP scheme is one that has a centralised plant room containing the heat pump(s). The heat pumps are typically either in heating or in cooling mode and provide space heating or cooling to the entire building or a specific zone of a building. A domestic heat pump scheme is an example of a small-scale centralised scheme.

Space heating can be provided in centralised schemes by means of warm air circulation (in a water-to-air heat pump) of by means of a building loop containing warm water and feeding an underfloor heating system or a wall-mounted radiator system. In reverse (cooling) mode, such underfloor and wall-mounted radiators, through which a cool fluid is circulating, are not especially efficient at delivering a large cooling effect (furthermore, they may lead to problems of condensation). Cooling is much more efficiently achieved by means of high-level structural elements such as chilled beams or panels. Thus, for GSHP systems that are designed to be reversed, we are faced with potentially two different delivery systems within the building.

This can be overcome, however, by utilising forced convection of air. Here, we circulate the hot or cold fluid in the building loop via a network of metal heat exchange elements. Air is forced (by a fan) through these heat exchangers and is heated or cooled by the building loop fluid. It is this warm or cold air that performs the active space heating or cooling. Fan coil units, mounted on a building loop, are perhaps the most common solution of this type.

6.6.3 Simultaneous heating and cooling

In large buildings, we must recognise that different parts of the building may have different heating and cooling requirements. In a University department, a south-facing computer laboratory may contain tens of sweaty students and computers (all generating heat) and may receive full sunlight for much of the day. It may thus have a significant cooling requirement, even in spring. On the top floor, a philosophy professor may have an office on the north side of the building. He may not know how to use a computer and may spend hours sitting immobile, pondering the significance of Platonic idealism in politics in sixth century Byzantium. In other words, he might get a bit chilly and have a requirement for heating. How can a GSHP system supply both heating and cooling at the same time? One solution is, of course, to divide the building into separate zones and have a heat pump array (in either heating or cooling mode) supply each zone independently.

More elegant solutions are available to us: even with a single, centralised heat pump array, we can provide both heating and cooling. We should remember that every heat pump has a 'hot' and a 'cold' side. Let us imagine a heat pump array that is providing a net heating load to the majority of a building. There may, however, be a computer facility that requires cooling. We can simply couple a heat exchanger (e.g. fan coil units) in the computer facility to the 'cold' side of the heat pump. In a closed-loop system, we could access the cold carrier fluid after it has passed through the heat pump but before it enters the borehole array. 'Waste' heat from the cooled building zone would be dumped to the carrier fluid. But the heat would not be 'wasted' – it would ultimately be re-extracted by the heat pump on the next circuit through the closed loop and be used to heat other zones of the building. Ultimately, the amount of ground source heat needing to be extracted from the boreholes would be reduced, potentially saving capital cost in terms of drilled metres of borehole! With some nifty valvework and automated controls, different building zones can be connected to the 'hot' or 'cold' sides of a heat pump array in response to changes in heating and cooling demand.

With a little imagination, we can likewise see that, for a building with a net cooling demand, heating can be provided to some zones by connecting them to the 'hot', ground-coupled side of the heat pump. This is essentially the same principle as that shown in Figure 6.23b for the production of hot water in a building with a net cooling demand.

6.6.4 Distributed systems

Another elegant solution to fluctuating heating and cooling demands in different zones of the same building is to install a so-called distributed system. Here, the ground loop does not simply enter a heat pump array in a centralised plant room. Rather, the ground loop actually becomes the building loop: it circulates around the building and feeds a network of small individual heat pumps in each zone or even each room. Each heat pump can be controlled either by thermostat or by the occupants of the room directly. On cold days, the individual heat pumps can be set to heating mode. They extract heat

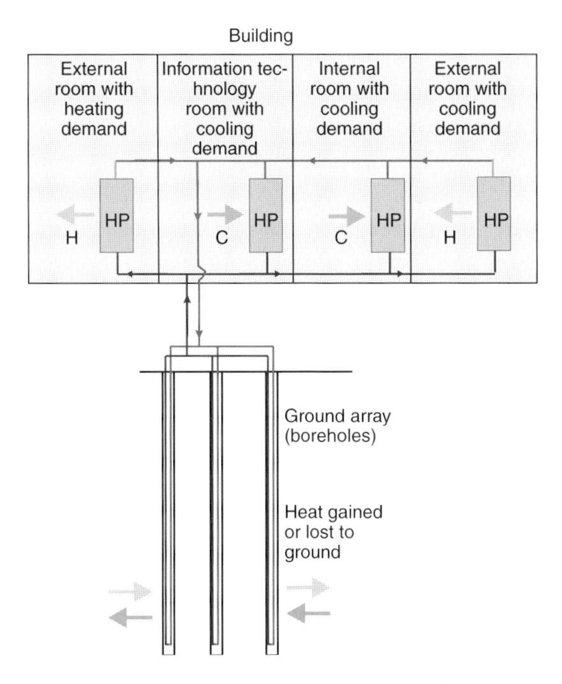

Figure 6.24 A schematic diagram of a distributed heat pump scheme.

from the building/ground loop's fluid and deliver it to the room (the heat pump may be a small wall-mounted water-to-air console unit of the type shown in Figure 4.8b, p. 77). On hot days, the heat pump extracts heat from the room and dumps it back into the building/ground loop. On some days, heat pumps in different parts of the building may be either dumping heat to, or extracting heat from, the building/ground loop, according to demand. Only the net imbalance between heating and cooling requirements needs to be rejected to or absorbed from the ground (Figure 6.24).

We can use some basic equations to assess the amount of net heat transfer from such a loop to the ground:

$$\text{Net waste heat from building loop} \approx C\left(1+\frac{1}{\mathrm{COP}_C}\right)-H\left(1-\frac{1}{\mathrm{COP}_H}\right) \tag{6.15}$$

where H and C are the heating and cooling loads delivered to the building, and COP_H and COP_C are the coefficients of performance of the heat pumps in heating and cooling mode, respectively. The electrical energy (E) required by the heat pumps is given by

$$E=\frac{C}{\mathrm{COP}_C}+\frac{H}{\mathrm{COP}_H} \tag{6.16}$$

Note that both these equations neglect any energy consumption by circulation pumps, and other inefficiencies in the system.

6.7 Heat from ice

Heat pumps can be utilised in extreme climatic conditions – IEA (2001b) document a scheme in Norwegian Lappland, at Kautokeino, where 16 deep coaxial closed-loop boreholes, to depths of up to 145 m, support 290 kW of installed heat pump capacity to heat a health centre at a location where undisturbed shallow ground temperatures may be as low as 2°C.

But what about the truly extreme corners of the planet, where the ground is permanently frozen? Here, heat pumps can be used to deliver a solution to an unusual geological problem. Significant parts of northern Asia and America are underlain by permafrost; that is, frozen ground, where the average annual air temperature is so low that the upper tens or hundred of metres of the geological column are permanently below 0°C. Pore space in rocks and sediments is saturated not with groundwater but with ice. Constructing viable dwellings in such terrain is not easy: they need to be heated, of course. But if a paraffin stove or electric heating system is installed in house on permafrost, the heat generated tends to melt the permafrost beneath the house and to cause subsidence. One way around this is to have a very well insulated floor, such that heat exchange between the lounge and the ground is minimised. The inhabitants of Longyearbyen, on Svalbard, build their houses on pillars in order to avoid contact between the house and the ground.

An alternative solution, described from Yukon and eastern Siberia (Perl'shtein *et al.*, 2000; Environment Canada, 2002), is to install a GSHP with the closed loop beneath the building. Heat is thus extracted from the permafrost beneath the house, keeping the ground frozen and stable, and delivered as usable space-heating energy to the building. Bingo ... heat from ice!

7

The Design of Groundwater-Based Open-Loop Systems

> Darcy's Law governs the motion of ground-water under natural conditions and ... is analogous to the law of the flow of heat by conduction, hydraulic pressure being analogous to temperature, pressure gradient to thermal gradient, permeability to thermal conductivity, and specific yield to specific heat.
>
> Charles V. Theis (1935)

The establishment of an open-loop heating or cooling system using groundwater depends on the existence of an aquifer. An aquifer is a body or stratum of rock or sediment that has adequate hydraulic properties (hydraulic conductivity and storage) to permit the economic exploitation of groundwater. In our specific case, the aquifer must yield enough water to support the required heating or cooling load of our system.

We have already encountered open-loop systems in Chapter 6. Here we will consider them in greater detail. We will look briefly at the different types of aquifer that exist in nature. We will look at conceptual designs of wellfields in various aquifers and at means of estimating the likely yields of wells. Means of obtaining information about aquifer properties are briefly covered. Thereafter, the sustainability of the wellfield will be examined – how long will it take water to travel between an injection and an abstraction well? Moreover, how rapidly will a heat signal from an injection well break through into an abstraction well? What can we do to improve the sustainability and design life of open-loop systems?

First, let us look at some of the common design flaws in open-loop groundwater-based systems.

7.1 Common design flaws of open-loop groundwater systems

- Lack of specialist design input from a hydrogeologist or groundwater engineer.
- Overoptimism regarding the hydraulic properties of aquifers. In particular, the common misconception that, if one well yields 5 L s^{-1}, a wellfield of 10 wells will yield 50 L s^{-1}. This is not necessarily the case. If no re-injection occurs, the total yield will typically be significantly less than simple multiplication would suggest. On the other hand, if spent water is re-injected it will serve to support the abstraction and it may be possible to abstract more water than would be the case with an 'abstraction only' scheme.
- Lack of consideration given to disposal of rejected (waste) water.
- Lack of appreciation that injection wells require specialist design, may under-perform compared with abstraction wells, and will need careful operation and maintenance to ensure a long life.
- Lack of appreciation that water chemistry and microbiology may affect the long-term performance of the system (Bakema, 2001).
- Lack of consideration of hydraulic breakthrough: that is, the possibility that warm (or cold) wastewater may flow back into the abstraction well, compromising the efficiency of the system (or even its long-term sustainability).

7.2 Aquifers, aquitards and fractures

Aquifers (from Latin: *water-bearing*) are strata or bodies of rock or sediment that yield economically useful quantities of groundwater. The critical hydraulic properties of an aquifer are the hydraulic conductivity (K) and the storage (S) of the material of which it is composed. We have already met hydraulic conductivity in Darcy's law (Chapter 1; Equation 1.2). Storage refers to the amount of water (in m^3) that a unit of aquifer material can release (or take up) in response to a 1 m decline (or increase) in head. Sediments such as sands and gravels, and rocks such as porous sandstones, are usually quite good aquifers, and are termed *porous-medium* or *intergranular flow* aquifers. They have a high K (Table 7.1) and a respectable porosity, implying a relatively high value of S. In such aquifers, water flows through the pore spaces between the grains. The wider the necks between the pores, the higher will be the value of K.

Fine-grained sediments, such as silts and clays, may have a high porosity, but the pore spaces are only very small. As K exhibits a very strong dependence on the aperture of the pore spaces (or, strictly speaking, the necks connecting the pores), the hydraulic conductivities of such sediments are low: wells yield poorly and the sediment is referred to as an *aquitard* (from Latin: *water + slow/late*).

Some rocks – in particular, crystalline rocks, such as granites, slates or gneisses – consist of tightly interlocked silicate crystals, with practically no intergranular pore space. One might think that such rocks would be wholly impermeable, but in fact, wells drilled in such aquifers commonly yield several hundred litres of water per hour,

Table 7.1 A summary of the typical hydraulic properties of geological formations (based on Domenico and Schwartz, 1990; Allen *et al.*, 1997; Fetter, 2001; Olofsson, 2002; Misstear *et al.*, 2006).

	Porosity (n) %	**Specific yield (S_Y) %**	**Hydraulic conductivity (K) m s^{-1}**
Clay	30–60	1–10	10^{-12}–10^{-8}
Silt	35–50	5–30	10^{-9}–10^{-5}
Sands	25–50	10–30	10^{-7}–10^{-3}
Gravel	20–40	10–25	10^{-4}–10^{-1}
Sandstone	5–30	5–25	10^{-9}–10^{-4}
Most unweathered crystalline silicate rocks (granites, schists, gneisses)	<1 0.1** <0.05*	<1	10^{-13}–10^{-5} depending on degree of fracturing
Basalt	<1–50	<1–30	10^{-13}–10^{-2}
British Chalk	10–45	0.5–5	10^{-10}–10^{-6}# (median 7×10^{-9})#

* Olofsson (2002) cites an effective (kinematic) porosity of crystalline rock aquifers of <0.05%.
** Domenico and Schwartz (1990) cite values as low as 0.1% for porosity and 0.0005% for effective porosity in granite.
# These figures refer to Chalk matrix hydraulic conductivities. Bulk transmissivities of the Chalk aquifer are typically in the range 10–10 000 m^2 day^{-1} (10^{-4}–10^{-1} m^2 s^{-1}), which provides some impression of the importance of fracture flow in the Chalk (Allen *et al.*, 1997).

and sometimes several thousands per hour. Such rocks have been subject to the imposition and release of enormous stresses throughout geological time, which have caused them to fracture and crack. The resulting fractures and joints are able to transmit groundwater. If a well intersects such interconnected fractures, it will yield a supply of groundwater. Such aquifers are termed *fracture-flow* aquifers.

Finally, some types of aquifer, such as limestone and dolomite, typically have a rather low intergranular permeability. Limestones often consist of interlocking carbonate crystals with minimal intergranular pore space, or tiny fossilised organisms (Chalk, e.g., is made of tiny *coccolith* microfossils) providing a large overall porosity but with very small pore apertures (Table 7.1). In these lithologies, too, most groundwater flow takes place via fractures and joints. However, unlike silicate rocks, the calcite or dolomite minerals constituting these lithologies are rather soluble in groundwater. Thus, over a period of thousands of years, the fractures become wider and more permeable. They can reach several mm or cm in aperture and transmit huge amounts of groundwater flow (Figure 7.1). In extreme cases, cave systems result and the flow regime is typically referred to as karstic. Such fissure and cave development was particularly intense in periglacial conditions towards the end of the last ice age, as CO_2 is more soluble in water at cold temperatures, and it is this carbon dioxide that is responsible for carbonate dissolution:

$$CaCO_3 + CO_2 + H_2O = Ca^{2+} + 2HCO_3^- \tag{7.1}$$

Limestone + Carbon dioxide + Water = Dissolved calcium and bicarbonate

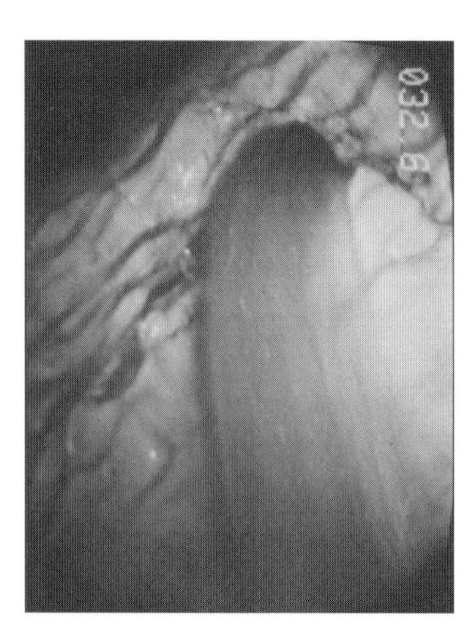

Figure 7.1 A photograph taken in a well, showing groundwater cascading out of a solution cavity (fissure) in a pumped borehole in the Chalk aquifer of southern England. The solution pipe is probably around 4 cm across. *Public domain information, provided by and reproduced with the permission of the Environment Agency of England and Wales (Thames Region).*

Aquifers such as limestones or Chalk, where groundwater flow has enhanced hydraulic conductivity by dissolution of fractures, are often referred to as *karstified* aquifers (Banks *et al.*, 1995; Waters and Banks, 1997), much to the disgust of many speleologists who wish to retain this term to describe very specific landforms and geomorphologies!

7.3 Transmissivity

We need one further term to complete our description of the hydraulic properties of aquifers. By considering Darcy's Law (Equation 1.2), it should be possible to see that an aquifer comprising a layer of sand 5 m thick will be able to transmit only half as much groundwater (per unit width) as a similar layer 10 m thick, assuming the same head gradient in both cases. Whereas hydraulic conductivity (K) is an intrinsic property of the material comprising the aquifer, the transmissivity (T) is an extrinsic property. Transmissivity describes the ability of the aquifer, as a geological unit, to transmit groundwater flow; it is defined as the product of the hydraulic conductivity (K) and the thickness (D) of the aquifer.

$$T = K \cdot D \tag{7.2}$$

Its units are $m^2\, day^{-1}$ or $m^2\, s^{-1}$. Consider Darcy's law (Equation 1.2):

$$Z = -K \cdot A \cdot \frac{dh}{dx} \tag{7.3}$$

where Z = flow of groundwater ($m^3\,s^{-1}$), $A = Dw$ = cross-sectional area of the block of material under consideration (m^2), w = width of aquifer under consideration (m), h = head (m), x = distance coordinate in the direction of decreasing head (m) and dh/dx = head gradient (dimensionless).

By substituting our expression for transmissivity we obtain

$$Z = -T \cdot w \cdot \frac{dh}{dx} \tag{7.4}$$

As a point of interest, we can also define the transmissivity of a fracture T_f (i.e. its ability to transmit water under a given head gradient). This value depends, of course, not only on the aperture of the fracture, but also on its roughness and its tortuosity. However, for an ideal fracture; smooth-sided, plane-parallel and of constant aperture b_a (Snow, 1969; Walsh, 1981; Misstear *et al.*, 2006):

$$T_f = \frac{\rho_w g b_a^3}{12\mu_w} \approx 629\,000 b_a^3 \tag{7.5}$$

where T_f is in $m^2\,s^{-1}$ and b_a is in m, ρ_w is the density of water ($\approx$1000 $kg\,m^{-3}$), g the acceleration due to gravity (9.81 $m\,s^{-2}$) and μ_w is the dynamic viscosity of water ($\sim$0.0013 $kg\,s^{-1}\,m^{-1}$). Hence, we can appreciate the huge dependence of fracture transmissivity on aperture.

7.4 Confined and unconfined aquifers

A confined aquifer is a transmissive stratum of sediment or rock that is overlain by a low-permeability aquitard and where the groundwater head is higher than the top of the aquifer. An example would be a layer of sand sandwiched between two layers of clay (Figures 2.5, p. 21, and 7.2). The groundwater wholly saturates the aquifer and is, in a sense, under excess pressure. Thus, when a borehole is drilled into the aquifer, water rises up the borehole to a level above the top of the aquifer – a level corresponding to the groundwater head in the aquifer. We can imagine a surface, called the piezometric surface, describing the locus of the head (h) at any point in the aquifer. The gradient of this surface is the hydraulic gradient or head gradient. This, of course, controls the direction and rate at which groundwater flows (down the hydraulic gradient), according to Darcy's Law.

If the piezometric surface is above ground level, a borehole drilled into the aquifer will be artesian – the water will overflow under its own pressure at ground level (Figures 2.5, p. 21, and 13.3, p. 284).

An unconfined aquifer is not held under pressure by a low-permeability cap (Figures 3.2, p. 41, and 7.2). Its upper boundary is a free water surface – the water table. Rainfall (or other recharge) can usually enter from the top of the aquifer. The unsaturated portion of strata is called the vadose zone, while the saturated portion of

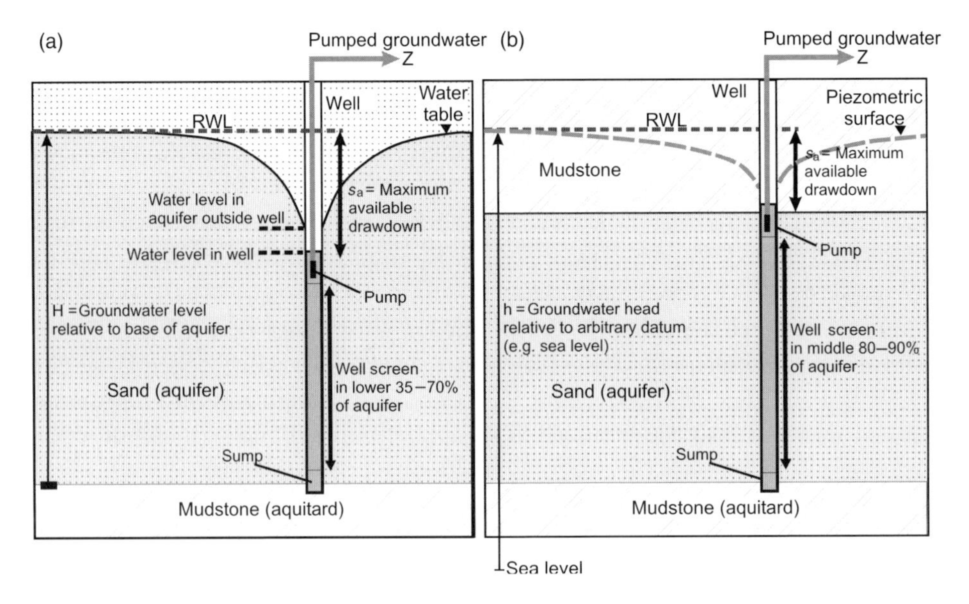

Figure 7.2 Schematic diagrams of (a) an unconfined and (b) a confined aquifer, illustrating typical placements of well screens. s_a = the available drawdown. RWL = rest water level.

the aquifer is the phreatic zone. The transmissivity of the unconfined aquifer is given by Equation 7.2 but, be careful! The thickness of the aquifer (D) can change as the water table rises and falls with the seasons. It can also change if we start abstracting water from the aquifer and drawing down the water table. Thus, whereas the transmissivity of a confined aquifer is usually constant, the transmissivity of an unconfined aquifer can vary seasonally and with abstraction. Especially in an unconfined aquifer such as the Chalk (Owen and Robinson, 1978), many of the most transmissive fissures are situated in the shallow part of the aquifer, around the water table. Thus, as the water table declines, these transmissive fissures become dewatered and the transmissivity of the aquifer can drop dramatically.

Finally, the storage coefficient of an unconfined aquifer is termed the specific yield (S_Y). If the water table falls by 1 m, a quantity of water will be released from storage that will be related to the porosity (n) but will be somewhat less as we can never fully drain a geological material by gravity. There will always be some small amount of water that is retained – adhering to mineral grains or tucked away in 'blind' pore spaces. If we consider 1 m^2 of aquifer, we can say that

$$S_Y < n \tag{7.6}$$

The units of S_Y are m^3 of water per m^2 aquifer area per m decline in head – that is, dimensionless. In granular porous media, S_Y is often a few percent (i.e. >1% or >0.01) and may exceed 10% (0.1 as a fraction). In fissured or fractured media, porosity (and hence S_Y) is very low. Indeed, porosities of <1% (and usually <0.1%) are likely in fresh crystalline silicate rocks (Table 7.1).

In confined aquifers, the amount of water released from every cubic metre of aquifer as head declines by 1 m is termed the specific storage (S_S). Here, the aquifer is not actually dewatered as the head drops; water is merely released because of the very small elastic responses in the water and the aquifer matrix. S_S in confined aquifers is thus very small (of the order 10^{-5} to 10^{-6} m^{-1}). Its units are m^3 water per m^3 aquifer per m decline in head, or m^{-1}. We can integrate S_S over the entire confined aquifer thickness to get an overall aquifer storage coefficient or storativity (S) – this is the amount of water in m^3 released per m^2 area of aquifer per m decline in head (i.e. dimensionless). Whereas S_S does not depend heavily on aquifer lithology, S is dependent on aquifer thickness and is given by

$$S = S_S \cdot D \tag{7.7}$$

7.5 Abstraction well design in confined and unconfined aquifers

In this brief section, we will consider some very basic elements of design of groundwater abstraction wells. For more detailed information, see Misstear *et al.* (2006). Re-injection wells can be similar to abstraction wells, but are often a wholly different kettle of fish – see Section 7.9.3.

7.5.1 Confined aquifers

In a confined aquifer comprising poorly lithified porous materials (sands, gravels, poorly cemented sandstone), it is wise to screen as much of our aquifer as possible to maximise the water yield from an abstraction well (Figure 7.2). Indeed, Driscoll (1986) suggests screening around 90% of the aquifer thickness, leaving small amounts of blank casing overlapping at the top and bottom of the aquifer to ensure that no fine particles from the overlying and underlying clayey aquitards migrate through the pore spaces of the aquifer to enter the well. Of course, if our required yield is small and our aquifer is very transmissive and relatively thick (>100 m, say), we may not choose to drill to the base of the aquifer, but to place a well screen in the upper portion of the aquifer only. This will reduce the hydraulic efficiency of the well, but this may be a price worth paying to save on the capital cost of a deep well to the base of the aquifer.

If the confined aquifer is well lithified (e.g. a lithified limestone, a well-cemented sandstone, a granite; Figure 6.6a, p. 114), the well may be completed 'open hole' (no well screen) within the aquifer itself. A section of blank casing would usually be installed in the upper part of the well (corresponding to the confining aquitard).

When drilling into a confined aquifer, especially if there is a suspicion that it may be artesian, it is good practice to grout a string of casing into the confining stratum (see Section 13.3, p. 281) before the aquifer is encountered. Drilling continues through the grout plug in the base of the cased section of borehole at a narrower diameter. This practice ensures that there is no hydraulic communication between the surface (or any overlying aquifer) and the target confined aquifer. It also allows any artesian overflow to be controlled. If a strong artesian overflow is encountered and a string of casing has not be been securely grouted into the overlying aquitard, it can be very difficult to bring the overflow under control (Figure 13.3, p. 284).

When pumping a well in a confined aquifer, it is generally recommended as good practice that the confined aquifer is not dewatered. In other words, the pumping water level in the abstraction well should remain above the top of the aquifer or top of the well screen. If we start to dewater the aquifer, we introduce air into a previously anaerobic environment. This can bring about all sorts of unwanted chemical reactions, such as the oxidation of sulphide minerals, oxidation and precipitation of iron and stimulation of bacterial growth. We can thus say that the available drawdown (s_a) is the difference between the undisturbed piezometric surface and the top of the aquifer (or top of the pump, whichever is higher). In all fairness, it should be stated that this rule is often 'bent' by practitioners, especially in the case of 'open' boreholes drilled into lithified confined aquifers. We break it at our own risk, however, in the awareness that doing so may shorten the life of our well and/or impair its performance.

7.5.2 Unconfined aquifers

In an unconfined aquifer, we will draw down the water table as a result of a groundwater abstraction (Figure 7.2). It is a general guideline that this drawdown will seldom exceed 50% of the aquifer thickness, but may be in the range 30–65% of aquifer thickness.

In designing a well screen for an abstraction well in an unconsolidated aquifer comprising poorly lithified, relatively homogeneous sediments or rocks, there are two schools of thought. Some will recommend placing a well screen in the bottom 35–70% of the aquifer, and operating the well such that the water level never falls below the top of the well screen. This means that the well screen always remains under water and is not exposed to oxygen that can promote corrosion, bacterial fouling and chemical incrustation. In this case, the available drawdown is the difference between the rest water table (in its undisturbed state) and the top of the well screen. Others will simply install a well screen in the entire saturated section of the aquifer and accept that pumping will cause the uppermost section of the screen to become alternately dewatered and re-saturated. They will accept that the performance of this upper screen section may deteriorate over time. The former philosophy makes sense if the cost of well screen is much greater than that of plain casing; the latter pragmatic philosophy makes sense if the costs are not too dissimilar.

If the aquifer is not homogeneous, but contains specific, high-transmissivity flow horizons, the zones that are installed with well screen will be selected to coincide with those horizons. Furthermore, as in the case of the confined aquifer (Section 7.5.1), if our required yield is small and the aquifer is very transmissive (or if the aquifer is very thick), we may choose not to drill to the aquifer's base.

In a lithified, well-cemented or crystalline aquifer, we will typically construct our well 'open hole'. A section of blank casing would usually be installed and grouted into the upper part of the well to exclude any superficial materials, badly weathered, unstable strata and potentially contaminated shallow groundwater. In the particular example of limestones, such as the Chalk, where transmissivity is concentrated at specific horizons (in the few metres or tens of metres below the water table in the Chalk), we will often aim not to pump at such a high rate that we wholly dewater these especially transmissive horizons. This consideration may restrict still further our available drawdown.

7.6 Design yield, depth and drawdown

Using the formulae discussed in Chapter 6 and Equations 4.7 and 6.7–6.11, we will usually have defined a groundwater yield that we require to satisfy a given load of ground-source heating or cooling. When we pump a well in a confined aquifer (Figure 7.2), we draw down the piezometric surface around the abstraction well, creating a radial flow field (and hydraulic gradient) towards the pumping well. The area where the groundwater levels are drawn down is called the *cone of depression*, and the vertical difference between the original piezometric surface and the new piezometric surface as a result of pumping, is termed the *drawdown*.

The American hydrogeologist Charles V. Theis (one of the key figures of modern hydrogeology) deduced in 1935 a formula for the transient (time-dependent) radial flow of groundwater towards a pumped well (Box 7.1). The formula provides the relationship between aquifer transmissivity (T), groundwater abstraction rate (Z) from the well, dimensionless storage (S) and drawdown (s) at a given radial distance (r) from the pumping well at any given time (t) after pumping started:

$$s = \frac{Z}{4\pi T} W(u) \tag{7.8}$$

where $W(u)$ is a function known as the Theis well function:

$$W(u) = -0.5772 - \ln u + u - \frac{u^2}{2.2!} + \frac{u^3}{3.3!} - \frac{u^4}{4.4!} + \frac{u^5}{5.5!} - \cdots \tag{7.9}$$

and

$$u = \frac{r^2 S}{4Tt} \tag{7.10}$$

The polynomial term $W(u)$ converges quickly and can usually be calculated by considering the first, say, five power terms (after checking that adequate convergence has

BOX 7.1 Charles V. Theis.

C.V. Theis. From public domain USGS document by White and Clebsch (1994).

The American hydrogeologist Charles Vernon Theis was born in Kentucky in March 1900. He studied civil engineering at the University of Cincinnati in 1917 and subsequently gained a post in the University's Geology Department. Later, in 1927, Theis became a junior geologist at the US Geological Survey (USGS), based in Moab, Utah. In 1929, he completed his doctoral degree (in geology) from the University of Cincinnati. He joined the Ground Water Division of the USGS in 1930 and stayed there for the rest of his career. Much of his early work was based on the assessment of groundwater abstractions for irrigation in the arid region of New Mexico. He was astute enough to realise that existing equations (such as that of Thiem) were not wholly adequate for describing the impact of an abstraction on regional groundwater levels. Indeed, he realised he needed a good, time-variant solution to the fundamental groundwater flow equations (Darcy's Law and the conservation of mass). It would seem that Theis's maths was not up to the job of solving this puzzle. Fortunately, he was not too proud to ask for advice and he sought it from his old University friend, Clarence Lubin. Lubin steered Theis in the direction of some earlier work by the physicist H.S. Carslaw (published in the 1921 book *Introduction to the Mathematical Theory of the Conduction of Heat in Solids*). Lubin and Theis figured that a non-equilibrium solution to groundwater flow to a well would probably be analogous to heat conduction through a solid towards a thermal sink. Theis published the non-equilibrium groundwater solution in his 1935 paper (of which Lubin was reputedly offered co-authorship, but modestly refused). In this paper, Theis explicitly draws attention to the analogy between heat conduction and groundwater flow (see Table 1.1, p. 7).

Theis enjoyed many further years with the USGS, and later worked on issues as diverse as artificial recharge, aquifer inhomogeneity and anisotropy, and radioactive waste disposal (White and Clebsch, 1994).

occurred), allowing manual calculation of drawdown for a given r, T, S and t. We can already see from this, admittedly complex, formula that

- well yield (Z) is approximately proportional to transmissivity (T);
- drawdown (s) is approximately proportional to abstraction rate (Z);
- drawdown is inversely proportional to transmissivity;
- drawdown increases with time of pumping;
- drawdown decreases with distance from abstraction well.

Cooper and Jacob (1946) realised that Theis's equation could be dramatically simplified if we assume that u is small (less than 0.01 or, some would say, less than 0.05). In this case, all the higher terms of the polynomial expansion become negligibly small and we are left with

$$s = \frac{Z}{4\pi T}\left[-0.5772 - \ln\left(\frac{r^2 S}{4Tt}\right)\right] = \frac{2.30Z}{4\pi T}\log_{10}\left(\frac{2.25Tt}{r^2 S}\right) \tag{7.11}$$

or

$$s = \frac{2.30Z}{4\pi T}\log_{10}\left(\frac{2.25T}{r^2 S}\right) + \frac{2.30Z}{4\pi T}\log_{10} t \tag{7.12}$$

Here we see that

- when u is small (r is small or t is large), drawdown increases in proportion to $\log_{10}(t)$;
- when u is small, drawdown decreases in proportion to $\log_{10}(r)$.

In the hypothetical, infinite, homogeneous aquifer considered by Theis, the cone of drawdown continues to expand forever. In a real aquifer, it continues to expand (Theis, 1940; Bredehoeft *et al.*, 1982) until it has

- induced sufficient recharge (i.e. induced vertical head gradients causing water to flow from wetlands, streams or lakes into the aquifer) or
- captured sufficient groundwater discharge (i.e. decreased the amount of spring flow overflowing as excess water from the aquifer)

to balance the abstracted quantity, at which point the aquifer stabilises to a steady-state condition.

Many years before Theis, Adolph Thiem (1887) had found a steady-state solution for radial groundwater flow towards an abstraction well. The following differential equation (derived from Darcy's Law in radial coordinates) describes this situation:

$$Z = -2\pi r T\frac{\mathrm{d}h}{\mathrm{d}r} \tag{7.13}$$

where h = groundwater head and r = radial distance from the abstraction well. We can integrate between two radial distances: r_1 and r_2.

$$h_2 - h_1 = s_1 - s_2 = \frac{Z}{2\pi T} \ln \frac{r_2}{r_1} \tag{7.14}$$

where h_1 and h_2 are the groundwater heads at these radii, and s_1 and s_2 are the drawdowns. If we consider our groundwater abstraction well as one of these two points, we can say that the expected drawdown in this well (for our yield Z) is s_w, where r_w is the radius of the abstraction well:

$$s_w - s_2 = \frac{Z}{2\pi T} \ln \frac{r_2}{r_w} \tag{7.15}$$

Logan (1964) devised the most famous approximation in hydrogeology when he dared to set r_2 to a certain distance r_e from the abstraction well, where the drawdown s_2 is effectively zero. Moreover, he said that (r_e/r_w) is often approximately equal to 2000. Thus, he obtained the Logan Approximation relating abstraction well yield (Z) to drawdown (s_w) in a pumped well.

$$T = \frac{Z}{2\pi s_w} \ln \frac{r_e}{r_w} = 1.22 \frac{Z}{s_w} \tag{7.16}$$

We should be rightly suspicious of the chain of dodgy assumptions leading to the Logan Approximation: Adolph Thiem assumed a steady state that Charles Theis demonstrated would never be reached in an ideal aquifer; Logan plucked a figure of 2000 from thin air! Yet, in fact, the Logan formula can prove a useful 'back of the envelope' first estimate of the relationship between transmissivity and well yield.

What does this mean for well design? If we know the transmissivity of the aquifer, and we have defined a maximum drawdown that is allowable in our well (e.g. to the top of the well screen), we can estimate the maximum yield we can expect from the well. Conversely, if our aquifer is thick and quite transmissive, we may wish to estimate how deep (D) we need to drill into our aquifer to achieve a given yield (Z). As transmissivity is the product of hydraulic conductivity (K) and effective aquifer thickness, and assuming that our well will be hydraulically efficient (see Section 7.7.1), we could say that

$$KD = 1.22 \frac{Z}{s_w}, \text{ thus } D = 1.22 \frac{Z}{K s_w} \tag{7.17}$$

In this case, our well of reduced depth would not perform as hydraulically efficiently as a well through the entire aquifer thickness – the drawdown (and hence the pumping costs) for a given yield would be greater. We are thus sacrificing running costs (electricity or fuel for the pump) for a lower capital cost (shallower well).

7.6.1 *A more sophisticated approach*

If the Logan approximation seems a little bit too primitive (and it should really only be used for an initial estimate of well yield), we can always utilise the Theis or Cooper–Jacob equations, if we can provide estimates of transmissivity and storage. To predict

BOX 7.2 Assessing the maximum yield of a drilled well.

We can use the Theis (Equation 7.8) or Cooper–Jacob (Equation 7.11) approaches to predict the drawdown (s) for any given yield (Z) at a given time (t) after pumping starts, provided we know the aquifer transmissivity (T) and storage (S). But what value of t should we use? Let us consider a proposed well, where the top of the well screen is at 27 m below ground level (bgl) and the pump is just below 25 m bgl. The natural groundwater level is 15 m bgl. Thus, we have an available drawdown of around 10 m. Let us assume that our aquifer has a transmissivity of 500 m^2 day^{-1} and a storage coefficient of 0.08.

If our well, of radius 0.1 m, is going to be pumping at a rate of 10 L s^{-1} (=864 m^3 day^{-1}) for 8 h day^{-1} throughout a 5 month winter heating season, it would probably be sensible to solve the Theis or Cooper–Jacob equation twice. First, we might set Z equal to the average winter abstraction rate of 3.3 L s^{-1} = 288 m^3 day^{-1} and use a value of t = 150 days. We apply Equation 7.11:

$$s_w = \frac{2.30Z}{4\pi T} \log_{10}\left(\frac{2.25Tt}{r_w^2 S}\right) = \frac{2.30 \times 288}{4\pi \times 500} \log_{10}\left(\frac{2.25 \times 500 \times 150}{0.1^2 \times 0.08}\right) = 0.9 \text{ m}$$

Second, we would set Z = 864 m^3 day^{-1} and t = 8 h = 0.33 days in order to simulate the peak abstraction condition on a single day.

$$s_w = \frac{2.30Z}{4\pi T} \log_{10}\left(\frac{2.25Tt}{r_w^2 S}\right) = \frac{2.30 \times 864}{4\pi \times 500} \log_{10}\left(\frac{2.25 \times 500 \times 0.33}{0.1^2 \times 0.08}\right) = 1.8 \text{ m}$$

Thus, using the average abstraction rate over an entire winter season, we would not expect the drawdown to exceed 0.9 m. Even if we superimpose a pulse of peak abstraction onto this long-term drawdown, we would not expect drawdown to exceed 3 m in total. This is well within our available drawdown of 10 m, even allowing for hydraulic inefficiency.

If we apply the Logan approximation (Equation 7.16) to the problem, we obtain 2.1 m drawdown for the peak condition and 0.7 m for the long-term average drawdown.

the drawdown for a given yield in our abstraction well, we would set r to the radius of abstraction well (r_w) and then solve the equation for a value of pumping duration (t). A worked example is given in Box 7.2.

7.7 Real wells and real aquifers

All of the above equations presuppose that our abstraction well is 100% hydraulically efficient and that our aquifer is 'ideal' – that is, confined, infinite, homogeneous, isotropic and generally well behaved in the way that real geology is not!

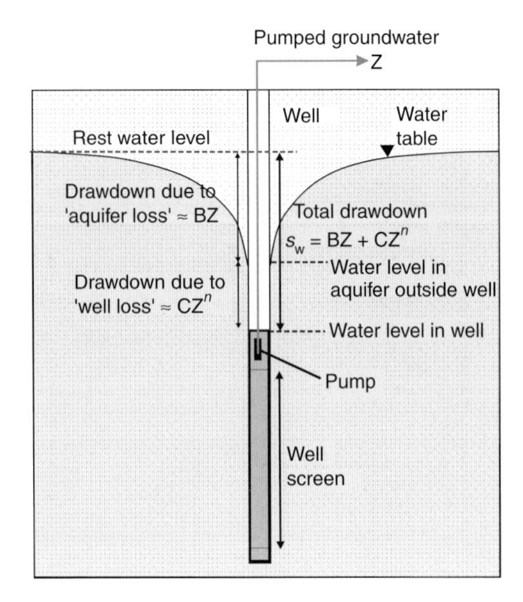

Figure 7.3 A real abstraction well, showing components of aquifer loss and well loss.

7.7.1 *Real wells*

No well is 100% hydraulically efficient. In a real well, turbulent flow, hydraulic resistance caused by the well screen/filter pack and several other factors all contribute to additional losses in head called well losses (Figure 7.3). These generally increase as abstraction rate (Z) increases, but in a non-linear way. While the Logan Approximation (Equation 7.16) predicts a simple proportionality between drawdown and yield, a more realistic equation might be

$$s_W = BZ + CZ^n = \left(\frac{1.22}{T} + B'\right)Z + CZ^n \tag{7.18}$$

where B represents linear head losses. These are dominantly natural head losses in the aquifer due to the hydraulic resistivity of the aquifer, but they may include a minor component of linear well losses, B'. C represents non-linear losses (typically well losses). n is a power law coefficient, often taken to be around 2 (Bierschenk, 1963; Hantush, 1964).

We have made so many assumptions already, what harm can another do? In fact, to make a first estimate of expected well yield, it is common practice (Misstear *et al.*, 2006) to use a modified version of Logan's Approximation, with the coefficient 1.22 increased (rather arbitrarily) to 2, to take into account possible well losses:

$$Z = \frac{Ts_W}{2} \tag{7.19}$$

Remember, Logan's approach is an approximation – use it only to gain a first estimate of well yield and drawdown, not for detailed design.

7.7.2 Unconfined aquifers

The equations derived hitherto have assumed that we are dealing with a confined aquifer. In fact, we can usually use them with a good conscience in unconfined aquifers, too, provided the drawdown is small in comparison with the total aquifer thickness. If this is not the case, the transmissivity of the aquifer may reduce significantly due to the decrease in its saturated thickness caused by drawdown. We thus run the risk of underestimating our expected drawdown or overestimating our yield. Modified versions of some of the standard equations may exist for unconfined aquifers. For example, the analogue to the Thiem equation (Equation 7.14) for an unconfined aquifer is as follows:

$$H_2^2 - H_1^2 = \frac{Z}{\pi K} \ln\left(\frac{r_2}{r_1}\right) \tag{7.20}$$

where H_1 and H_2 are the water table elevations at radii r_1 and r_2 relative to the base of the aquifer (and *not* to some arbitrary datum as is normally the case when we are considering groundwater head – Figure 7.2).

7.8 Sources of information

The formulae we have looked at are very fine, but they presuppose some knowledge of the properties of the aquifer. The best means of making a prognosis of the likely yield of a new well is to examine the performance of any nearby wells in the same aquifer. Most countries will have a database of wells and boreholes: the host organisation may be the Ministry of Water, the Geological Survey or the Environmental Authority. In the UK, the body charged with maintaining the database is the British Geological Survey (BGS). It is usually possible, on request, to obtain copies of drilling logs and pumping tests from boreholes and wells within, say, a 1 km radius of our proposed drilling location.

If there are no existing wells in the immediate vicinity upon which to base our prognosis, we may be able to interpolate or deduce aquifer properties from more distant sources of information. In many aquifers, especially sedimentary rock aquifers, properties will often vary gradually in space, such that maps can be made showing how properties such as transmissivity vary. These spatial variations may be related to systematic changes in grain size or sedimentological facies, or they may be more related to structural geological factors or even to recent weathering or dissolution history (the latter being relevant for limestone aquifers). In the UK, the Environment Agency and BGS (Allen *et al.*, 1997; Jones *et al.*, 2000) have published comprehensive summaries of data for the major and minor British aquifers. These publications provide an invaluable starting point for the prospective well owner: they contain maps of hydraulic properties, statistical distributions of well and aquifer properties and details of pumping tests. If we can estimate transmissivity from such maps and data sources, we can start to make prognoses for the yield from our proposed well from the equations discussed above.

If the available published information is inadequate and/or if we are planning a very expensive wellfield comprising a number of wells, we may wish to drill a pilot borehole, upon which we can carry out a pumping test. This pilot borehole will provide us with site-specific information on the aquifer and, if successful, can be incorporated into the final ground source heating or cooling scheme.

7.8.1 Pumping tests

Whole books have been written on the subject of pumping tests, of which the definitive volume is by Kruseman *et al.* (1990). A concise summary of test pumping is given in many hydrogeological textbooks, such as Misstear *et al.* (2006). In a pumping test, we typically pump our well at a constant rate (Z), and measure the drawdown (s) as it evolves with time (t), either in the abstraction well itself or in some observation borehole at a distance (r) from the abstraction well. With these data, we can perform an inverse solution to one of the equations described above to derive values for transmissivity (T) and storage (S), for example:

- the Theis equation – Equations 7.8–7.10;
- the Cooper–Jacob approximation – Equation 7.12;
- the Hantush (1964) and Bierschenk (1963) method – Equation 7.18, assuming $n = 2$;
- the Logan approximation – Equations 7.16 or 7.19;
- or some, more complex, variant of the Theis equation taking into account deviations from Theis's ideal confined aquifer.

A typical pumping test on a well will normally comprise several phases (Misstear *et al.*, 2006):

1. *Step testing* – where the well is pumped at a low rate (Z_1) for a short period (typically 2 h) and the drawdown (s_{w1}) measured in the abstraction well. After this first step, the groundwater is allowed to recover to the rest condition and a second step is commenced at a higher rate (Z_2). The new drawdown after 2 h for step 2 (s_{w2}) is measured. This process is typically repeated for four 'steps', the rates of pumping bracketing our design yield (Z_d), such that rates Z_1, Z_2, Z_3 and Z_4 will typically be equal to $0.33 \times Z_d$, $0.67 \times Z_d$, Z_d and $1.33 \times Z_d$. Alternatively, we can perform 'continuous' step testing, where the pumping rate is simply increased to the next rate at the end of the previous step, with no period of recovery. This cuts down the time required for step testing but the results are slightly more difficult to interpret. In either case, we end up with four well yields $Z_1 \ldots Z_4$, with corresponding drawdowns $s_{w1} \ldots s_{w4}$, enabling us to construct a curve of s_w versus Z (Figure 7.4), and allowing us to predict the 2 h drawdown for any yield. Note that the shape of the curve is often convex upwards, implying that for ever higher pumping rates, we develop disproportionately high drawdowns. This implies that well efficiency decreases as pumping rate increases: this non-linear relationship is predicted by Equation 7.18. Hantush (1964) and Bierschenk (1963) proposed a simple and elegant method for estimating well efficiency from such data. Remember that, in thin unconfined aquifers, a non-linear,

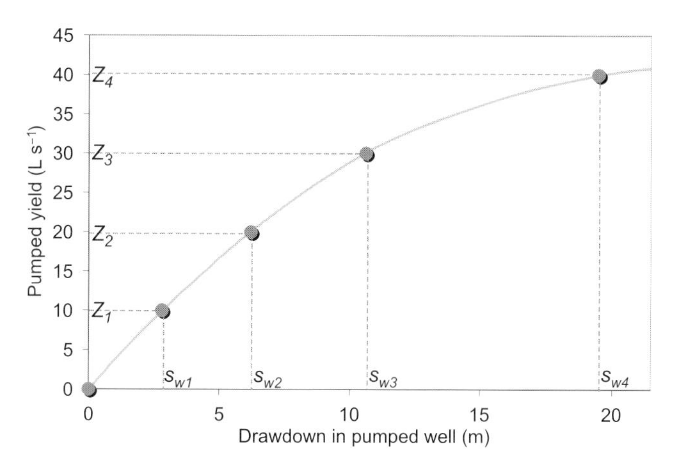

Figure 7.4 Plot of yield versus drawdown for step testing. For any yield Z, we can predict a corresponding drawdown (s) after 2 h of pumping.

convex-upwards curve may also be due to the declining transmissivity of the aquifer as it is dewatered due to increasing pumping rate.

2. *Constant rate testing.* Here the borehole is pumped at a constant rate (usually Z_d) for a period of typically between 24 and 72 h. The evolution of drawdown is monitored either in the abstraction well or an observation borehole (or both). From this data set, we are able (through inverse solution of the Theis or Cooper–Jacob equations) to derive a value of transmissivity. We can also mathematically derive a value of storage coefficient (S), although this is seldom reliable if the drawdown data are only measured in the abstraction well rather than in a dedicated observation borehole.
3. We may then carry out an 'environmental' *or* 'sustainability' test. This is typically a longer-term test, running over periods of several days to several months, depending on the size and importance of the wellfield scheme. If the ground source heating scheme is designed as a doublet (with abstraction and re-injection wells), we will probably choose to re-inject the abstracted water during this phase of testing, to simulate operational conditions. During this test, one will ensure that the design yield can be sustained over longer periods, and that there are no adverse impacts of drawdown caused by the abstraction (or of groundwater mounding caused by re-injection) on other users of the aquifer, or on environmental features that are supported by groundwater. Such environmental features may include groundwater-fed rivers and streams, lakes, ponds and springs. In the context of ground source heating and cooling schemes, we may also use this period of testing to ensure that thermal breakthrough of a heat signature is not occurring from our re-injection point to our abstraction well (i.e. we monitor temperatures). We could also monitor temperatures in any nearby rivers/streams, adjacent wells or observation boreholes to ensure that we are not causing any unacceptable heat pollution to these features. In addition, we

might also monitor the ground surface around any buildings or structures for signs of ground movement related to groundwater abstraction or re-injection.

Throughout the program of test pumping, we would typically also take samples of groundwater for physical, chemical and microbiological analysis. The following typical set of analytical parameters would be considered a minimum for hydrogeochemical interpretation and for assessing the potential that the groundwater has for corrosion, incrustation and biofouling.

To be determined in the field, using portable meters or kits
pH, temperature, electrical conductivity, dissolved oxygen (or redox potential, Eh), alkalinity, turbidity (possibly also cold acidity or dissolved CO_2).

To be determined in the laboratory
Major cations: calcium, magnesium, potassium, sodium, ammonium.
Major anions: chloride, sulphate, nitrate, alkalinity (bicarbonate).
Other metals: iron, manganese, aluminium (these three should be sampled and analysed as both total and dissolved metals), barium.
Microbiology: faecal coliforms, total heterotrophic plate counts.
Other: total organic carbon, turbidity, colour, total suspended solids.

Environmental authorities may insist that other parameters are also analysed, especially if wastewater is being discharged to a sensitive recipient, or if the site is potentially contaminated. Sampling and analysis should be undertaken according to strict protocols, which are discussed by USGS (2004) and Misstear *et al.* (2006).

7.8.2 Statistical data for hard rock aquifers

In some aquifers, especially crystalline rock aquifers (e.g. slates, granites, quartzites, gneisses and some limestones and marbles), hydraulic properties do not vary geographically in a systematic or predictable manner. This is because the hydraulic conductivity depends on secondary structures (fractures and joints) rather than primary ones. We cannot predict either yield or water quality in a deterministic manner in such aquifers. We can, however, use large data sets from such aquifers to describe the distributions of yield and water quality statistically (Banks and Robins, 2002; Banks *et al.*, 2005). For example, Figure 7.5 shows the distribution of well yields in three different aquifer lithologies in Norway. This diagram enables us to say that the median yield (50% chance of achieving this yield) is around 600 $L\,h^{-1}$ in the Iddefjord granite and nearer 750 $L\,h^{-1}$ in the Precambrian gneisses. It also enables us to say that, if we wish to achieve a yield of 5000 $L\,h^{-1}$ in the Caledonian metasediments, we have a less than 5% chance of doing this with a single randomly drilled borehole (5000 $L\,h^{-1}$ corresponds to >95 percentile on the cumulative frequency diagram). Be aware that the power of this technique depends on the quality and representativity of the underlying data set.

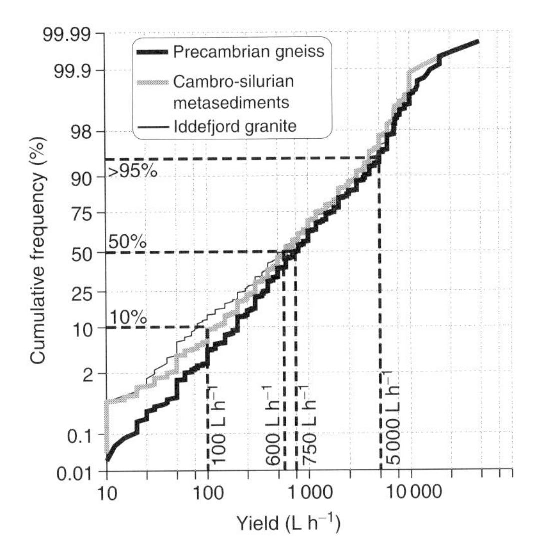

Figure 7.5 Cumulative frequency distribution curve showing distribution of well yields in three lithologies (Iddefjord Granite, Precambrian gneisses and Caledonian metasediments/metavolcanics) in Norway. The diagram enables us to predict that we have a 50% chance of obtaining a short-term yield of 600 L h^{-1} from the granite and 750 L h^{-1} from the gneisses, but only a <5% chance of obtaining 5000 L h^{-1} from the aquifers. We typically have ≈90% chance of obtaining >100 L h^{-1}. *Modified after Banks and Robins (2002) and reproduced by kind permission of Norges geologiske undersøkelse, Trondheim.*

Note also that yields cited in such data sets are often short-term yields reported by drillers. Long-term, 'sustainable' yields may be somewhat less.

Banks (1998) and Gustafson (2002) have also demonstrated methods of using such statistical data from single wells to predict the probability of achieving a given total yield from a wellfield comprising multiple wells.

7.9 Multiple wells in a wellfield

While some open-loop ground source heating and cooling schemes are based on a single well, many will require more than one well to achieve the required yield. It is a common fallacy amongst engineers that, if one well yields 5 L s^{-1}, then 10 wells will provide a yield of 50 L s^{-1}. This will not usually be the case for several (inter-related) reasons:

- the aquifer may simply not receive enough recharge to support such a large total abstraction;
- the hydraulic properties of the aquifer may not be adequate to transmit a total flow of 50 L s^{-1} to a compact wellfield;
- the wells within the wellfield will hydraulically interfere with each other.

On the other hand, if the scheme utilises injection wells to return the wastewater to the original aquifer, the achievable, sustainable yield may be higher than it would have been than with an 'abstraction only' scheme.

7.9.1 Multiple abstraction wells

Let us imagine an abstraction wellfield comprising three identical wells (A, B and C) of radius r_w in a line at spacing 30 m. For simplicity, we will assume no injection wells. Drawdown is an additive property: thus, the total drawdown in the middle well (B) at any time is not merely the drawdown predicted by the Theis or Cooper–Jacob equations, using the yield from the middle well (Z_B) and its radius (r_w). We must add to this the drawdowns caused by the abstraction from neighbouring wells A and C. Thus, the total drawdown (s_{wB}) in well B at a time t after pumping commences is given by

$$s_{\mathrm{wB}} = \frac{Z_{\mathrm{A}}}{4\pi T}W(u_{\mathrm{A}}) + \frac{Z_{\mathrm{B}}}{4\pi T}W(u_{\mathrm{B}}) + \frac{Z_{\mathrm{C}}}{4\pi T}W(u_{\mathrm{C}}) \quad (7.21)$$

where $u_{\mathrm{A}} = u_{\mathrm{C}} = r^2S/4Tt = (30m)^2S/4Tt$ and $u_{\mathrm{B}} = r_{\mathrm{w}}^2S/4Tt$.

If the values of u are small, we can apply the Cooper–Jacob approximation, resulting in

$$s_{\mathrm{wB}} = \frac{2.30Z_{\mathrm{A}}}{4\pi T}\log_{10}\left(\frac{2.25T}{(30m)^2S}\right) + \frac{2.30Z_{\mathrm{C}}}{4\pi T}\log_{10}\left(\frac{2.25T}{(30m)^2S}\right) + \frac{2.30Z_{\mathrm{B}}}{4\pi T}\log_{10}\left(\frac{2.25T}{r_{\mathrm{w}}^2S}\right) + \frac{2.30(Z_{\mathrm{A}}+Z_{\mathrm{B}}+Z_{\mathrm{C}})}{4\pi T}\log_{10}t \quad (7.22)$$

Similarly, for the drawdown in well A:

$$s_{\mathrm{wA}} = \frac{2.30Z_{\mathrm{B}}}{4\pi T}\log_{10}\left(\frac{2.25T}{(30m)^2S}\right) + \frac{2.30Z_{\mathrm{C}}}{4\pi T}\log_{10}\left(\frac{2.25T}{(60m)^2S}\right) + \frac{2.30Z_{\mathrm{A}}}{4\pi T}\log_{10}\left(\frac{2.25T}{r_{\mathrm{w}}^2S}\right) + \frac{2.30(Z_{\mathrm{A}}+Z_{\mathrm{B}}+Z_{\mathrm{C}})}{4\pi T}\log_{10}t \quad (7.23)$$

In other words, the drawdown in any given well in a wellfield comprising multiple abstraction wells is greater than it would be for the same well, pumping at the same rate, on its own. Thus, the yield from each well in a multiple wellfield, for a given design drawdown, will be less than if the well was alone. Note that the above equations assume ideally efficient wells: real drawdowns may be larger due to well inefficiency (see Section 7.7.1).

7.9.2 Abstraction and injection wells

Injection wells can be considered as 'negative' abstraction wells, where a groundwater pumping rate – Z can be substituted into the Theis, Cooper–Jacob or Logan equations. This results in a negative drawdown, or 'upconing', of the water table or piezometric surface (Figure 7.6).

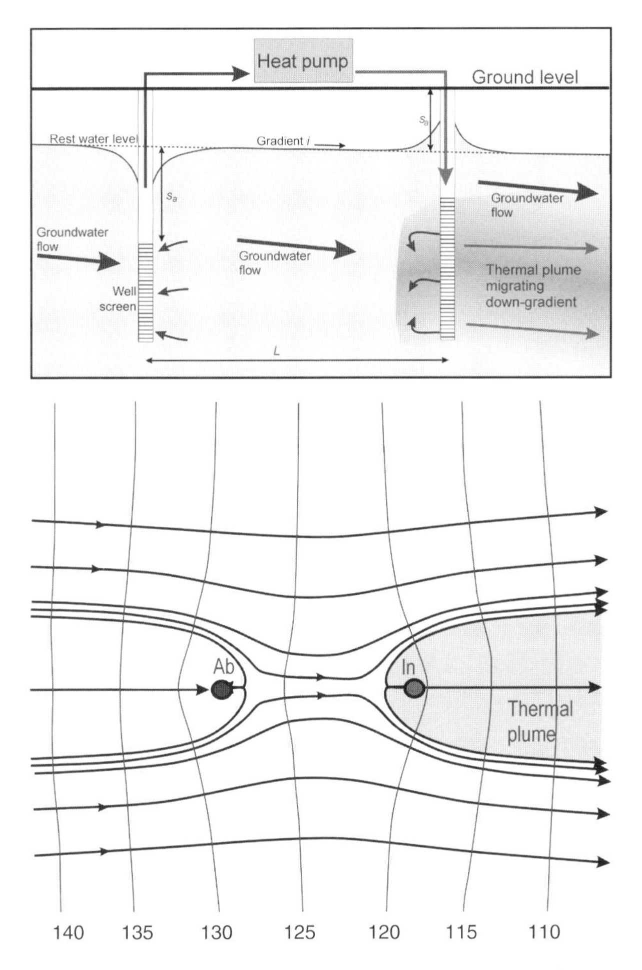

Figure 7.6 A well doublet system, comprising an abstraction well (Ab), situated immediately up the hydraulic gradient from a re-injection well (In). The upper diagram shows a schematic cross-section and the lower diagram a plan view. The distance L between the wells in greater than $2Z/T\pi i$ (see Equation 7.28) and there is no hydraulic (or thermal) feedback between the wells. In the lower diagram, black arrows show groundwater flow lines, thinner numbered lines are groundwater contours, with arbitrary head values (groundwater levels decline from left to right). s_a shows the available drawdown/upconing for each well.

The same technique can be applied as in Section 7.9.1, but this time subtracting the terms relating to the injection wells. Let us consider a simple 'doublet' arrangement, with one abstraction well (rate Z) and a single injection well (rate $-Z$) at a distance L apart, both with radius r_w. The relevant equation for drawdown in the abstraction well (s_w), for small values of u (i.e. large values of t), is

$$s_w = \frac{2.30Z}{4\pi T}\log_{10}\left(\frac{2.25T}{r_w^2 S}\right) - \frac{2.30Z}{4\pi T}\log_{10}\left(\frac{2.25T}{L^2 S}\right) + \frac{2.30(Z-Z)}{4\pi T}\log_{10} t \quad (7.24)$$

Note here that the time-dependent term becomes zero and a steady-state drawdown and flow field should arise. The drawdown in the abstraction well in the doublet is lower than it would have been in case of a single abstraction well, and is given by Gringarten (1978):

$$s_W = \frac{2.30Z}{4\pi T} \log_{10}\left(\frac{L^2}{r_W^2}\right) = \frac{Z}{2\pi T} \ln\left(\frac{L}{r_W}\right) \tag{7.25}$$

The upconing in the injection well should also be $-s_W$, if our injection well is 100% hydraulically efficient. Reality is never so obliging.

7.9.3 Injection wells

In theory, injection wells are the negative image of abstraction wells. Re-injection of water causes an upconing of groundwater levels instead of a drawdown. The available upconing may prove to be a major constraint on the system: the available upconing is the difference between the rest water level and the ground surface (Figure 7.6). If we wish to inject quantities of water that will result in an upconing greater than this value, then we must either inject water under excess pressure into a sealed well, or use a larger number of conventional injection wells.

Injection wells require specialist design and construction. As a trivial example, wire-wrapped well screens in abstraction wells are designed to prevent clogging by particles by using a V-shaped aperture (Figure 6.7, p. 115). If we choose to use such a well screen in an injection well, the groundwater will flow into the 'V' rather than out of it, an ideal situation to promote clogging if particles are present in the water.

Injection wells also require specialist operation for several reasons. The re-injected water must be particle-free to prevent clogging of the well screen or aquifer. It should also be microbiologically inactive to prevent bacterial biofilm growth on the well-screen or borehole wall: it is common practice to apply UV disinfection to injected water. The re-injected water should not contain gas bubbles or concentrations of dissolved gas that are likely to exsolve in the aquifer. Bubbles of dissolved gas in pore spaces or fractures can clog an aquifer as effectively as particles or biofilms! Finally, one must also be aware of the possibility of chemical precipitation and clogging of the injection well's screen. Contact between water and atmospheric oxygen may increase the risk of precipitation of iron and manganese oxyhydroxides. Degassing of excess carbon dioxide, especially in ground source cooling schemes where the re-injected water has been heated, may promote precipitation of calcite. Chemical analyses of the groundwater are useful to predict this risk. The operation of an abstraction–re-injection well doublet as a pressurised, sealed system minimises contact between water and atmosphere and may help to minimise the risk of chemical clogging (Bakema, 2001).

In summary, re-injection wells should not be expected to behave ideally: there is usually some risk of deterioration of the performance of a re-injection well for a variety of reasons. Putting it very bluntly, we should usually expect to drill more re-injection wells (or metres of injection well) to accept a given flux of groundwater back into an aquifer, than we would require to abstract it.

7.10 Hydraulic feedback in a well doublet

7.10.1 Hydraulic feedback

Let us, for the purposes of Sections 7.10–7.12, assume that we are designing a ground source cooling scheme, where we abstract cold water at a rate Z from an aquifer at temperature θ_{gout}, and re-inject it to the same aquifer at the same rate at a higher temperature θ_{ginj} via an injection well at a distance L down the hydraulic gradient of the aquifer. Let us call such an arrangement an *open-loop well doublet*. The heat rejected to the groundwater (G) is given by

$$G = (\theta_{ginj} - \theta_{gout}) \cdot S_{VCwat} \cdot Z \tag{7.26}$$

where S_{VCwat} is the specific volumetric heat capacity of water. The cooling load (C) delivered to the building is then estimated by

$$C = \left(\theta_{ginj} - \theta_{gout}\right) \cdot S_{VCwat} \cdot \frac{Z}{(1 + (1/\mathrm{SPF}_C))} \tag{7.27}$$

where SPF_C is the seasonal performance factor for the cooling system. This should be very high ($1/\mathrm{SPF}_C \approx 0$) for free cooling systems, while SPF_C may be around 2–3 for heat-pump-based systems.

In a fit of optimism, we might hope that, if we site the abstraction well up the hydraulic gradient from the injection well, all the waste heat will be carried away (to some unspecified destination) by groundwater flow (Figure 7.6) and our cooling scheme will function sustainably. Of course, the plume of waste heat from the re-injection well may impact groundwater users or environmental features further down-gradient and may prove unacceptable to environmental authorities. However, if our injection well is situated too close to the abstraction well (Figure 7.7), we may start to induce a component of warm groundwater flow back towards the abstraction well. Clyde and Madabhushi (1983) have in fact demonstrated that this will occur if

$$L < \frac{2Z}{T\pi i} \tag{7.28}$$

where i is the natural hydraulic gradient (m per m; dimensionless) and T is the aquifer transmissivity. If the spacing of the well doublet is shorter than this critical value, there is a possibility that a proportion of the warm, re-injected wastewater will find its way back into the abstraction well. This will increase the temperature of the abstracted water and render the ground source cooling scheme less efficient or even, in the worst case, unsustainable. Let us plug some typical values into Equation 7.28: assuming $T = 150\ \mathrm{m^2\,day^{-1}}$, $Z = 10\ \mathrm{L\,s^{-1}} = 864\ \mathrm{m^3\,day^{-1}}$ and $i = 0.01$, then we find that the

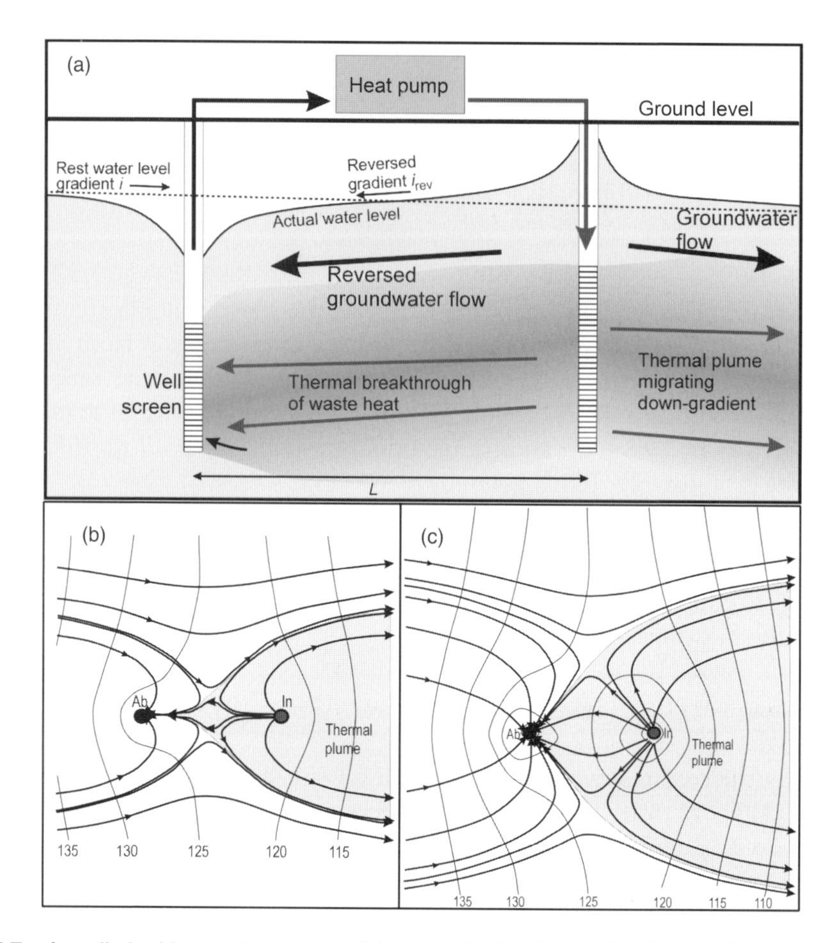

Figure 7.7 A well doublet system, comprising an abstraction well, situated immediately up the hydraulic gradient from a re-injection well. (a) Shows a schematic cross-section. The distance L between the wells is less than $2Z/T\pi i$ and there is hydraulic (and thermal) feedback between the wells. (b) Shows a plan view of the situation where L is slightly less than $2Z/T\pi i$ and there is a minimal amount of hydraulic feedback. (c) Shows the situation where L is significantly less than $2Z/T\pi i$. Symbols as for Figure 7.6.

spacing L must be at least 367 m for there to be no risk of hydraulic or thermal feedback. In many cases, it is impossible to achieve such a large spacing between abstraction and recharge wells, and we thus often have to accept a risk of hydraulic and thermal feedback. Just because the risk is present, however, does not mean that the system is doomed to failure. In fact, such a scheme can be sustainable or can, at least, have a very long lifetime, because

i. Breakthrough of heat in the abstraction well does not happen immediately. In fact, it may take many years (although it may only take weeks or months in some fissured or karstified aquifers).

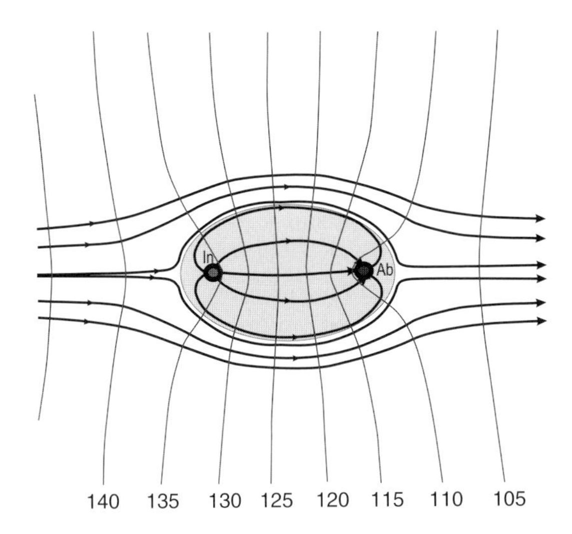

Figure 7.8 A well doublet system, comprising an abstraction well, situated immediately down the hydraulic gradient from a re-injection well. This plan view shows that the thermal 'plume' forms a closed cell, with all the re-injected water being relatively rapidly returned to the abstraction well. Symbols as for Figure 7.6.

ii. Only a small portion of the water abstracted from the abstraction well may consist of re-circulated water from the injection well.
iii. If we have a heating demand in winter and a cooling demand in summer, we may be able to operate the scheme reversibly, effectively recovering the summer's warm wastewater during the winter heating season without any thermal breakthrough occurring.

To assess the risk of thermal breakthrough, we thus need to consider the speed with which groundwater travels between the injection and abstraction wells. Before we do this, however, we should note that, if we are crazy enough to locate our injection well directly up-gradient of the abstraction well, it is likely that we will develop a closed thermal cell (Figure 7.8), with all our re-injected warm water feeding back to our abstraction well.

7.10.2 Rate of hydraulic breakthrough

The ideal way of assessing hydraulic breakthrough time (i.e. the time it takes for groundwater to travel from the injection well to the abstraction well) is to use a computer-based numerical model, which has a so-called particle tracking capability. This will allow us to simulate the motion of large numbers of water molecules along different flow paths (and often to consider dispersion effects) and to calculate an average hydraulic breakthrough time in what is rapidly becoming a complex hydrogeological problem. There are two analytical methods we can use to estimate this breakthrough time, however, (1) by simple application of Darcy's Law and (2) by the Doublet breakthrough method.

Darcy's Law

We can calculate, from Equation 7.25, the expected steady-state drawdown and upconing of groundwater levels in the aquifer adjacent to the abstraction and injection wells. In our example here ($T = 150\ \mathrm{m^2\,day^{-1}}$, $Z = 10\ \mathrm{L\,s^{-1}} = 864\ \mathrm{m^3\,day^{-1}}$), if we assume that $r_w = 100$ mm and the well separation $L = 100$ m, we calculate that the drawdown adjacent to the abstraction well is 6.3 m and the upconing adjacent to the injection well is 6.3 m. Thus, we have a head difference (Δh) between the injection and abstraction well of 12.6 m. We can, if we are feeling sufficiently daring, make three dubious assumptions: (a) we can forget that the flow field around the wells will not be linear; (b) we can assume that the hydraulic gradient between the two wells is constant; and (c) we can assume that the natural hydraulic gradient (i) is negligible. We can thus say that the reverse hydraulic gradient (i_{rev}) between the two wells is 12.6 m/L or 0.126. Darcy's Law states that the flow rate per unit cross-sectional area of aquifer (the so-called Darcy velocity v_D) is given by

$$v_D = K i_{rev} = \frac{K\Delta h}{L} \tag{7.29}$$

But we must remember that this is not the actual linear velocity of groundwater flow. As flow only takes place through pore space and not through solid mineral grains, we also have to divide by the effective porosity (i.e. the interconnected porosity through which flow takes place, n_e) to obtain the actual linear flow velocity (v):

$$v = \frac{K i_{rev}}{n_e} = \frac{K\Delta h}{L n_e} \tag{7.30}$$

Thus, the hydraulic breakthrough time (t_{hyd}) can be estimated as follows:

$$t_{hyd} = \frac{L}{v} = \frac{L^2 n_e}{K\Delta h} \tag{7.31}$$

This probably represents a worst-case estimate of minimum breakthrough time along the most direct flow path between the injection and abstraction wells.

Double breakthrough time for zero hydraulic gradient

If we are troubled by the dubious assumptions involved in using Darcy's Law (which we should be!), we may choose to apply a somewhat more sophisticated equation derived from a geometric consideration of the flow paths in a well doublet. In fact, if we assume that the natural hydraulic gradient (i) is zero, the flow paths will all be arcs of circles (Figure 7.9). The doublet formula for the travel time along the shortest flow path (a straight line) between the injection and abstraction well is found to be (Grove, 1981; Güven *et al.*, 1986; Himmelsbach *et al.*, 1993)

$$t_{hyd} = \pi n_e D \frac{L^2}{3Z} \tag{7.32}$$

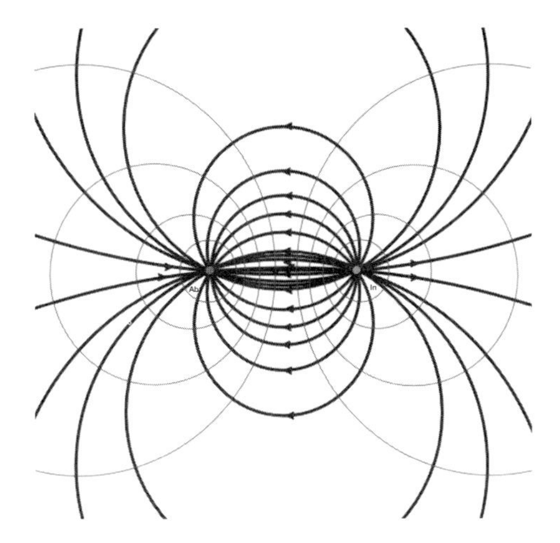

Figure 7.9 A well doublet system, comprising an abstraction well (left) and a re-injection well (right), in this case there is no natural hydraulic gradient. Both the flow lines between the wells (thicker arrowed lines) and groundwater head contours (thinner lines) all form arcs of circles, making the scenario amenable to geometric analysis.

where D is the effective aquifer thickness (m). This equation is probably most appropriate for simple dipole systems of one abstraction and one injection well. The Darcy approach may be better justified where we have a row of several abstraction wells separated from a row of injection wells, and where flow conditions may approach linear flow rather than radial flow.

The Lippmann–Tsang modification for finite hydraulic gradient

Equation 7.32 presupposes a negligible initial natural hydraulic gradient (i). If this gradient is not negligible, Lippmann and Tsang (1980) and Clyde and Madabhushi (1983) proposed a modified version for the case where the injection well is situated down the regional (natural) hydraulic gradient from the abstraction well (as in Figure 7.7).

$$t_{\mathrm{hyd}} = \frac{Ln_{\mathrm{e}}}{Ki}\left[1 + \frac{4\alpha}{\sqrt{-1-4\alpha}}\tan^{-1}\left(\frac{1}{\sqrt{-1-4\alpha}}\right)\right] \tag{7.33}$$

where

$$\alpha = \frac{Z}{2\pi KDiL} = \frac{Z}{2\pi TiL} \tag{7.34}$$

and where i and α are negative quantities. When using these equations, note that

1. Hydraulic breakthrough times are proportional to effective porosity. In fissured and fractured limestones or crystalline rocks, this effective porosity may be <0.01 and we can expect potentially rapid breakthrough.
2. The equation does not account for hydrodynamic dispersion. In real aquifers, some fractures or pore channels are wider than others; some water molecules flow down

the middle of channels while others drag along the side of pore spaces. In other words, dispersion means that some water molecules arrive before the theoretical breakthrough time while others arrive after. The equations give the average breakthrough time along the most rapid flow path. In karstified or fractured aquifers, we may get a surprisingly rapid breakthrough if we have encountered a transmissive flow pathway (large-aperture fracture or fissure) in our wells.

7.11 Heat migration in the groundwater environment

The astute reader will have noted that, hitherto, we have only talked about the breakthrough of groundwater and water molecules from the injection to the abstraction well. Can we assume that heat travels at the same speed as groundwater and is merely carried along passively? In fact, we cannot. In a groundwater aquifer, heat travels by three mechanisms:

1. conduction through mineral grains and water filled pores;
2. advection with bulk groundwater flow;
3. exchange between moving groundwater and the matrix of the aquifer (mineral grains and any immobile pore water).

Many hydrogeologists assume that heat exchange is very rapid and that the aquifer matrix equilibrates with groundwater temperature speedily. De Marsily (1986) estimated that thermal equilibration between a sand grain (1 mm diameter) and surrounding groundwater takes less than 1 min, while equilibration between a 10 cm cobble and groundwater may take 2 hours. This means that a single point in an aquifer system can be characterised by a single temperature (θ) at any given time t (which applies both to the matrix and to the mobile groundwater). This is called the *assumption of instantaneous thermal equilibration*. The mathematically inclined can write an equation describing heat transport in such an aquifer (de Marsily, 1986; Shook, 2001a) – in one dimension only it is

$$\lambda^* \frac{d^2\theta}{dx^2} - S_{VCwat} \frac{d(v_D\theta)}{dx} = S_{VCaq} \frac{d\theta}{dt} \qquad (7.35)$$

where λ^* is the effective thermal conductivity of the saturated aquifer (see Box 7.3), S_{VCwat} and S_{VCaq} are the specific volumetric heat capacities of the groundwater and the saturated aquifer (i.e. mineral matrix plus pore water), respectively. v_D is the Darcy velocity of groundwater flux through the aquifer ($= v/n_e$). The first term in Equation 7.35 describes heat transport by conduction only; the second term describes heat transport by convection and the third term the change in heat stored in a unit volume of aquifer with time.

Let us imagine a 'front' of warm groundwater entering an aquifer of cross-sectional area A and effective porosity n_e at a rate Z. After a time t, the influent water will have

BOX 7.3 The analogy between heat transport and solute transport in groundwater.

In its full three-dimensional glory, Equation 7.35 can be written (de Marsily, 1986) as follows:

$$\mathrm{div}\left(\frac{\lambda^*}{\rho_{wat} \cdot S_{Cwat}}\mathrm{grad}\theta\right) - \mathrm{div}\,(v_D \cdot \theta) = \frac{\rho_{aq} \cdot S_{Caq}}{\rho_{wat} \cdot S_{Cwat}} \cdot \frac{d\theta}{dt}$$

where the following (in addition to gradθ) are tensors:

λ^* = effective thermal conductivity of the saturated aquifer material, modified to take into account effects of hydrodynamic dispersion;
v_D = Darcy velocity (rate of flow of groundwater per metre cross section).

The following are scalars;
θ = temperature at a given point in the aquifer–groundwater system;
S_C = specific heat capacity ($\mathrm{J\,kg^{-1}\,K^{-1}}$);
ρ = density.

The subscripts 'wat' and 'aq' refer to the groundwater and the bulk saturated aquifer (pore water plus matrix), respectively. We can compare this to the equation for the migration of a non-adsorbed solute in groundwater, where C is a dimensionless concentration of that solute (relative to natural background).

$$\mathrm{div}(D^* \cdot \mathrm{grad}C) - \mathrm{div}(v_D \cdot C) = n_e \cdot \frac{dC}{dt}$$

where D^* is a dispersion term reflecting both molecular diffusion and hydrodynamic dispersion. By analogy then, we can now say that

- $\lambda^*/(\rho_{wat} S_{Cwat})$ is a term for 'thermal dispersion', reflecting both thermal conduction and hydrodynamic dispersion. It *may* be comparable, but will not necessarily be the same as D^*.
- The velocity of the non-sorbing chemical v (which travels with bulk groundwater flow) is related to v_D by a factor n_e.
- The velocity of the heat front v_{the} is related to v_D by a factor $(\rho_{aq} S_{Caq})/(\rho_{wat} S_{Cwat})$

Thus, $v_D = vn_e = v_{the}\rho_{aq} S_{Caq}/(\rho_{wat} S_{Cwat})$ and $v_{the} = vn_e S_{VCwat}/S_{VCaq}$.

filled a volume of aquifer V_{hyd} and penetrated a distance x_{hyd}, where

$$V_{hyd} \cdot n_e = A \cdot x_{hyd} \cdot n_e = Z \cdot t \tag{7.36}$$

However, if the influent water is at a temperature $\Delta\theta$ higher than the ambient aquifer temperature, the amount of heat entering the aquifer in time t will be $ZtS_{VCwat}\Delta\theta$. If the injected heat is immediately absorbed by and equilibrates with the aquifer matrix, then the heat will only fill a volume V_{the}:

$$V_{the} S_{VCaq} \Delta\theta = Zt S_{VCwat} \Delta\theta \tag{7.37}$$

Thus, in that given time t, the groundwater front will travel a distance x_{hyd} and the heat front will only travel a distance x_{the} ($=V_{the}/A$), where

$$R = \frac{x_{hyd}}{x_{the}} = \frac{S_{VCaq}}{n_e S_{VCwat}} \tag{7.38}$$

In other words, the absorption of heat by the aquifer matrix has retarded the progress of the heat front relative to groundwater velocity, by a retardation factor R. We should remember that this is not strictly speaking a constant, as the volumetric heat capacities will vary slightly with temperature.

7.11.1 *Thermal breakthrough*

Thus, if we wish to estimate the time for thermal breakthrough in an open-loop well doublet, we can simply multiply the hydraulic breakthrough time by a factor R. Thus, if the natural hydraulic gradient is zero (Figure 7.9), the doublet equation (Equation 7.32) becomes (Gringarten, 1979; Clyde and Madabhushi, 1983)

$$t_{the} = \pi D \frac{S_{VCaq} L^2}{3 S_{VCwat} Z} \tag{7.39}$$

Lipmann and Tsang (1980) and Clyde and Madabhushi (1983) also provide a formula for predicting how temperature in the abstracted water (θ_{gout}) will evolve, following thermal breakthrough (for times $t > t_{the}$):

$$\begin{aligned}\frac{\theta_{gout} - \theta_{ginj}}{\theta_0 - \theta_{ginj}} &= 0.34 \exp\left(-0.0023\frac{t}{t_{the}}\right) + 0.34 \exp\left(-0.109\frac{t}{t_{the}}\right) \\ &\quad + 1.37 \exp\left(-1.33\frac{t}{t_{the}}\right)\end{aligned} \tag{7.40}$$

where θ_{ginj} is the temperature of the injected water (which is assumed to be constant, although in a real operation it may increase as the temperature of the abstracted water increases), θ_0 is the initial ambient groundwater temperature, t is time following commencement of abstraction/injection and t_{the} is the thermal breakthrough time. Equation 7.40 does, however, underestimate the temperature evolution, compared with the purely geometric approach of Güven *et al.* (1986) – see Figure 7.10a. Again, equation 7.40 assumes no initial natural hydraulic gradient.

If the initial natural hydraulic gradient is finite, the Lippmann–Tsang Equation 7.33 becomes (Clyde and Madabhushi, 1983)

$$t_{the} = \frac{S_{VCaq} \cdot L}{S_{VCwat} \cdot K \cdot i}\left[1 + \frac{4\alpha}{\sqrt{-1-4\alpha}} \tan^{-1}\left(\frac{1}{\sqrt{-1-4\alpha}}\right)\right] \tag{7.41}$$

It is important to realise that these systems of equations build upon a number of simplifying assumptions:

- That groundwater flow obeys Darcy's Law and that the aquifer can adequately be simulated as a homogeneous, saturated porous medium (although even heterogeneity can be tackled – Shook, 2001a).
- The equations do not account for dispersion effects. Some thermal breakthrough will inevitably occur ahead of the calculated mean travel time. If there are major open fractures present or if the aquifer is heterogeneous with rapid flow pathways, this may result in macrodispersive effects and the very rapid breakthrough of a significant portion of the thermal signature.
- That thermal equilibration is instantaneous between groundwater and aquifer matrix. If flow is through a limited number of widely spaced, discrete fractures, rather than through an extensive network of pores and small fractures, thermal equilibration between water and matrix (i.e. the blocks of rock between fractures) will not be instantaneous. This will lead to overestimation of thermal travel times.
- That there is no vertical conductive transfer of heat to strata overlying or underlying the aquifer (although, if this occurs, it will typically improve reservoir lifetime and can be tackled mathematically – Gringarten, 1978).

In short, the equations are likely to be adequate in sedimentary aquifers such as sands, gravels and even some sandstones. In fractured or fissured aquifers, such as limestones, Chalk or crystalline rocks, they are likely to overestimate thermal breakthrough times, especially over short distances. Some progress towards developing thermal breakthrough models for fractured rock aquifers has been made by Shook (2001b) and Law (2007).

7.12 Theoretical and real examples

7.12.1 *A theoretical example*

Let us return to our example of a doublet scheme of separation $L = 100$ m, pumping at a rate $Z = 10\ \mathrm{L\,s^{-1}} = 864\ \mathrm{m^3\,day^{-1}}$ in a sand aquifer of transmissivity $T = 150\ \mathrm{m^2\,day^{-1}}$, thickness 75 m, hydraulic conductivity $K = 2\ \mathrm{m\,day^{-1}}$ and effective porosity $n_e = 0.15$. Let us further assume that we abstract groundwater at 11°C and re-inject it at 17°C.

Our simple 'Darcy's Law' approach (Equation 7.31) yields a minimum hydraulic breakthrough time of 60 days, while the more sophisticated doublet equation yields a better estimate of 136 days. If we assume a natural hydraulic gradient $i = -0.01$ and that the injection well is down-gradient of the abstraction well, we obtain a hydraulic breakthrough time of 175 days from Equation 7.33.

We know that the specific heat capacity of water is around 4180 $\mathrm{J\,L^{-1}\,K^{-1}}$. Let us assume that our aquifer is composed of quartz-dominated sand grains and has a

BOX 7.4 Volumetric heat capacity of aquifers.

Let us consider an aquifer of saturated quartz sand, with a total porosity of 23% ($n = 0.23$). If quartz has a specific heat capacity of 740 J kg^{-1} K^{-1} (Ward, 1992) and a density of 2620 kg m^{-3}, we can estimate the volumetric heat capacity of quartz as 1.9 MJ m^{-3} K^{-1} (de Marsily, 1986). The volumetric heat capacity of the quartz sand aquifer is then given by

$$S_{VCaq} = nS_{VCwat} + (1 - n)\, S_{VCmat} = (0.77 \times 1.9\ \text{MJ m}^{-3}\,\text{K}^{-1}) + (0.23 \times 4.18\ \text{MJ m}^{-3}\,\text{K}^{-1}) = 2.4\ \text{MJ m}^{-3}\,\text{K}^{-1}$$

Where S_{VCmat} is the volumetric heat capacity of the aquifer matrix. If the sandstone is dry, we can substitute the volumetric heat capacity of air (0.0012 MJ m^{-3} K^{-1} – Eskilson *et al.*, 2000) for that of water. This yields a specific heat capacity of the dry sand of 0.77 × 1.9 MJ m^{-3} K^{-1} = 1.5 MJ m^{-3} K^{-1}.

saturated volumetric heat capacity S_{VCaq} = 2.4 MJ m^{-3} K^{-1} (Box 7.4). The retardation factor for the aquifer can then be estimated from Equation 7.38 as follows:

$$R = \frac{2.4}{(0.15 \times 4.18)} = 3.83 \qquad (7.42)$$

The doublet equation (Equation 7.39, no hydraulic gradient) then predicts a thermal breakthrough time of 522 days or 1.43 years, and a temperature evolution in the abstraction well as shown in Figure 7.10a. Note that the predicted temperature in the abstraction well only exceeds 14°C after 4 to 9 years' continuous operation. The Lippmann–Tsang equation (7.41), with a hydraulic gradient of 0.01, predicts a thermal breakthrough at 671 days or 1.84 years.

7.12.2 *A real example from Manitoba, Canada*

Ferguson and Woodbury (2005) provide one of very few documented case studies of the breakthrough of hot water from an open-loop doublet system into an abstraction well, at Winnipeg, Canada. The system is a pseudo-'doublet' comprising two closely spaced abstraction wells in a limestone aquifer and a single injection well some 90 m away. The transmissivity of the aquifer is very high and estimated as around 620 m^2 day^{-1}, while the depths of the wells are in the range 68–90 m. The system was operated at a maximum rate of 13 L s^{-1} for 8 h every working day, giving a long-term average of 3.9 L s^{-1}. Such a fissured (probably karstified), high transmissivity aquifer, with a low effective porosity (n_e) and short doublet spacing is clearly a recipe for rapid feedback of water and heat breakthrough. Indeed, plugging these values into equation 7.32 (setting n_e arbitrarily at 0.02) yields a hydraulic breakthrough time of only 1–2 months. A thermal

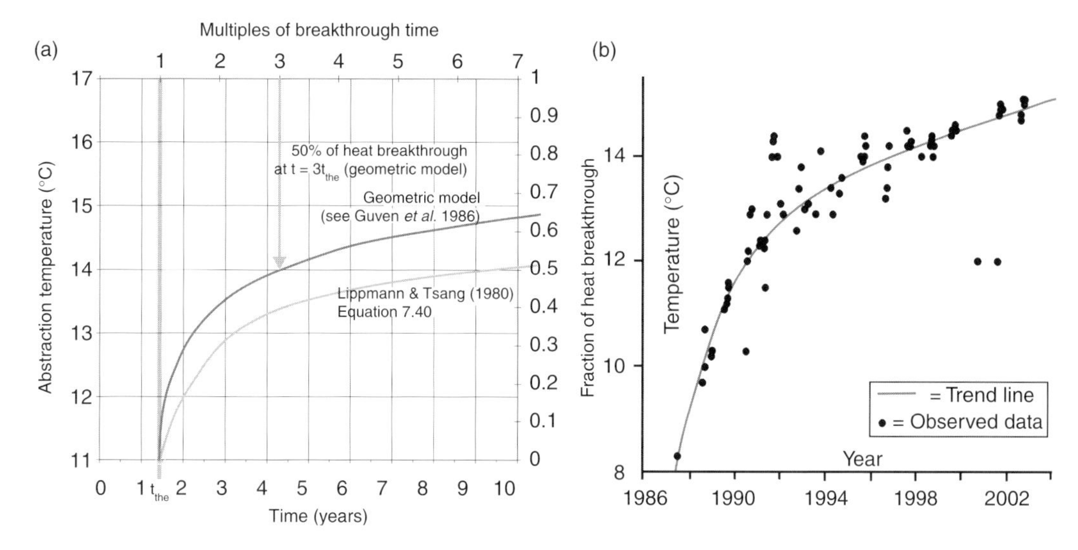

Figure 7.10 (a) The calculated evolution of temperature of water from an abstraction well in a well doublet. The re-injection well is operating at a constant temperature of 17°C and the ambient groundwater temperature is 11°C. Thermal breakthrough is assumed to occur after 1.43 years of operation; (b) the real evolution in temperature in an abstraction well of an open-loop doublet system in Winnipeg, Canada, described in Section 7.12.2. *Diagram (b) is redrawn after Ferguson and Woodbury (2005), and reproduced by kind permission of the National Research Council of Canada.*

breakthrough of around 3 years is calculated but, for reasons discussed in Section 7.11.1, this is likely to be significantly overestimated in fissured limestone aquifers. In reality, thermal breakthrough occurred with very little delay (less than a few months) and, over the course of the subsequent 8 years, the temperature of the abstracted water rose by almost 6°C (Figure 7.10b), presumably compromising the scheme's long-term viability.

7.13 ATES: thermally balanced systems and seasonal reversal

In our theoretical example of Section 7.12.1, we calculated a potential thermal breakthrough of warm water in the abstraction well after less than 2 years. The predicted temperature in the abstraction well exceeds 12°C after 2 years but has only reached 14°C after 9 years of continuous operation. Thus, the potential for thermal breakthrough does not mean that the system fails immediately, although its lifetime may ultimately be limited. In such assessments we should remember that

- We can significantly decrease the risk, speed and magnitude of thermal breakthrough by increasing the separation (L) of the wells. The thermal breakthrough time increases with L^2.

- We may not be pumping the doublet continuously at a given rate Z. There may be seasons of the year where the scheme is dormant.
- We may have a so-called thermally balanced scheme, where the amount of heat rejected during the summer is approximately equal to that abstracted during winter. In such a case, we are injecting successive warm and cold pulses of water into the aquifer at (approximately) six monthly intervals. If the separation of the abstraction and injection wells is long enough and the thermal breakthrough time long enough, it is likely that the thermal storage properties of the aquifer will 'smear' out these temperature pulses during their passage through the aquifer, resulting in little net change in temperature at the abstraction well.

Even better, in a thermally balanced scheme, we may be able to reverse the polarity of our well doublet. Thus, we may abstract cold water from our first well and reject warm water to a second injection well in the summer. In winter, we can convert the second well to a pumping well to re-abstract the warm water. After passage through a heat pump system, the chilled water is then rejected to the first well. The following summer the polarity is reversed again. For such a scheme to function, we need to ensure that

- the heating and cooling loads are approximately balanced;
- the thermal breakthrough time is less than one heating or cooling season (say, 6 months);
- the well construction allows recharge and abstraction functions to be reversed. We have already seen that re-injection wells may be specialised, dedicated structures, although downhole valve devices are available that allow functions to be reversed (Cla-val, 2007).

Indeed, Kazmann and Whitehead (1980) have developed tables of minimum well doublet separations, based on exactly this assumption. If we wish to 'store' the summer's waste heat in the aquifer for re-abstraction the following winter, we ideally also want an aquifer with a relatively low natural throughflow of groundwater, so that the heat is not advected away from the doublet system. Such a reversible scheme, with dedicated 'hot' and 'cold' wells, is termed an 'Aquifer Thermal Energy Storage' (ATES) scheme (Andersson, 1998; Bakema and Snijders, 1998; Vos, 2007). Oslo's Gardermoen Airport (Box 3.4, p. 41) is an example of such a scheme. As a good balance between heat dumped to the aquifer and heat abstracted is a prerequisite for the sustainable operation of ATES systems, Vos (2007) advocates continuous monitoring of groundwater flows and temperatures into and out of the aquifer. She also notes that some Dutch provinces require an energy balance, over a period of 5 years, to be demonstrated in order for such an ATES scheme to be permitted. Furthermore, in Holland, ATES schemes are typically not permitted to operate at temperatures $<5°C$ or $>25°C$ or $30°C$ and they should not be operated in the vicinity of drinking water wells. In reality, it is seldom that a building will have exactly balanced annual heating and cooling loads: in order to balance the

Figure 7.11 Gringarten (1978) argued that, where multiple abstraction and injection wells are necessary, the most efficient pattern is an alternating grid of abstraction (Ab) and injection (In) wells – a so-called five-spot pattern.

heat rejected to or abstracted from the ground, we may need to employ conventional heating or cooling solutions to supply the 'unbalanced' portions of the loads.

7.14 Groundwater modelling

Naturally, with larger schemes, things soon get complicated. The magnitude of our scheme may require us to drill multiple injection and abstraction wells (Figure 7.11). Our aquifer may not be homogeneous or isotropic. Our heating and cooling loads may be variable and complex, as may our re-injection rates and temperatures. There may be other groundwater abstractors nearby that perturb the flow dynamics of our scheme. Furthermore, it may not be possible to neatly align our abstraction well directly up-gradient of our injection well. To tackle problems of greater complexity, it may be necessary to invoke a numerical computer model that can simulate coupled groundwater flow and heat transport, using finite element or finite difference algorithms based on the equations of Box 7.3. As heat transport in groundwater is directly analogous to contaminant transport (Box 7.3), it may be possible to utilise (with careful thought) models that are designed to simulate groundwater pollution. There are, however, several models that are developed with the explicit capability to simulate heat transport in aquifers. These currently include (but are not limited to)

- HST3D (Heat and Solute Transport in 3-Dimensional Groundwater Flow Systems). This is a finite difference code in the public domain for simulation not only of groundwater flow and heat transport, but also of contaminant transport, produced by the United States Geological Survey (Kipp, 1997).

- SHEMAT (Simulator for HEat and MAss Transport) is a German program with similar functionality to HST3D (Clauser, 2003).
- FEFLOW® (Finite Element subsurface FLOW system) is a commercial (and expensive) finite element programme, with a large capability to simulate variable-density groundwater flow problems, coupled with heat or solute transport.

7.15 Further reading

In a single chapter, you have just been subjected to a crash course in some aspects of quantitative hydrogeology. If you desire more information, there are several, less quantitative introductory books on hydrogeology that I can recommend:

- Mike Price's (1996) concise introductory volume *Introducing Groundwater*, focussing to some extent on British geological conditions.
- Paul Younger's (2006) recent *Groundwater and the Environment* – blessedly equation-free.
- C.W. Fetter's (2001) established *Applied Hydrogeology* – gives you the equations, but in a painless way!

For groundwater engineering, try

- *Water Wells and Boreholes* by Bruce Misstear, Dave Banks and the late Lewis Clark (2006). This is an updated and expanded version of Clark's (1988) *Field Guide*.
- *Groundwater and Wells* (1986) by Fletcher. D. Driscoll – groundwater engineering from an American perspective.

8

Horizontal Closed-Loop Systems

> A rapid increase in heat pump sales ... is likely as long as energy costs increase faster than wealth.
>
> R.D. Heap (1979)

From Chapters 4 and 6, it should have become apparent that heat pump technology provides us with many ways of extracting useful heat from our environment: from the air, from cows' milk, from wastes and even from permafrost. In Chapter 7, we have seen that, if we have a reservoir of groundwater in the subsurface, we can extract the heat directly from this water. To do this, however, we need an aquifer, an expensive water well and (usually) a submersible pumping system. There is, fortunately, a far simpler way of extracting heat from the ground, which requires none of these things. All we need is a buried closed loop to act as a heat exchanger between the ground and our heat pump. This sounds complicated but, in its simplest form, such a ground heat exchanger is very basic – a pipe buried in a trench. This is what we mean by a horizontal closed-loop system.

We have, in fact, already encountered horizontal ground loops in Chapter 6. To recap, these comprise some form of buried pipe, conveying a refrigerant (in direct circulation systems) or a carrier fluid (indirect circulation systems) that extracts heat from or dumps heat to the shallow subsurface. In the rest of this chapter, we will focus on the more commonly used indirect circulation systems.

The pipe is usually buried in a trench at a depth of some 1.2–2 m and the carrier fluid is usually a solution of antifreeze, circulating under turbulent flow conditions (Tables 8.1 and 8.2; Box 8.1), in order to ensure efficient heat transfer. Clearly, with such a trench-based system, we are not drawing on a deep 'block' of rock in the same way that we are with a vertical borehole-based closed-loop system (Chapter 10). Rather, we are extracting heat in the winter from the shallow soil around the trench and relying on this depleted heat reservoir to be replenished during the summer season. We are

Table 8.1 Properties of common antifreeze solutions at around 0°C. For comparison, solution concentrations have been chosen that yield a freezing point of −15°C (modified after Rosén *et al*., 2001 and including data from Energi-spar, 2007).

Fluid	Freezing point (°C)	S_C ($kJ\,kg^{-1}\,K^{-1}$)	λ ($W\,m^{-1}\,K^{-1}$)	Density ($kg\,m^{-3}$)	Dynamic viscosity (cP)
30.5% Ethylene glycol	−15	3.67	0.445	1046	4.38
32.9% Propylene glycol	−15	3.86	0.417	1034	8.12
24.4% Ethanol	−15	4.29	0.426	972	5.85
19.9% Methanol	−15	4.09	0.462	973	3.26
18.8% Sodium chloride	−15	3.41	0.549	1146	2.57
24% Freezium™	−15	3.33	0.51	1152	2.21
24.0% Potassium acetate	−15	3.36	0.492	1130	3.36

Table 8.2 Properties of water and various antifreeze solutions at varying temperatures, and the necessary flow velocities (F_{turb}) for $Re > 3000$ (turbulent flow) for pipes of 35.4 mm ID and 26 mm ID (corresponding to 32 mm OD pipe of wall thickness 3 mm). Fluid properties derived from Eskilson *et al.* (2000) and from Table 8.1.

	Freezing point (°C)	Viscosity ($kg\,m^{-1}\,s^{-1}$)	Density ($kg\,m^{-3}$)	F_{turb} 35.4 mm ($L\,min^{-1}$)	F_{turb} 26 mm ($L\,min^{-1}$)
Water at 5°C	0	0.00152	1000	7.6	5.6
Water at 10°C	0	0.001308	999.8	6.5	4.8
Water at 15°C	0	0.001139	999.2	5.7	4.2
Water at 20°C	0	0.001003	998.3	5.0	3.7
Water at 25°C	0	0.000891	997.2	4.5	3.3
Water at 30°C	0	0.000798	995.8	4.0	2.9
Water at 35°C	0	0.00072	994.1	3.6	2.7
30.5% Ethylene glycol at 0°C	−15	0.00438	1046	21.0	15.4
32.9% Propylene glycol at 0°C	−15	0.00812	1034	39.3	28.9
24.4% Ethanol at 0°C	−15	0.00585	972	30.1	22.1
19.9% Methanol at 0°C	−15	0.00326	973	16.8	12.3
18.8% Sodium chloride at 0°C	−15	0.00257	1146	11.2	8.2
24% Freezium™ at 0°C	−15	0.00221	1152	9.6	7.1
24.0% Potassium acetate at 0°C	−15	0.00336	1130	14.9	10.9

using the earth's surface as a solar collector and its subsurface as a temporary storage. The question then becomes: how can we most effectively harvest this absorbed solar energy input? The following factors will need to be considered:

- The area overlying the loop will need to be large enough to receive sufficient replenishment of solar and atmospheric heat during the summer season.
- The ground should be conductive enough to transmit heat efficiently to the loop.
- The contact between the ground and the pipe should be thermally efficient.
- The pipe should be constructed of a material that is durable, tough and sufficiently thermally conductive.

BOX 8.1 Turbulent flow in closed-loop arrays.

The Reynolds number (Re) is named after Osborne Reynolds (1842–1912) of the University of Manchester. For circular pipes, carrying a fluid flow F

$$Re = \frac{2\rho F}{\pi r \mu}$$

where μ = the dynamic viscosity of the fluid in $[M][L]^{-1}[T]^{-1}$, ρ = the fluid density $[M][L]^{-3}$ and r = radius of pipe [L].

The Reynolds number allows us to predict whether the fluid flow is laminar ($Re < 2000$) or turbulent ($Re > 3000$). In practice, there is also a transition stage between the two states, at intermediate Re values. For circular pipes, the critical value for the onset of turbulence is believed to be in the range $Re = 2000$–2300. In terms of maximising heat transfer, turbulent flow is desirable. If we wish to minimise heat transfer, laminar flow is desirable. Thus, for closed-loop ground source heat schemes, we often attempt to achieve turbulent flow in the ground heat exchange elements (i.e. the borehole U-tube, the slinky or the lake coils) but laminar flow in the header pipes between the ground array and the building. We do this by selecting a pipe of large radius (such that Re is less than the lower critical value) for the header pipe, and a smaller radius for the ground array itself. Of course, turbulent flow optimises heat exchange but carries a penalty in as much as it results in greater head losses and greater expenditure of energy in pumping (i.e. higher pumping costs).

We see, furthermore, that Re increases (and turbulent flow becomes more likely) as density increases and viscosity decreases. With water, for example, dynamic viscosity decreases with increasing temperature (Table 8.2). Density also decreases, but more slowly. Adding antifreeze to water also changes the density and viscosity characteristics of a carrier fluid. Design programs such as EED (Eskilson *et al.*, 2000) take these factors into account and flag up a warning if carrier fluid flow in a heat exchanger becomes non-turbulent.

We can estimate the flow (F_{turb}) of water at 25°C required in a 26 mm ID polyethene pipe to achieve a value of $Re > 3000$ (turbulent flow) from

$$F_{turb} = \frac{3000 \times \mu \times \pi \times r}{2 \times \rho}$$

By setting $\mu = 0.000891\ \mathrm{kg\,m^{-1}\,s^{-1}}$, $\rho = 997\ \mathrm{kg\,m^{-3}}$ and $r = 0.013$ m, we obtain

$$F_{turb} = 0.000055\ \mathrm{m^3\,s^{-1}} = 0.055\ \mathrm{L\,s^{-1}} = 3.3\ \mathrm{L\,min^{-1}}$$

Table 8.2 shows the required flow rates to provide turbulent conditions in pipes of different diameter for water at various temperatures and for a selection of

continued

BOX 8.1 (*Continued*)

antifreeze agents. We will note that viscosity increases dramatically as temperature approaches freezing point (Figures 8.1 and 8.2). Viscosity also increases significantly with the addition of antifreeze. Thus, although addition of antifreeze to a closed-loop carrier fluid allows us to run our loop at much lower temperatures, we should be aware that

i. the performance of the heat pump will diminish with lower carrier fluid temperatures;
ii. pumping costs for the closed loop will increase as the viscosity increases, both due to the lower temperatures and the antifreeze itself;
iii. we run the risk of flow reverting from turbulent to laminar flow as viscosity increases. This results in a decrease in heat transfer efficiency in the closed loop and corresponds to an increase in borehole thermal resistance (R_b) for a closed-loop, vertical borehole array (see Section 10.6.1, p. 227).

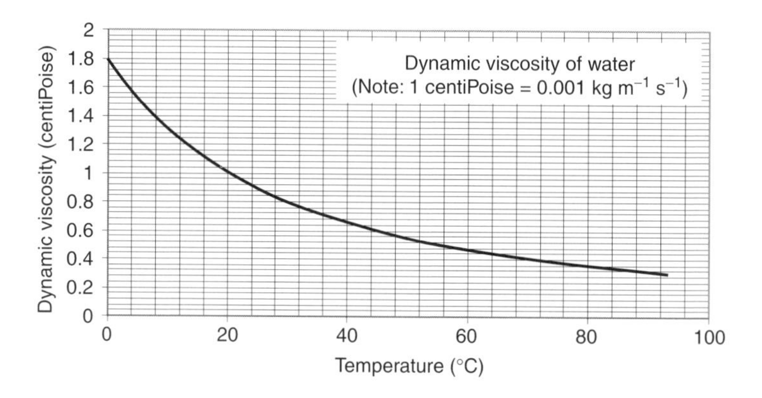

Figure 8.1 The dynamic viscosity of pure water as a function of temperature. Note that 1 cP = 0.001 kg m^{-1} s^{-1}.

- The carrier fluid should efficiently exchange heat with the loop wall, should not be too viscous, should be of low toxicity (in case of leaks), should have a freezing point below the minimum operating temperature of the system and should ideally not be flammable.

8.1 Depth of burial

The depths of burial for horizontal loop systems recommended by various practitioners (1.2–2 m) ensure

- that the pipe is isolated from diurnal fluctuations in temperature (which only penetrate a few tens of cm – Figure 8.3);

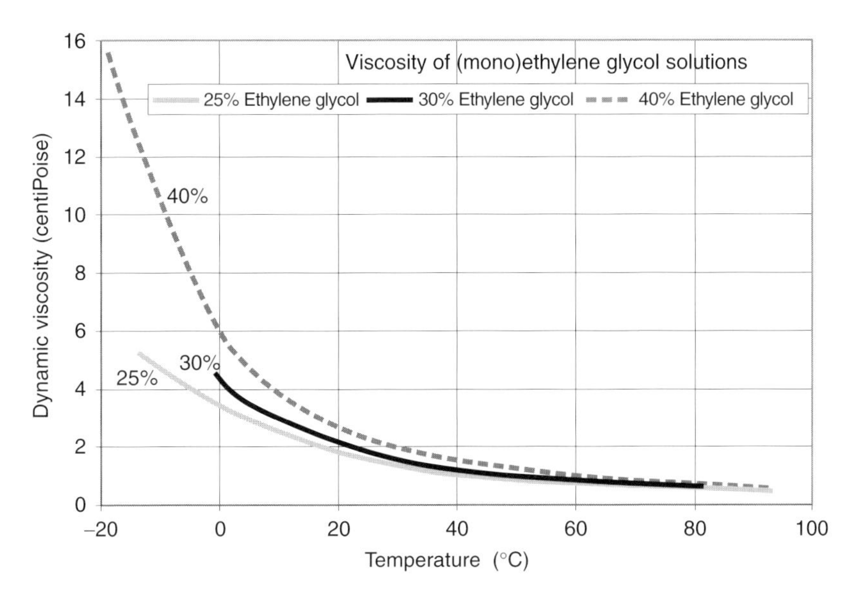

Figure 8.2 Viscosity of ethylene glycol solutions as a function of temperature. Note that 1 cP = 0.001 $kg\,m^{-1}\,s^{-1}$. Concentrations are as vol./vol. % values. *Based on data compiled from Eskilson et al. (2000), Rosén et al. (2000) and http://www.engineeringtoolbox.com/.*

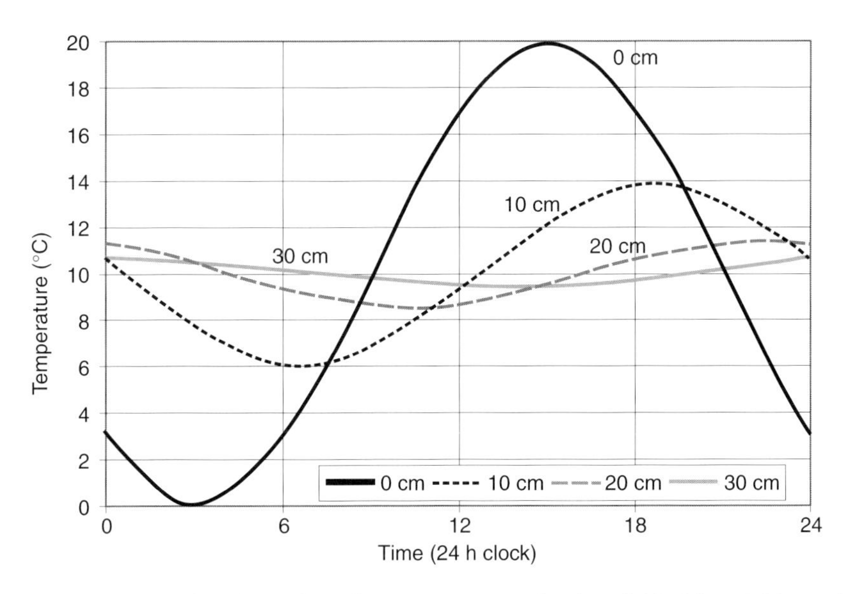

Figure 8.3 The simulated variation in soil temperature at depths of 10, 20 and 30 cm depth in a clayey soil, in response to a diurnal variation in surface temperature of 0–20°C (i.e. temperature at zero cm depth). Note both the rapid 'damping' of amplitude with depth and the progressive time lag. *Based on data provided by Rosén et al. (2001).*

- that the pipe is below the usual depth of frost formation in the subsurface due to longer periods of subzero surface temperature (although ice may form around the collector due to heat extraction!);
- that the pipe is close enough to the surface for summertime solar radiation and induced heat flow from the surface to replenish the shallow soil's heat reservoir each year.

With a trenched pipe system, in the short term (scale of months), during the heating season, we draw heat out of the soil reservoir immediately surrounding the ground loop. During the summer, we would anticipate that this heat in the upper 1–2 m of the soil would be replaced by solar energy. In the long term, we would thus hope to achieve some form of dynamic equilibrium with solar/atmospheric energy input. Note that some authorities recommend a depth of burial at the shallower end of the range suggested above: VDI (2001a) suggest a depth of 1.2–1.5 m to ensure complete replenishment of the heat reservoir during summer.

8.2 Loop material

The ground loop can be made up of a number of materials. Copper has, of course, a very high thermal conductivity, but it is rather expensive and is not regarded as being particularly resilient: it may be subject to corrosion and also to damage by, for example, later excavation works. It is therefore normal for the ground loop to be constructed of a plastic. Of the common varieties, polyethene has one of the highest thermal conductivities, while being tough and durable (Table 8.3). Both medium-density polyethene (MDPE) and high-density polyethene (HDPE) can be used in shallow ground loops, as they tolerate temperatures of up to 60°C. As ground loops are commonly pressurised under operation to 2–3 bar, pipe materials with a pressure rating of at least PN6 (6 bar)

Table 8.3 Thermal conductivities of common pipe materials (after the databases of [1]Eskilson *et al.*, 2000; [2]Rosén *et al.*, 2001; [3]VDI, 2001b; and [4]Engineering Tool Box, 2007).

Material	Thermal conductivity ($W\,m^{-1}\,K^{-1}$)
High-density polyethene (HDPE)[2]	0.45
Polyethene (generic)[1]	0.42
Medium-density polyethene (MDPE)[2]	0.4
Polypropene (PP)[1]	0.22
Polybutene[3]	0.22
Polyvinyl chloride (PVC)[1]	0.23
Steel	20[3]; 60[1]; 16–54[4]
Copper[1,2,4]	390–401

Table 8.4 Approximate dimensions of high-density (PE80-100) polyethene pipes (data derived from www.engineeringtoolbox.com, KWH pipes and Vylon 2007). The PN number is the pressure rating in bars. SDR = standard dimension ratio = the ratio of outside diameter to wall thickness. Specifications may vary according to material and manufacturer and the reader is advised to consult specific manufacturer's literature for design purposes.

Nominal outer diameter (OD) (mm)	Internal diameter (ID) (mm)			
PN system	*PN6.3*	*PN10*	*PN16*	
20		16	14.4	
25	21.0	20.4	18.0	
32	28.0	26.2	23.2	
40	35.2	32.6	29.0	
50	44.0	40.8	36.2	
SDR system	*SDR 9*	*SDR11*	*SDR13.6*	*SDR17*
Pressure rating (bar)	*13.8*	*11.0*	*9.0*	*6.9*
33.4 (Nominal 1″ ID)	26.0	27.4	28.5	
42.2 (Nominal $1\frac{1}{4}''$ ID)	32.8	34.5	36.0	
48.3 (Nominal $1\frac{1}{2}''$ ID)	37.5	39.5	41.1	42.6

are used, although higher pressure ratings (SDR11 or PN10) are typical (Table 8.4). The internal diameter (ID) is typically nominally between $\frac{3}{4}''$ and $1\frac{1}{4}''$ (19–32 mm). Low-density polyethene is seldom used, as it does not tolerate temperatures in excess of 20°C (Rosén *et al.*, 2001). Ground-loop pipes are usually delivered in coils, ready to be laid out as straight lengths or as coiled 'slinkies' in a trench.

8.3 Carrier fluid

The most obvious type of carrier fluid for a closed-loop system is water; however, this is only suitable in systems where heat is being rejected to the ground (i.e. cooling systems). Where we wish to abstract heat from the ground, the carrier fluid will usually be chilled by the heat pump to temperatures below 0°C (in order to induce heat flow from the ground to the fluid). Thus, we will usually add some form of antifreeze to the carrier fluid. Rosén *et al.* (2001) recommend that its freezing point should be no higher than −8°C. This antifreeze may be a solution of an inorganic salt (sodium chloride, magnesium chloride, calcium chloride or potassium carbonate), an organic salt (potassium acetate) or an alcohol or glycol (methanol, ethanol, ethylene glycol or propylene glycol). In selecting a suitable carrier fluid one must consider various factors, including viscosity, flammability, freezing point, thermal properties (Figure 8.4) and stability. Rosén *et al.* (2001) also urge us to consider the consequences of a leakage and recommend that the fluid should have a low toxicity and preferably be biodegradable. The environmental toxicity of a carrier fluid will be more important in a shallow closed-loop system (as it will be more prone to accidental damage than a deep system), especially if the loop is installed in or near a lake or other surface water body (Chapter 9).

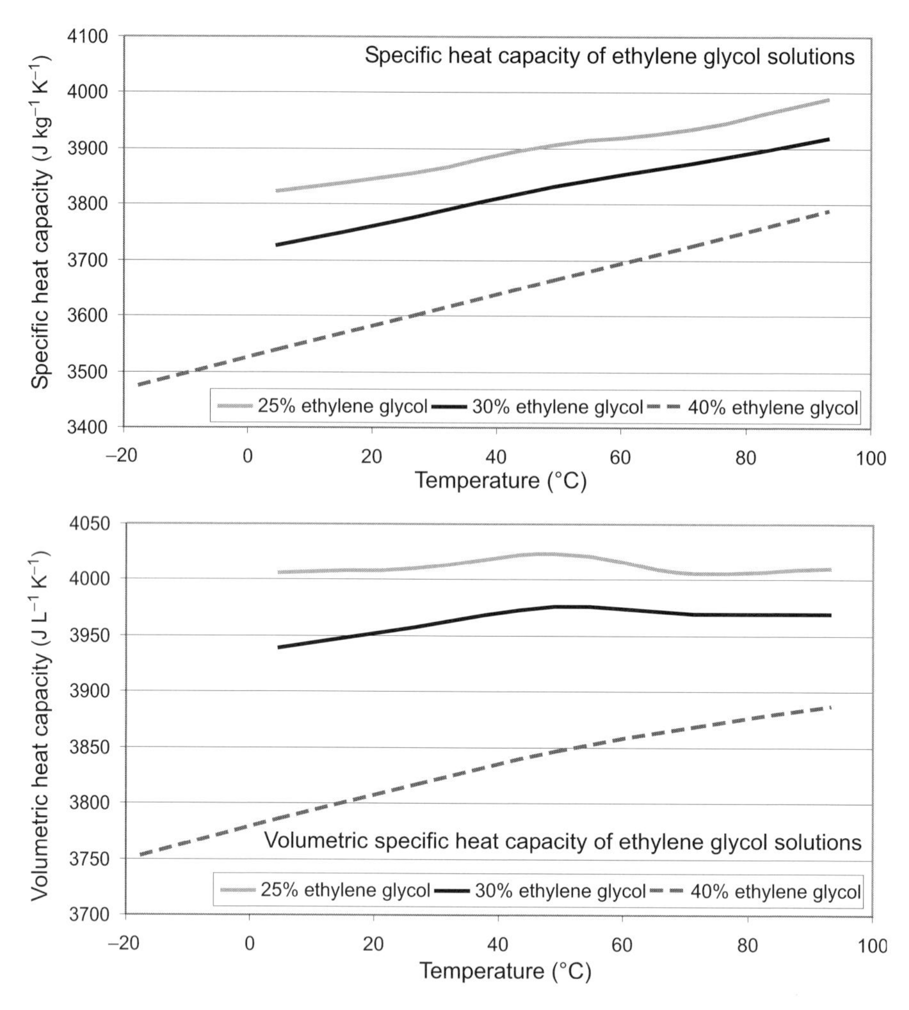

Figure 8.4 Specific and volumetric heat capacity calculated for ethylene glycol solutions as a function of temperature. Concentrations are as vol./vol. % values.

Table 8.1 summarises the selected properties of several common carrier fluids. Ethylene glycol is a popular choice in European systems, as it has good thermal properties and is biodegradable in nature. It is, however, somewhat toxic at high concentrations and is banned in ground loops in a number of US states (Den Braven, 1998). Propylene glycol is less toxic and more environmentally benign but, as can be seen from Table 8.1, it has a high viscosity, leading to larger energy losses in pumping and making it more difficult to achieve turbulent flow conditions. Ethanol, in low concentrations, is not especially toxic and is biodegradable. It too has a relatively high viscosity, however, and, in higher concentrations, can be both toxic and flammable. The salt solutions may have corrosive properties in contact with some metals. Proprie- tary carrier fluids (such as Freezium™, based on potassium formate) have also been developed, claiming to have optimised combinations of low viscosity, high specific heat capacity and low toxicity.

8.4 Carrier fluid flow conditions

When a ground loop has been installed, it will typically be pressure-tested at somewhere between 6 and 10 bar (although the exact regime of pressure testing may be specified by national legislation) to ensure that the loop has no leaks. This compares with typical operational pressures of 2–3 bar (or sometimes less). The heat pump is usually supplied with a pressure sensitive trip switch. Thus, if a leak does develop in the ground loop such that pressure falls and circulation ceases, the heat pump should cut out, preventing freeze damage to the evaporator (and hopefully minimising leakage, too).

When designing a flow rate (F) for the carrier fluid, we need to achieve two objectives. First, the heat transfer rate (Q) should result in flow and return temperatures that are suitable for the heat pump in question. This is governed by the equilibrium between heat transfer at the loop–ground interface and at the evaporator of the heat pump. At the heat pump

$$Q = S_{\mathrm{VCcar}} \cdot \Delta\theta \cdot F \tag{8.1}$$

Ground source heat pumps often operate with a temperature differential across the evaporator ($\Delta\theta$) of around 4–5°C. If we assume that the volumetric heat capacity of the carrier fluid is around 3.8 kJ L^{-1} K^{-1}, then the flow rate will be around 3.2 L min^{-1} for every kW of heat transfer for $\Delta\theta = 5$°C.

Second, for efficient transfer of heat from wall of the pipe to the fluid, the flow should be turbulent (Box 8.1). For example, if we consider a typical 25 m slinky trench, designed to extract 2 kW of ground source heat, utilising 26 mm ID pipe, we will see that a fluid flow of 6.4 L min^{-1} (2×3.2 L min^{-1}) will not be enough to produce turbulent flow at temperatures of around 0°C, if 30% ethylene glycol is the coolant (Table 8.2). Thus, in this case, our first estimate of the carrier fluid flow rate must be modified upward to produce turbulent flow. We must therefore consider carefully the interplay of the diameter of the pipe, the carrier fluid, the operational requirements of the heat pump and the length of ground exchanger in order to design an optimally functioning system.

The temperature of the carrier fluid during operation is typically in the interval 0°C to −5°C (although it can be lower under very brief peak loading conditions). In Germany, VDI (2001a) make the more specific recommendation that the temperature of the fluid returning to the heat pump from the ground loop should not deviate by more than ±12°C from the undisturbed ground temperature under base load (i.e. weekly average) conditions, or by more than ±18°C under peak conditions. If we assume that the undisturbed ground temperature is 10°C in temperate Europe (including the UK), this implies typical acceptable base-load temperatures of −2°C in heating mode and 22°C in cooling mode, with extremes under peak loading not exceeding −8°C and 28°C, respectively.

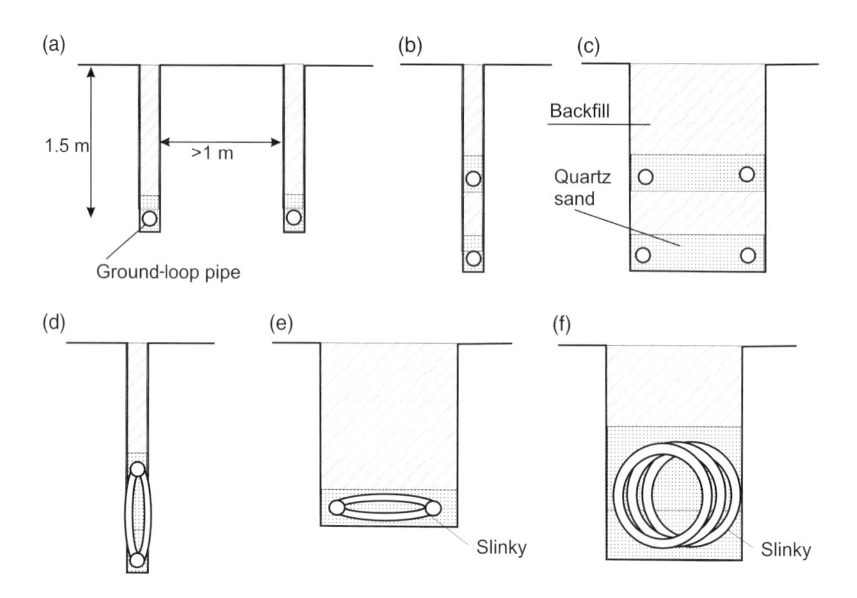

Figure 8.5 Possible configurations of horizontal ground loops in trenches: (a) single pipes in parallel trenches; (b) vertically installed double pipe system (flow and return); (c) 'square' four-pipe system (two flow, two return); (d) vertically installed slinky; (e) horizontally installed slinky; and (f) three-dimensional spiral slinky. Note that pipes are bedded in conductive quartz sand.

8.5 Geometry of installation

There is clearly an almost unlimited number of ways of burying a pipe in the ground; so, in this section, we will consider just a few (Figure 8.5). In all cases (at least in temperate European and North American climes), it would appear that, if the pipe is buried much more than 2 m deep, the time for 'recharge' of the heat extracted during the heating season may take more than one summer. If the pipe is too shallow, however (<0.8 m according to data cited by Rosén *et al.*, 2001), it is suggested that root systems of vegetation may be damaged by the installation. Whatever the geometry of the installation, it seems that backfilling around, immediately above and below the pipe or slinky with fine quartz-rich sand will ensure good thermal conductivity in the zone around the pipe, and a good thermal contact between pipe and soil.

8.5.1 Areal constraints

Some international guidelines for dimensioning horizontal ground-loop systems are provided in this section. Although it might appear that we obtain a lot more ground source heat per metre of trench from a 'slinky' than from a single pipe buried in a trench, the amount of ground source heat that can be removed from a given area will be limited ultimately by typical net solar/atmospheric radiation rates of several tens of W m^{-2} (for temperate in Europe, see Box 3.5, p. 43). In fact, there is a clear inverse

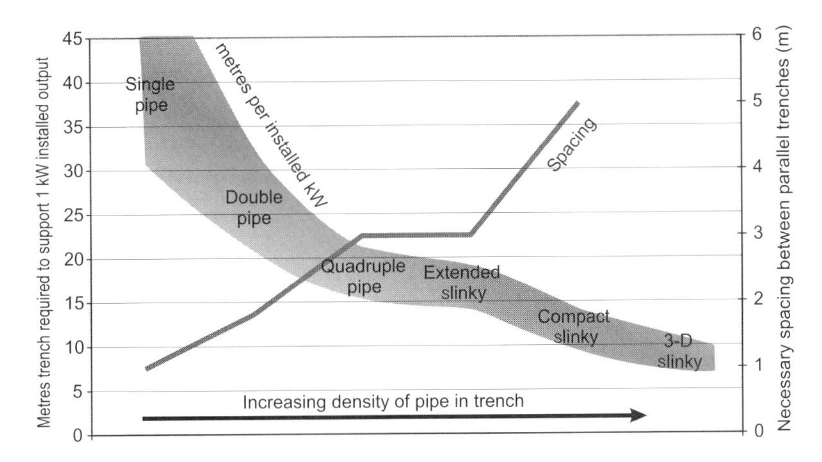

Figure 8.6 The relationship between specific installed thermal output (number of metres of trench required to support 1 kW installed heat pump capacity) and necessary separation between parallel trenches, for different configurations of horizontal ground loop. This figure is indicative only and should not be used for design purposes.

relationship between the density of the ground-loop pipe that we can install in a trench (i.e. density of heat extraction) and the necessary spacing between parallel trenches (Figure 8.6). Thus, if we have a given land area available to us, we can choose to bury

- single pipes (which may support 15–30 $W\,m^{-1}$ installed heat pump capacity) in parallel trenches at 1 m spacings, or
- slinkies in trenches (which may support 100 $W\,m^{-1}$) at 3–5 m spacings.

In the case of slinkies, we are still limited by the available area: we simply require fewer metres of trench to be excavated to access the energy (although we install a greater linear length of pipe, which is, fortunately, cheap). Indeed, some guidelines for dimensioning operate on the basis of the maximum heat extraction per m^2 of ground area. VDI (2001a) make suggestions (based on the assumption of simple horizontal collector pipes in parallel) that

- for normal damps soils, with heating seasons equivalent to 1800 h full operation, the maximum specific heat extraction rate should not exceed 20–30 $W\,m^{-2}$. For dry, friable soils, a figure of 10 $W\,m^{-2}$ is used and for water-saturated sands and gravels, as high as 40 $W\,m^{-2}$.
- for a heating season of 2400 h, these figures are reduced to 8 $W\,m^{-2}$ for dry soils, 16–24 $W\,m^{-2}$ for normal soils and 32 $W\,m^{-2}$ for saturated sands and gravels.

These German guidelines further suggest that total annual extraction of heat from the soil should not exceed 50–70 $kWh\,m^{-2}$ each year (180–250 $MJ\,m^{-2}\,yr^{-1}$) to ensure

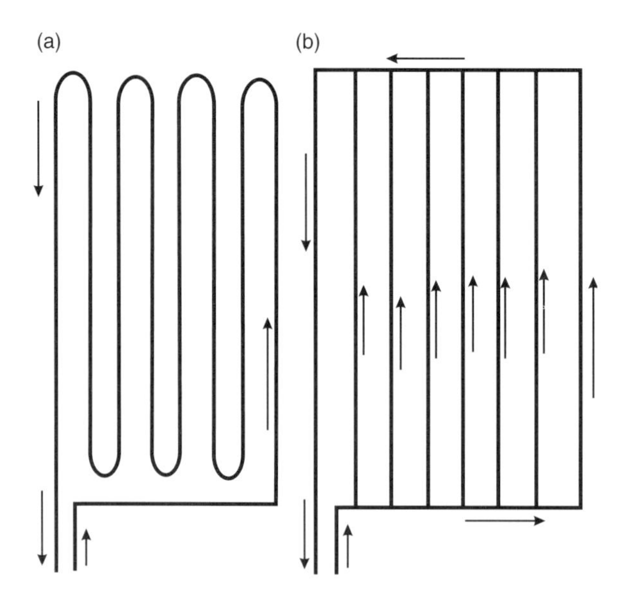

Figure 8.7 Two possible ways of installing parallel trenches of single pipe; (a) in series, (b) in parallel.

long-term sustainability. If waste heat is re-injected during a cooling season, the total annual availability of heat will be increased.

8.5.2 Single horizontal pipe

Rosén *et al.* (2001) cite data from the USA suggesting that each metre of buried single horizontal pipe (Figure 8.7) will support 15 W of installed GSHP capacity in dry soils and up to 30 W in 'normal' soils. They also cite a typical value for 'heating-only' European conditions of 19 $\mathrm{W\,m^{-1}}$ pipe. Rosen *et al.* (2001) also note the generally higher values cited from American practice and suggest that this may reflect a typically shorter GSHP heating season in the USA, and their frequent use for cooling as well as heating (in other words, there will be artificial recharge of 'waste' heat during the summer), which will hasten the regeneration of heat in the soil. Note that the figures referred to here are 'specific installed thermal outputs': the installed peak capacity of the system divided by the metres of pipe (or borehole or trench). Other authors cite such 'rules of thumb' as 'specific heat absorption' or 'specific heat extraction' rates, that is, the peak rate of heat transfer from the ground to the ground loop. In heating schemes, this value is less than the 'specific installed thermal output' by a factor of $[1 - (1/COP_H)]$.

Often, an installation will consist of parallel rows of buried horizontal pipe at spacings of not less than 1 m (Figure 8.5a). The areal constraints cited in Section 8.5.1 are based on parallel trenches of single pipes, and form the basis for Figure 8.8, which allows us to estimate the area required to support a given peak installed output, for given soil conditions and heating season durations. Swiss data support these figures,

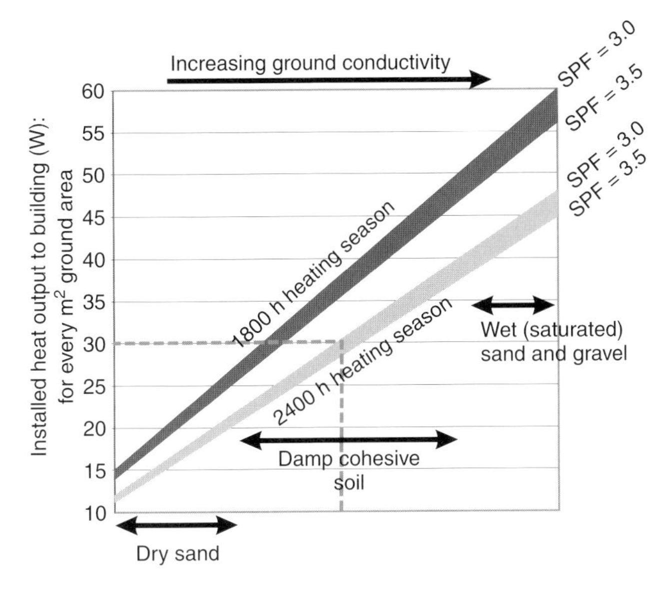

Figure 8.8 Recommended specific installed thermal outputs for differing ground conditions and different lengths of heating season. *Based on data cited for parallel single horizontal pipes by VDI (2001a) and Rosén et al. (2001).* The dashed line shows an example where, in a damp silty soil, around 30 W of heat pump capacity can be installed per m^2 of ground underlain by parallel trenches, for 2400 h equivalent running hours per heating season and a seasonal performance factor of 3.0. If the trenches contain single pipes and are 1 m apart, this equates to 30 W m^{-1} of trench.

with specific heat extraction rates ranging from as low as 10 W m^{-2} for dry soils to 40 W m^{-2} for high thermal conductivity substrates (Rosén *et al.*, 2001).

8.5.3 *Multiple pipes per trench*

Instead of burying a single pipe per trench, we can bury two pipes per trench: for example, with the upper pipe at depth 0.9–1.2 m and the lower at 1.5–1.8 m (Figure 8.5b). American studies tend to be unanimous in concluding that around 50–60% (and even as much as 80%) more energy per metre of trench can be extracted compared with a single pipe (Rosén *et al.*, 2001). Thus, if a single-pipe system has a specific heat extraction rate of 20–30 W m^{-1}, 1 m of a two-pipe trench might yield 30–50 W m^{-1}. The reason for this appears to be that, early in a heating 'episode' or season, there is no thermal interference between the two pipes and they effectively extract heat at double the rate of a single pipe. Later, interference occurs as the temperature fields of the upper and lower pipes begin to overlap – this ultimately reduces the heat available per metre of trench. To maintain a long-term equilibrium with atmospheric/solar heat input, Rosén *et al.* (2005) suggest that such double-pipe trenches should be spaced at correspondingly larger intervals in parallel systems (around 1.8 m). The advantage of a double-pipe system over a single-pipe system is that it reduces the total length of ditch required. It increases the length of pipe required, and the volume of collector fluid, however. Double-pipe systems would be particularly effective in situations with a short heating season, such that interference between the two pipes is minimal.

Clearly, many other geometric variants are possible, including four pipes per trench either stacked vertically or in a 'square' configuration (Figure 8.5c). Rosén *et al.* (2001) suggest that four-pipe systems yield even greater specific heat extraction rates (per metre trench), some 120% to >150% higher than with single-pipe systems. This implies that trench lengths can be reduced to 40–45% of those required single-pipe systems. The spacing between the trenches must also increase, however, with values of 2.5–3.6 m being cited by Rosén *et al.* (2001).

8.5.4 Coiled collectors – 'slinkies'

Clearly, stuffing more and more pipe into a single trench appears to yield dividends in terms of specific heat extraction rate per metre of trench, at least in the short term. The downside of this is

- that there is an inevitability of the 'law of diminishing returns' kicking in;
- the necessity for increased spacing of trenches (as the heat available will ultimately be constrained by available area and insolation);
- that our gains in terms of reduced trench length will be offset by the disproportionately large amounts of pipe and carrier fluid that we require (although, admittedly, polyethene pipe is very cheap).

The apparent culmination of this tendency is the 'slinky' or Svec Coil, which we have already encountered in Chapter 6. Here, large lengths of HDPE or MDPE pipe are installed in trenches in the form of overlapping coils, which are typically supplied at diameters of around 1 m. There is no wholly standardised design for slinky geometry: slinky lengths and diameters can vary. Successive coils may overlap to varying degrees (a so-called compact slinky) or may not overlap at all (an extended slinky), if space is available. If we consider Figure 8.9, we will see that

- if the original slinky coil diameter is D^o;
- if the slinky is stretched out to fit in a trench, such that each successive coil is displaced by a pitch (P);

then the original coil diameter will reduce slightly such that the *in situ* (stretched) diameter D^* is approximately given by

$$\pi D^* = \pi D^o - P \tag{8.2}$$

Furthermore, neglecting the length of the return pipe, the length of trench (L) and pitch (P) are related by the formula:

$$D^* + P(n - 1) = L \tag{8.3}$$

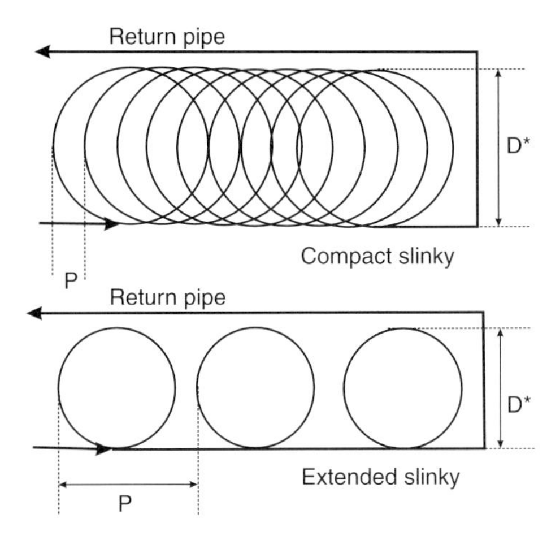

Figure 8.9 Two methods of installing 'slinkies': in compact and extended modes. *P* is the pitch between successive coils, while D^* is the *in situ* loop diameter.

where *n* is the number of coils in the trench. Thus, if the supplied diameter of a slinky is 0.8 m and if it is installed as a compact slinky, with a pitch of only 250 mm, we can say that the 'stretched' diameter D^* is around 0.72 m. If it is installed in a 25 m long trench, there will be around 98 loops of pipe, containing a length of pipe of 246 m. In other words, every metre of trench contains around 10 m of pipe (or 11 m if we count the return pipe). The same length (246 m) of slinky coil, in extended form, with a pitch of, say, 1 m, would have a coil diameter of around 0.5 m and would fit in a total trench length of 97.5 m. Here, every metre of trench would contain 2.5 m of pipe (or 3.5 m of pipe if we include the return pipe). The extended slinky extracts a larger amount of heat in total than the compact one, as there is less thermal interference between overlapping coils, but requires almost 4 times the length of trench.

Slinkies can be installed vertically in a narrow slot trench (Figure 8.5d) or horizontally in a broader trench (Figure 8.5e), around 1 m wide. Either way, the 'rules of thumb' for dimensioning do not appear to vary greatly. The depth of installation is typically between 1.2 and 2 m. The spacing between parallel slinky trenches depends on the density of heat transfer: for extended slinkies, a spacing of 3 m between adjacent trenches may be sufficient, while a spacing of up to 5 m may be required for compact (overlapping) slinkies (Alliant Energy, 2007).

Typically, individual slinkies contain around 150–240 m pipe (and lengths greater than this are not recommended, due to head losses in very long pipes and difficulties in achieving the flow rates necessary for turbulent flow). Installed diameters are typically in the range 0.6–1 m.

According to Alliant Energy (2007), 230–240 m of $\frac{3}{4}''$ (19 mm, presumed ID) HDPE slinky pipe, in compact horizontal configuration, with a pitch of 43 cm, installed in a 30 m long, 1.8 m deep, 90 cm wide trench is adequate to support 1 ton of heat pump capacity (3.5 kW). With this American recommendation, we should be aware that the postulated system may be dominated by cooling demand. For the (usually heating-dominated) systems designed for the British Isles, an average of 10.5 m of slinky trench is typically installed per kW of installed peak output (the values range from 10 to 14 m – Figure 6.10.b, p. 123). Installers such as GeoWarmth (*pers. comm.*) typically operate on the basis of 10–12 m slinky trench per installed kW. Rosen *et al.* (2001) suggest that a compact slinky-based system requires around 25% of the trench length calculated for a single-pipe system, while an extended slinky requires around 45% of the length (see Figure 8.6).

It should be noted that we can also install a slinky as a three-dimensional spiral, if we are prepared to construct a large enough trench (Figure 8.5f). According to information cited by Rosén *et al.* (2001) for a climate similar to Ontario, 150 m of slinky pipe can be packed into a 20–25 m long, 60 cm wide, 1.8 m deep trench in this manner and can be used to support an installed heat pump capacity of 3.5 kW. This corresponds to only 7 m trench per installed kW.

Computer-aided design tools are available to assist in the design of horizontal ground-loop installations, although they are not as sophisticated as those available for vertical systems (Spitler, 2005).

8.6 Horizontal ground collectors and soil properties

Such 'rules of thumb' discussed above should not blind us to the fact that system performance will also depend on the soil's thermal properties. These will depend, not just on lithology, but also on seasonally varying conditions: the soil's moisture content and whether it is frozen or not. Generally, both the thermal conductivity and the volumetric heat capacity of geological materials increase as they become wetter, due to the fact that these properties are significantly higher for water than for air. As moist geological materials freeze, their thermal conductivity increases again (due to the conductive nature of ice), but their volumetric heat capacity falls.

Purely theoretically, we could consider a quartz sand, with 25% porosity. If it is perfectly dry, we could calculate the volumetric heat capacity as the volume-weighted arithmetic mean of its constituents: quartz (1.9 $MJ\,m^{-3}\,K^{-1}$) and air (negligible). This gives us a value of 1.4 $MJ\,m^{-3}\,K^{-1}$. If we allow that its thermal conductivity can be roughly estimated as the geometric mean of the constituents (using values from Table 3.1, p. 35), we obtain 1.8 $W\,m^{-1}\,K^{-1}$. If the sand becomes saturated with water, these values increase to 2.5 $MJ\,m^{-3}\,K^{-1}$ and 4.1 $W\,m^{-1}\,K^{-1}$, respectively. If the saturated sand then freezes such that pores become filled with ice, the values are recalculated as 1.9 $MJ\,m^{-3}\,K^{-1}$ and 5.3 $W\,m^{-1}\,K^{-1}$. These are high values, but remember that we have selected an unrealistically pure quartz sand for this example. Furthermore, the geometric mean is probably not a good way of estimating thermal conductivity of a composite material. A preferred method (the Russell equation) is presented by Russell (1935) and discussed by Spencer *et al.* (2000) – and even they admit that this rather complex formula has its limitations. Further discussion of the thermal properties of porous materials is provided by Clauser and Huenges (1995) and Waples and Waples (2004b).

As we extract or reject heat through our horizontal ground loop, the ground may heat and cool to the extent that its properties change. In heat extraction mode, we may cause the ground to freeze. This is not necessarily a bad thing (as long as there are no sensitive structures or foundations above the loop that might be sensitive to frost heave). Indeed Scandinavian thermogeologists are rather proud of the 'ispølser' (ice sausages) that may form around ground loops, as was John Sumner (1976a) in his day. The formation of ice results in an elevated thermal conductivity of the ground. It also releases heat in the form of latent heat of freezing (fusion) to the tune of 0.34 MJ kg^{-1} of pore water (see Box 3.2, p. 37). Horizontal ground loops are seldom installed below the water table. Thus, it often happens that the extraction of heat causes condensation of water vapour near the loop, decreased water vapour pressure and a tendency for water vapour to migrate towards the loop. The water vapour condenses and freezes on and near the loop (in addition to the direct freezing of existing pore water). If excessive accumulation of ice occurs around the loop, it may result in visible soil 'heave' at the surface. In designing our system, one of our criteria should be that any ice lens around the ground loop should not grow to such an extent that it joins up with that around a parallel loop or that it merges with any front of frozen ground extending down from the surface during winter frost conditions. If this does occur, we will effectively be trying to extract heat from ice. Moreover, we might doubt whether such an extensive ice lens could be melted by solar warmth penetrating down into the soil during the summer.

When we reject waste heat in cooling mode, we will be heating the air within the pore spaces. We will also increase the vapour pressure of water, causing it to migrate by convection away from the loop. The net effect of this might be a drying out of the ground around the loop, a decrease in thermal conductivity and a decline in the system's efficiency.

From Box 8.2, we will note that extensive horizontal closed-loop systems can be used for both heating and cooling large building spaces. We should be aware, however, that we are relying on 'leakage' of heat from the surface to the loop (and *vice versa*) to achieve this. Horizontal closed-loop systems are not especially good at 'storing' surplus heat from the summer for subsequent use in the winter. In order to use the subsurface as a thermal 'store', we probably need to construct carefully designed, deep, vertical borehole systems (see Section 10.9, p. 246).

8.7 Earth tubes: air as a carrier fluid

As a parting thought, we should be aware that we limit ourselves in Chapters 8–10 by considering only water-based solutions as carrier fluids. There is, of course, good reason for this: water has a huge specific heat capacity! We can, however, consider air as a carrier fluid. Indeed, we could conceive of a system of subsurface pipes (or even boreholes) through which air is circulated, gradually equilibrating towards the subsurface

BOX 8.2 The Green House, Annfield Plain.

The Green House is a major 3000 m^2 modern office complex in a former coal-mining area of County Durham, UK. At an early stage, a desire was expressed to provide space heating and cooling by ground source heat pumps, to be delivered by a combination of underfloor heating and fan-coil units.

The Green House, Annfield Plain. Photo by D Banks.

Initially, the intention was to employ an open loop system, using the abundant water that often occurs in the flooded, abandoned coal mine workings that are so prevalent in this part of the world (Younger, 2004). Such mine-water-sourced heat pump schemes are not without problems, due to the water's high iron content leading to the potential deposition of iron hydroxide ('ochre') in heat exchangers. Nevertheless, British and international experience demonstrates that mine sourced schemes are viable (Banks *et al.*, 2004a). Unfortunately, a feasibility study suggested (and two boreholes confirmed) that the Coal Measures strata beneath the Annfield Plain site were largely dry to at least 100 m depth! This is most likely due to the presence of an old underground regional mine drainage network that dewaters the abandoned coal mine workings and adjacent strata. The mine water in this drainage system flows northwards under gravity and is eventually discharged to the River Derwent, some 7 km away, at Hamsterley John.

The next possibility was to employ closed loop vertical boreholes as ground heat exchangers. Trial boreholes were constructed and tested (Fernández Alonso, 2005), although it proved inordinately difficult to grout the ground loop into a borehole that passed through open, abandoned mine workings – the grout seemed to slump into the workings, distorting the loop. The problem could have been overcome by 'casing out' the mine workings, but this would have added significantly to the cost of drilling.

BOX 8.2 (*Continued*)

Eventually, a plan was hatched to use most of the available land area around the building to install a huge network of horizontal slinky trenches. It proved difficult to integrate this work with the rest of the construction process, and to protect loops and manifolds from wandering excavators and bulldozers. Eventually, however, the system was commissioned, using fifteen 55 m long slinky trenches, containing a total of 6500 m pipe, to deliver peak loads of 80 kW heating and 30 kW cooling. This works out at just over 10 m trench per 1 kW installed capacity. The ground source scheme is integrated with photovoltaic cells and a wind turbine to generate electricity.

temperature. Such systems are probably most useful for summer air conditioning: clearly, a throughflow of air, which has been cooled by flowing through the ground (at, say, 11°C in Britain) could potentially be very useful in the summer for cooling a house or small office. Alternatively, in winter, the circulated and warmed air could be used for preheating a more conventional system, or the warmed air could be utilised in an air-sourced heat pump.

Such air-based ground circulation systems have been christened 'earth tubes'. Clearly, we need to power a fan to drive the air around the earth-tube system, which consumes power. VDI (2004) believe that such earth tubes are not energy effective unless they can achieve a temperature differential of at least ±2°C relative to the outside air temperature. Pipes are typically made of polyethene, PVC or concrete and installed in trenches of depths of between $1\frac{1}{2}$ and 3 m. Consideration must be given to problems of condensation and how such condensate will be collected.

VDI (2004) estimate that residential houses of floor area 100–250 m^2 typically require air throughflows of 100–500 $m^3\,h^{-1}$. Furthermore, flow velocities in pipes of 2–4 $m\,s^{-1}$ are typically sufficient to permit adequate heat exchange with the ground. The pipe diameter selected will depend on the flow rate: for small houses requiring 100 $m^3\,h^{-1}$ of air flow, some 25–28 m length (depending on the thermal properties of the ground) of 100 mm diameter pipe will be required to provide a minimum standard of thermal performance (VDI, 2004). The corresponding flow velocity here is 3.5 $m\,s^{-1}$ and the pipe heat exchange area is ≈8–9 m^2. For a larger house, with an air throughflow rate of 500 $m^3\,h^{-1}$, some 40–50 m of 300 mm diameter pipe will be required. This corresponds to 2 $m\,s^{-1}$ flow velocity and 38–47 m^2 of exchange area.

Earth tubes can be utilised to provide air conditioning and preheating to larger building spaces, although some form of computer-aided design is likely to be required.

9

Pond- and Lake-Based Ground Source Heat Systems

> The heat pump would appear to offer great possibilities in situations, such as the centre of London, where, for example, the large block of Government buildings now being erected in Whitehall could be warmed by heat extracted from the Thames, with a minimum of smoke and sulphur dioxide production in the centre of the city. Thus one might even recover some of the heat wasted in the cooling water of the Fulham and Battersea power houses.
>
> Mr T.H. Turner, in the discussion following the presentation of John Sumner's (1948) paper to the Institution of Mechanical Engineers in March 1947

In Chapter 8, we have seen how we can extract heat from shallow soils using closed horizontal loops or coils as a heat exchanger. We have also seen that the wetter the soil, the better it is as a source for a ground source heating scheme. We can also use heat-exchange coils to extract warmth from rivers, ponds and lakes, where the surrounding water provides an excellent heat transfer medium.

9.1 The physics of lakes

Many lakes are not simply homogeneous reservoirs of water and heat, nor will they be static. They will often be in some form of dynamic hydrological and thermal equilibrium with their environment. Our attempts to extract either water or heat (or both) will disturb that equilibrium in a manner which may or may not be acceptable to the ecological systems that depend on the lake.

Shallow lakes (less than, say, 4 m; Kavanaugh and Rafferty, 1997), where solar radiation can reach and warm the lake bed, are more likely to be homogeneous in their composition than deeper lakes. Convection processes in the former are likely to ensure that water is mixed throughout the lake depth.

In deep lakes, however, significant stratification can develop. Water is at its most dense at around 4°C. As temperatures drop during winter, cold water will sink to the base of the lake, as its temperature approaches 4°C, and it will tend to stay there. The water nearer the surface will continue cooling (but will be unable to sink, because density decreases as its temperature falls below 4°C) and it will remain near the surface, eventually forming a layer of ice. Thus, in climates where temperatures fall near or below freezing point, the water at the base of a deep lake will remain at around 4°C throughout the winter, unless there is a significant groundwater or surface water throughflow. In climates where temperatures do not approach freezing, the temperature of the lower part of the lake will, of course, remain above 4°C.

During summer, solar radiation and heat transfer from the atmosphere will warm the lake's water. If the lake is sufficiently deep, however, the sun's radiation will be unable to penetrate to the base of the lake and only the surface layers will be warmed. As the surface temperature approaches 4°C, the density of the deep and surface layers becomes similar and they may start convective exchange again resulting in an 'overturn' of the lake's stratigraphy (the 'spring overturn'). As the lake continues to warm, however, dense water at 4°C will tend to remain at the lake base, while the surface layers warm further and become less dense. A stratigraphy thus re-establishes itself, with dense, cold water below a so-called thermocline (i.e. the relatively steep temperature gradient where a layer of warm water overlies a layer of cold water). The depth at which the thermocline occurs varies from about 3 m (in turbid water, where solar radiation cannot penetrate deeply) to over 15 m in clear water lakes (University of Alabama, 1999). In autumn, the surface layers cool again, until they approach 4°C, when convective exchange with the bottom layers may restart, resulting in an 'autumn overturn'. We now see why deeper lakes in higher latitude climates are so attractive for ground source heating and cooling schemes – we have relatively warm (4°C) water below the thermocline in winter and relatively cold water (4°C) in summer!

In lakes with pronounced thermoclines, we may often see a chemocline as well. The dense, cool bottom layer of water may be relatively immobile (at least for several months of the year): it may become depleted in oxygen and other oxidised species, and become rich in reducing gases such as hydrogen sulphide (H_2S) produced by sulphate-reducing microbes in bottom sediments.

Of course, local factors can modify these generalised conceptual models significantly. Influxes of groundwater from the banks or bed of a lake can cool portions of the lake in summer and warm them in winter. Indeed, infrared imaging of lakes can be very useful in identifying groundwater influxes (Banks and Robins, 2002).

Figure 9.1 shows a summer thermochemical profile of a deep Siberian lake (Banks *et al.*, 2001, 2004b) with a thermocline between around 7 and 12 m depth and a redox gradient (a type of chemocline). Above 9 m depth, the water is close to saturation with dissolved oxygen. Below 13 m, dissolved oxygen is effectively absent and concentrations of hydrogen sulphide increase steadily with depth. In other respects, the lake is far from typical: the lake is saline due to high evaporation rates in the semi-arid southern Siberian climate, coupled with low surface water inflow and the somewhat saline

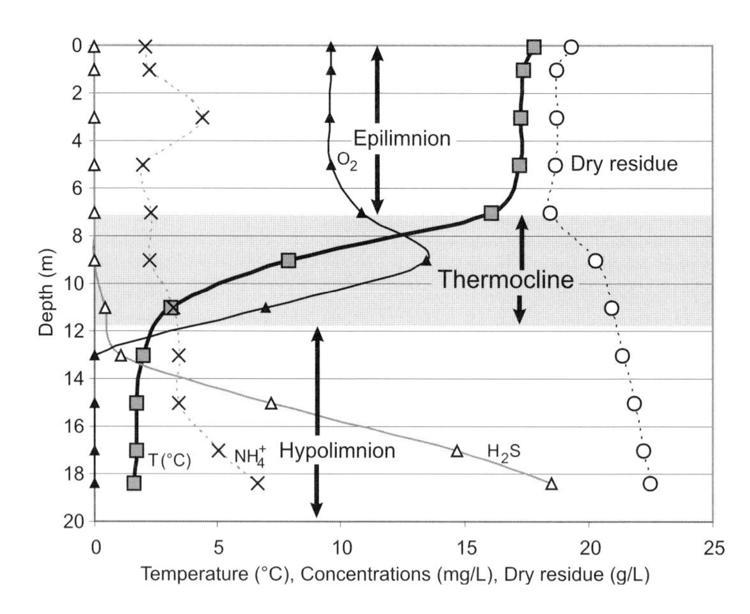

Figure 9.1 A typical summer thermochemical profile for a deep lake in a cold climate exhibits many of the features of this profile from Lake Shira, southern Siberia, measured in September 2000. Note the presence of a thermocline separating deep cold water from shallow warm water, and the redox gradient. Lake Shira is, however, saline (dry residue is a measure of salinity) and has extremely cold basal temperatures: in these respects, is not typical of lakes in less extreme, non-continental climates. *Based on data from Banks et al. (2001).*

nature of groundwater inflows. The salinity is a little over half that of seawater, as reflected in the recorded dry residue concentrations. Thus, it is not merely temperature that controls the density and the dynamics of the lake, one might expect saline water to form by evaporation in the upper layers of the lake during summer. During autumn and winter cooling, it is possible that this increasingly dense saline water sinks and accumulates near the lake base (note the more saline water at the base of the lake and in the uppermost sample). An added complicating feature is the belief that the lake is partly fed by submerged groundwater inflow, and in Siberia, this may be colder than 4°C. It is plausible that both these mechanisms may go someway to explaining the unusually low temperatures (<2°C) at the base of the lake. These complicating factors are speculative, but should serve to remind the reader that lake dynamics are complex.

9.2 Some rules of thumb

Our objective, when supplying heat or cooling from a pond or lake, is to ensure that we do not cause unacceptable change to the lake's temperature, its water level or through-flow (note that, if we increase a lake's temperature, we will also increase losses through evaporation), its chemistry or to its stratification. By 'unacceptable change', we usually mean change that will significantly damage the lake's ecology, amenity value or value as a resource. Clearly, if we are basing a large heating or cooling operation on a pond or lake, we need to thoroughly evaluate both the thermal and water budgets of the lake. However, with smaller schemes, Kavanaugh and Rafferty (1997) suggest that if

- the lake is deeper than 3–4 m,
  AND
- the peak cooling load (resulting in waste heat) is <174 kW Ha^{-1} (17.4 W m^{-2}),
- the peak heating load (extraction of heat from lake) is <87 kW Ha^{-1} (8.7 W m^{-2}),
  OR
- there is a substantial replenishment or throughflow of water (and thus heat) to the lake,

then a lake-sourced heating or cooling scheme is likely to be acceptable. If these criteria are violated, the scheme may still be plausible but will require a more detailed risk assessment.

9.3 The heat balance of a lake

If we need to perform a more detailed risk assessment of the impact of a heating/cooling scheme on a pond or lake, we need toconstruct not only a water balance, but also a

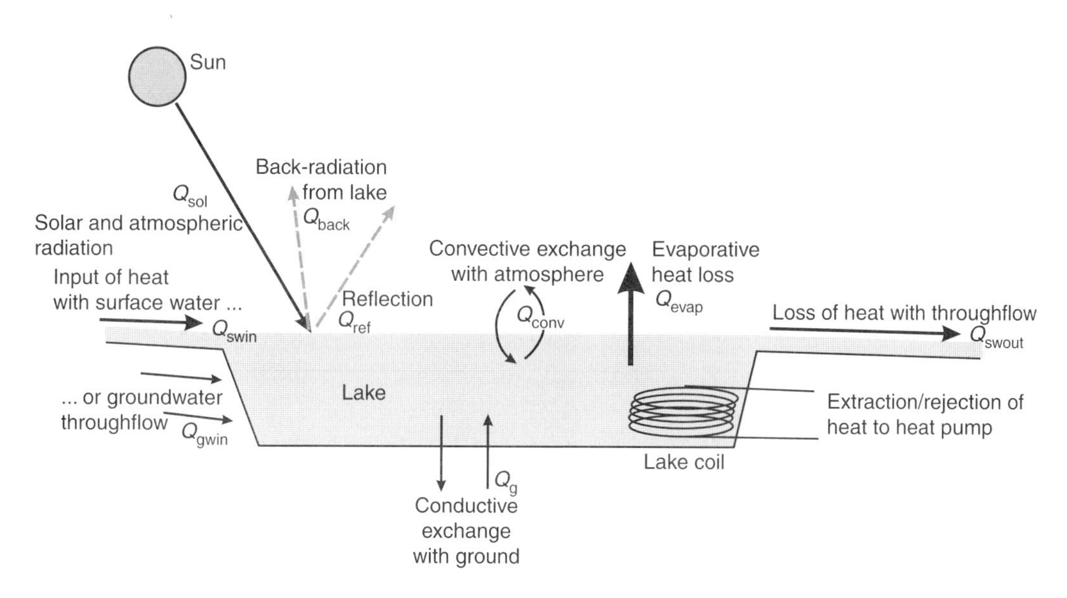

Figure 9.2 The main components of a lake's heat balance.

heat balance of the lake (Hostetler, 1995; Rouse *et al.*, 2005). Figure 9.2 illustrates the main heat transfer mechanisms for a natural lake; these include

- evaporative heat loss Q_{evap}: this is sometimes termed the latent heat flux and may be modified by plant cover;
- *sensible heat flux*: this is dominantly convective heat loss or gain to/from air Q_{conv};
- conductive heat loss or gain to/from ground Q_g;
- input of heat with surface water inflow Q_{swin};
- loss of heat with surface water outflow Q_{swout};
- input of heat with groundwater inflow Q_{gwin};
- loss of heat with groundwater outflow Q_{gwout};
- gains from solar and atmospheric radiation Q_{sol}. This term includes both insolation of short-wave solar radiation from the sun (Q_{sw}), and long-wave radiation from the warm atmosphere (Q_{lw}) – see Box 3.5 (p. 43);
- reflective losses of solar and atmospheric radiation Q_{ref}. This will be governed by the albedo of the lake with respect to the long- and short-wave components of incoming radiation (α_{lw} and α_{sw}), respectively.
- back-radiation (long-wave) of heat from lake Q_{back}.

As an example, let us consider a ground source cooling scheme, where, on the hottest day of the year, we expect to wish to dump Q_{gsc} MJ day^{-1} of heat energy to a lake. The University of Alabama (1999) argue that, for many relatively shallow, non-stratified lakes, the main natural modes of heat loss are (a) evaporative loss and (b) back-radiation losses (which become particularly important in the heat budget during cool nights), while the main mode of heat gain is from solar/atmospheric radiation. The dumping of heat raises the lake's temperature and evaporation increases. Eventually, evaporative (and back-radiated) heat losses become so great that (hopefully) a new thermal balance is achieved with additional heat loss via evaporation balancing the new heat input from the cooling scheme. Remember that increased evaporation implies not just additional heat loss, but additional water loss, which may be important for the water balance of the lake. In fact, 1 kg of evaporated water loss corresponds to 2.272 MJ of heat loss (Box 3.2, p. 37). In other words, an evaporation rate of 0.01 L s^{-1} corresponds to 22.7 kW.

The solar/atmospheric radiation input is location-specific and varies according to season and time of day. We have already considered it in Box 3.5 (p. 43) and Table 3.2 (p. 45). Remember that a proportion will be immediately reflected away from the water surface. According to Laval (2006) short-wave albedos (α_{sw}) of between 6% (summer) and 10% (winter) are typical for open water, with long-wave albedos (α_{lw}) being even less. However, we should remember that albedo may be modified if the lake is very turbid or if the lake bottom is shallow (i.e. reflection from the lake base as well as the water surface).

The back-radiation from the lake is largely a function of lake water temperature. According to Hostetler (1995), the relevant equation is related to the Stefan–Boltzmann Law (Equation 3.4):

$$Q_{back} = \varepsilon\sigma(\theta^{o}_{sur})^4 \qquad (9.1)$$

where ε = emissivity ($\approx$0.97), σ = Stefan–Boltzmann constant and θ°_{sur} = the temperature of the water surface in K. The University of Alabama (1999) state that, for a weighted-average dew point of 18°C in St Louis, the back-radiation rate for an 8 h night is approximately 3.4 MJ m^{-2} for a lake water temperature of 27°C and 4.5 MJ m^{-2} for a water temperature of 32°C.

Our simplified heat balance (assuming negligible groundwater and surface water throughput, and assuming convective 'sensible heat' losses to be negligible) for the lake is thus

$$Q_{gsc} + Q_{sol} - Q_{ref} - Q_{back} - Q_{evap} = 0 \tag{9.2}$$

or

$$Q_{evap} = Q_{gsc} + Q_{sw}(1 - \alpha_{sw}) + Q_{lw}(1 - \alpha_{lw}) - Q_{back} \tag{9.3}$$

In Equation 9.2, we have set the rate of change of heat stored to zero – this implies a long-term thermal balance. We need to be careful to consider which volume or layer of the lake our balance is valid for. In temperate and cold climates, as we have seen, it may only be the surface layer of the lake (above a thermocline) that actively absorbs solar radiation and exchanges heat with the atmosphere.

The rate of evaporation can also be calculated independently, from vapour pressure (which depends on temperature) and wind speed, using empirical equations from region-specific studies or by more theoretically based approaches such as that of Penman. Such equations typically relate the evaporation rate to wind speed and vapour pressure differential, and take the form:

$$E = a(e_0 - e_a) \times (b + cv_w) \tag{9.4}$$

where E = evaporation rate in mm day^{-1} (or L day^{-1} m^{-2}); e_0 = saturated vapour pressure of water at surface temperature (mmHg); e_a = vapour pressure of the surrounding air (mmHg). e_a is also equal to the vapour pressure of saturated air at the dew point temperature (Box 9.1), which can in turn be found from the wet bulb and dry bulb temperature measurements of the air. v_w is the wind speed (miles day^{-1}), and

BOX 9.1 Dew point.

The dew point is the temperature at which air, when cooled, first becomes saturated with water vapour. At that point, the vapour pressure of water in the air is equal to the saturated vapour pressure of water. The dew point temperature θ_d (°C) is given by manipulation of the Magnus–Tetens formula (Barenbrug, 1974):

$$\theta_d = \frac{b\Omega}{a - \Omega}$$

where $\Omega = (a\theta/(b + \theta)) + \ln(\text{RH})$, $a = 17.27$, $b = 237.7$°C, θ = actual temperature (°C) and RH = relative humidity at temperature θ on a scale of 0 to 1.

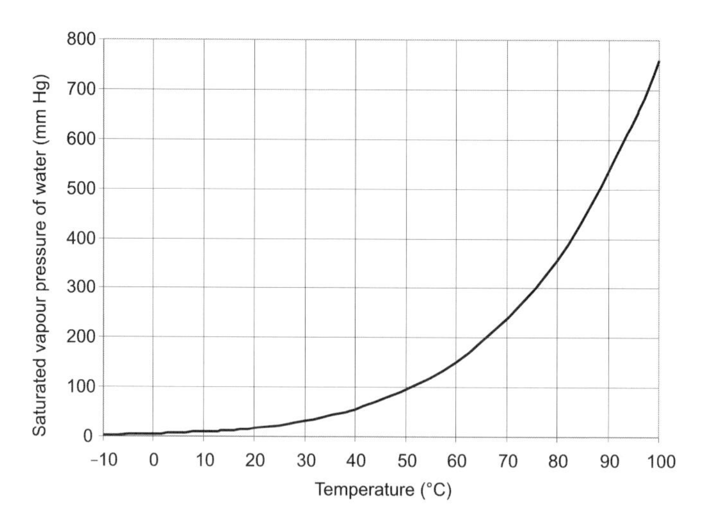

Figure 9.3 The dependence of the saturated vapour pressure of water on temperature. Formulae for calculating saturated vapour pressure are published by Tetens (1930).

a, b and c are constants, which may vary between empirical studies from different regions. Kavanaugh and Rafferty (1997) cite an equation where $a = 0.72$, $b = 0.417$ and $c = 0.004$ (for the set of units cited above).

We can then calculate the heat loss from the lake through evaporation, by

$$Q_{evap} = L_V \cdot \rho_w \cdot E \cdot A_{lake} \quad \text{in MJ day}^{-1} \tag{9.5}$$

where A_{lake} is the area of lake in m^2, L_V is the latent heat of vaporization of water (2.272 MJ kg^{-1}) and ρ_w is the density of water in kg L^{-1}.

The saturated vapour pressure of water at various temperatures can be obtained from Figure 9.3. We can thus consider Equations 9.3–9.5 and find the water temperature for which an equality (i.e. thermal equilibrium) is obtained. This may take several iterations, as we have already seen that Q_{back} depends on temperature, as well as Q_{evap}. The final equilibrium temperature represents the predicted maximum temperature of the lake during the operation of our ground source cooling scheme.

This type of simplified calculation ignores many elements of the heat balance shown in Figure 9.2. In other circumstances, the calculation may need to be modified to take account of these. For example, for a lake source heating scheme, where the peak loading will occur in winter, convective transfer of heat *from* air may become a significant mechanism. If the lake freezes, heat conduction through a layer of ice will need to be considered.

9.4 Open-loop lake systems

In principle, lake-water-based open-loop systems are similar to groundwater-based open-loop systems (Chapter 7). A prophylactic heat exchanger will often be used to prevent lake water directly entering the heat pump unit, thus restricting any problems of biofouling to the open loop and the heat exchanger. A filter will be placed on the lake inlet pipe to prevent entry of debris to the system and the intake area will need to be periodically checked to ensure that the intake is unobstructed.

The submersible pumps typically used for groundwater-based open-loop systems may also be used in lakes or ponds. However, if the head difference between the lake surface and the surrounding ground surface is less than around 5–6 m, surface-mounted suction pumps may alternatively be used: indeed, they may be preferable due to ease of access and maintenance.

Surface-water-based open-loop systems are suited to cooling schemes and to heating schemes in mild climates. In cold winters, however, water from natural water bodies may need to be abstracted at temperatures as low as 1–4°C (as was the case with John Sumner's (1948) pioneering heat pump system, based on the River Wensum in Norwich, UK). As no antifreeze can be added to the water flow, this leaves little margin for heat extraction and such schemes will run the risk of either low efficiency or ice formation in the heat exchanger or discharge pipe. Sumner (1948) noted the importance of ensuring that entry of debris to the heat pump is precluded, as this might constrict flow, increasing the likelihood of ice formation.

The locations of intake and outlet for the open loop will need to be carefully selected to prevent rapid 'short-circuiting' of water from the one to the other. It may be possible to use the thermal stratification of deep lakes to optimise system performance of a cooling scheme, with cold water being abstracted below the thermocline in summer and the waste warm water being rejected above the thermocline. Tempting though such an arrangement may sound, one must be aware of the chemical characteristics of the water below the thermocline. In the case of Figure 9.1, the deep water is oxygen-deficient and H_2S-rich. Rejection of such water above the thermocline could have drastic consequences (i.e. death!) for the aerobic organisms dwelling in the upper part of the lake.

9.5 Closed-loop surface water systems

In closed-loop systems, the antifreeze-based carrier fluid is normally circulated through a heat exchange element submerged in the surface water body. This heat exchanger can be a custom-built metal construction, but often simply comprises a network of high-density polyethene (HDPE) pipe. The polyethene should be tolerant of exposure to light (including UV). Closed-loop systems enjoy the advantage, over open-loop systems, of not being as susceptible to fouling. Moreover, they are also able to extract heat from relatively cold lakes (as antifreeze can be added to the closed-loop carrier fluid). The carrier fluid temperature will typically be some 2–7°C (or more) lower than the lake

Figure 9.4 Seven pond mats (loose coils of polyethene pipe, anchored to a steel frame) in the base of a newly constructed artificial lake in northern England. *Photo reproduced by permission of GeoWarmth Ltd.*

temperature and will thus absorb heat by conduction through the walls of the HDPE pipe. There are several means of installing closed-loop circuits of HDPE pipe in lakes:

- Parallel 'slinkies' (see Chapter 8): extended coils of pipe on the bed of the lake, which is a thermally efficient arrangement but relatively tricky to install.
- Loose coil bundles (i.e. unextended slinkies!) of HDPE pipe may be placed on the bottom of the lake. These will usually be anchored to some form of frame or fixed within a wire cage. In the case of artificial lakes (Figure 9.4), these can be installed during construction, before the lake is filled.
- A number of loose coils can be mounted on a 'raft' framework. The raft is floated into position and then sunk to rest on the base of the lake or pond.

The bundles of pipe emplaced on the pond bed, or attached to a raft or other framework, are sometimes referred to as 'pond mats'. The coil bundles should not be 'tight' (Figure 9.5), otherwise thermal 'short-circuiting' between adjacent loops of pipe will reduce the efficiency of heat absorption. Rather, each bundle should comprise loose coils of pipe. As rough rules of thumb (Kavanaugh and Rafferty, 1997; University of Alabama, 1999)

- Between 20 and 43 m of HDPE pipe length are typically required per peak kW of heating or cooling load depending on the coil configuration, mode (heating or cooling) and the temperature differential between the carrier fluid and the lake water. Kavanaugh and Raffery (1997) provide nomograms to calculate this.

Figure 9.5 Close-up of one of the 'pond mats' in Figure 9.4, comprising around 300 m of coiled HDPE pipe. Note the loose bundling of the pipe to reduce thermal interference. The coils are fastened to a steel frame, anchored with concrete blocks. *Photo reproduced by permission of GeoWarmth Ltd.*

- Coil bundles typically contain upwards of 150 m pipe and can be installed as rafts or as discrete bundles on a lake bed, with centres ideally spaced at least 6 m apart (but often less, especially if there is good water circulation within or through the lake).

In the UK, GeoWarmth (C. Aitken, *pers. comm.*) can document two case studies of lake-sourced systems that have employed (i) 2 × 200 m bundles of PE coil to support 6 kW heating capacity and (ii) 7 × 300 m bundles to support 60 kW heating and cooling capacity. These equate to 66 m and 35 m pipe length per installed kW, respectively. In the first case, however, the system was consciously over-dimensioned and the second bundle is effectively a 'back-up' – the true figure for specific installed capacity was thus nearer to 33 m per installed kW.

When dimensioning lake coils, we must remember that, if the building to be heated or cooled does not lie immediately on the shore of the lake or the bank of the river, we should take into account heat losses and gains to the carrier fluid during its passage (usually in a trench in the ground) between the building and the lake.

As with borehole-installed and trenched closed-loop systems, the total carrier fluid flow rate through the closed-loop array ($F_{carrier}$) should be high enough to achieve turbulent flow conditions within the lake exchange elements of the polyethene pipe (i.e. the bundled coils or slinkies), and thus to ensure efficient heat transfer (Box 8.1, p. 185). A flow rate of around 0.053 $\mathrm{L\,s^{-1}kW^{-1}} = 3.2\ \mathrm{L\,min^{-1}\,kW^{-1}}$ is a typical figure for $F_{carrier}$ for a water-based carrier fluid, with a temperature change of 4–5°C at the heat pump (see Section 8.4, p. 191). The corresponding figure in US literature is 3 US gallons per minute per ton (Kavanaugh and Rafferty, 1997).

Typically, each bundle of coil is connected in parallel although, if many parallel bundles are installed on a single circuit, a check should be performed to ensure that flow rates in each parallel circuit are adequate to ensure turbulent flow. The maximum number of parallel bundles (N_{max}) can be found by

$$N_{max} = \frac{F_{carrier}}{F_{turb}} \quad (9.6)$$

where $F_{carrier}$ is the total carrier fluid flow rate in the whole circuit under consideration (and is often, as a first approximation, around $G \times 3.2$ L min^{-1} kW^{-1}, where G is the peak rate of heat extraction from/rejection to the surface water) and F_{turb} is the flow rate (L min^{-1}) in each parallel circuit (pipe bundle) necessary to achieve turbulent flow. F_{turb} will depend on the internal diameter of the HDPE pipe (typically in the range 19–32 mm), and the viscosity and density of the carrier fluid. The viscosity will in turn depend on its chemical composition and temperature (Box 8.1, p. 185).

Although we can circulate the carrier fluid at a temperature below 0°C in an antifreeze-filled closed-loop system, we should be aware that significant build-up of ice on the lake coils could lead to two types of problem. First, a thick layer of ice, while improving heat conduction, will hinder convection of water around the coils and impede convective heat transfer, decreasing the overall efficiency of heat exchange. Second, large ice accumulations may lead to the coils becoming buoyant and floating to the surface, if not adequately anchored.

9.5.1 *Alternatives to polyethene pipe*

HDPE is clearly a cheap and robust means of installing a heat exchanger in a lake. However, it is not especially thermally conductive and for large schemes, we soon start running into km of pipe to satisfy our heating or cooling load. In this case, we would be well advised to consider prefabricated or custom-built metal heat exchange elements, which require a far lower exchange surface area with the water per kW of heat transfer. One alternative is to install flat heat exchangers ('lake plates'), prefabricated in stainless steel or titanium (depending on the salinity of the water). These look something like wall-mounted household radiators. One popular range, trading under name 'Slim Jim™', includes lake plates from 2 tons refrigeration capacity (nominal 7 kW, dimension 61 cm × 183 cm × 1 cm deep) to 8 tons (nominal 28 kW, dimension 122 cm × 366 cm × 1 cm). These exchangers weigh around 3.6 kg per kW capacity and require a carrier fluid flow of around 3.2 L min^{-1} kW^{-1}.

9.6 Closed-loop systems – environmental considerations

With surface-water-based closed-loop systems, particular consideration should be given to the type of any antifreeze employed in the closed loop. Closed HDPE loops in lakes and rivers, although fairly robust, will be far more vulnerable to damage (from boats, dredging, etc.) than borehole or trenched systems. In the event of a leakage, there is likely to be an ecosystem that will be directly exposed to the antifreeze. Thus, the potential eco-toxicity of any antifreeze is especially important. While ethylene glycol is an attractive choice in underground loops due to its low viscosity, it is relatively eco-toxic compared with, for example, propylene glycol (which has a significantly higher viscosity – see Table 8.1, p. 184). Thus, the latter is often preferable in lake-based systems, although other alternatives are available on the market today, including biodegradable, low-toxicity preparations derived from vegetable extracts.

10

Subsurface Heat Conduction and the Design of Borehole-Based Closed-Loop Systems

---

Heat is in proportion to the want of true knowledge.

Laurence Sterne, *The Life and Opinions of Tristram Shandy, Gentleman (1761)*

10.1 Rules of thumb?

In Chapter 8, we covered the basic design factors for horizontal closed-loop systems, which function largely as subsurface solar collectors. Chapter 6 also introduced us to vertical, borehole-based, closed-loop systems. We saw that for the (typically, though not exclusively, heating-dominated) ground source heat pump systems installed in the British Isles, each metre drilled length of borehole typically supports 50–104 W (average 75 W) of installed peak heat pump capacity, irrespective of lithology (Figure 6.10a, p. 123). This is referred to as the *specific installed thermal output*. If we assume that a heat pump scheme has a typical coefficient of performance of 3.4, then these figures equate to *specific heat absorption rates* (i.e. heat extracted from the ground) of 35–73 $W m^{-1}$ (average 53 $W m^{-1}$). This reflects the relatively low range of thermal conductivities in the geological environment (Tables 3.1, p. 35, and 3.3, p. 53). The thermal yield of ground source heat schemes contrasts with the groundwater yield of water wells, which can range from a few tens of litres per hour (or even less!) in hard crystalline bedrock to 200 $L s^{-1}$ or more in fissured limestones (Misstear *et al.*, 2006). This reflects the huge range of hydraulic conductivities in different lithologies, spanning nine or more orders of magnitude (Table 7.1, p. 149).

Many installers of small domestic heat pump systems will simply use this rule of thumb (i.e. around 60–100 W peak installed heating capacity per drilled metre) and add a small factor of safety for design purposes (drilling a few extra metres to provide a safety margin may be cheaper than indulging in a site-specific design assessment). They may consider thermal conductivity: they may, for example, assume 6 kW peak output from a 100 m closed-loop borehole in a quartz-poor lithology and maybe 8 kW in quartz-rich lithology.

The 'rule of thumb' derived from Figure 6.10a (p. 123) appears to be in line with international experience, which is summarised below on the basis of Rosén *et al.* (2001):

- In the United States, typical *specific installed thermal outputs* of 68–82 $W\,m^{-1}$ are reported for boreholes installed with single U-tubes.
- In Switzerland, *specific heat absorption rates* above 75 $W\,m^{-1}$ are not recommended.
- In Austria, recommended peak *specific heat absorption rates* range from 30 $W\,m^{-1}$ for dry sediments to 70 $W\,m^{-1}$ for granite, for a temperature difference of 10°C between carrier fluid and undisturbed ground.
- In Germany peak *specific heat absorption rates* of 20–25 $W\,m^{-1}$ are recommended for low conductivity ($<1.5\ W\,m^{-1}\,K^{-1}$) strata, 50–60 $W\,m^{-1}$ for medium-conductivity strata and 70–84 $W\,m^{-1}$ for high conductivity ($>3\ W\,m^{-1}\,K^{-1}$) strata. In each cited range, the lower value applies to systems with long operational usage (2400 $h\,yr^{-1}$) and the higher value to short operational usage (1800 $h\,yr^{-1}$). More detailed tables and nomograms are given for specific rock types, thermal conductivities, etc., by VDI (2001a).
- Across Europe in general, the average peak *specific heat absorption rate* is estimated at 62 $W\,m^{-1}$, (and taken over the entire heating season, 159 $kWh\,m^{-1}\,yr^{-1}$) for systems with operating times of 1600–2400 $hr\,yr^{-1}$ (Rosén *et al.*, 2001).

From these international experiences, we can begin to sense that a simple rule of thumb (a certain number of Watts per drilled metre) may be a little *too* simplistic. The performance of a closed-loop system will also depend on

1. The thermal conductivity, specific heat capacity and temperature of the ground. We have already seen that the conductivity does not vary too much from rock type to rock type (i.e. typically within the range 1.5–6 $W\,m^{-1}\,K^{-1}$) and heat capacity varies even less. Ground temperature is also relatively constant within a given region (although it may be significantly higher beneath urban areas – see Chapter 13).
2. The operational pattern of the scheme: for example, the number of equivalent full-load operational hours per heating season and the duration of peak operation on a daily basis. The number of equivalent full-load operational hours per heating season for domestic properties can range from 3000 to 4000 in Scandinavia to as few as 1000 in middle Europe.

3. The operating temperature of the ground loop. In heating mode, this will typically be around or just below 0°C (although it can fall significantly lower during times of peak demand). In cooling systems, however, there is no 'natural' upper limit. Indeed, Kavanaugh and Rafferty (1997) noted that, in one example, raising the loop's operating temperature from a value of 29.4°C to a value of 32.2°C could result in a 14% shorter ground loop – although this would be at the expense of the efficiency of the heat pump.

The 'rules of thumb' discussed thus far in Section 10.1 tend to be derived from consideration of relatively small, heating-only (or, at least, heating dominated) schemes in temperate and northern Europe, because the factors above tend to remain fairly constant. However, ground temperature will vary somewhat from country to country and we should remember that 'rules of thumb' from chilly Sweden or Canada may not necessarily be directly applicable to the United Kingdom: the ground temperature will be lower in Sweden and the achievable differential between carrier fluid and ground may also be less. This means that one might expect generally lower rates of heat absorption as well. Indeed, SVEP (1998) provide a set of nomograms for dimensioning of Swedish closed-loop systems where the climatic zone (ground temperature) is explicitly taken into consideration. The temperature difference between carrier fluid and undisturbed ground is assumed to be 7°C in the far north of Sweden and 12°C in the south.

With larger projects, cooling requirements become more dominant. Furthermore, in sizable schemes the number of boreholes begins to be large compared with available land area. In these circumstances, the following additional factors also begin to come into play:

4. Thermal interference between closely spaced boreholes.
5. Complex heating and cooling loads. Within a single day, one may have a heating demand in the early morning and cooling demand in the afternoon. Carrier fluid temperature differentials and COP_C values will likely be different in cooling mode to those in heating mode (possibly yielding different 'rules of thumb'). Seasonally reversible schemes may require shorter borehole arrays than comparable 'heating only' schemes because, in winter, heat is being abstracted from an aestifer that has been pre-heated by waste heat 'dumped' the previous summer.

Thus, as schemes become larger and larger, the 'rules of thumb' that are founded on a number of assumptions become less and less reliable (although they can still be a very useful starting point for design). It is no longer enough to 'guestimate' a specific heat absorption rate and add a margin of safety: the 25% difference between a rock thermal conductivity of 2.5 and 2.0 $W m^{-1} K^{-1}$ becomes significant in terms of drilling costs for a borefield comprising, say, 20 boreholes. Thus, for large ground source heating and cooling projects, we require a more sophisticated understanding of subsurface heat storage and transfer, site-specific input data and more nuanced design tools than mere 'rules of thumb'.

10.2 Common design flaws

The most common design flaws in closed loop heating or cooling systems can be summarised as follows:

- Using simplistic rules of thumb (e.g. 7 kW per 100 m borehole) for large systems, without appreciating that some of the assumptions underlying these 'rules' may be violated (e.g. we may have thermal interference between boreholes; or we may have a complex mixture of heating and cooling loads). Whether schemes are unidirectional or reversible and, in the latter case, whether the heating and cooling loads are balanced or not (see Glossary) may have a significant impact on the necessary drilled metres of borehole.
- Placing boreholes too close together such that thermal interference becomes significant. Some installers use borehole spacings as low as 4–5 m. While such short spacings may be necessary if land area is very limited, this author would suggest using a minimum spacing of 10 m as a starting point for design of schemes with a highly dominant heating or cooling load. For balanced, reversible schemes, smaller spacings will be acceptable.
- Lack of appreciation that GSHP cooling systems *may* require greater borehole lengths to deliver a given cooling load, than heating systems do to provide the same heating load (depending on the loop's operating temperatures). See Section 6.3 (p. 111).
- Using too short a design life for simulation of the performance of a closed-loop borehole array. Some designers use a simulation period as low as 10 years. We will see later that the system may not begin to approach a steady state until after some three decades or more have elapsed. We would certainly hope that most buildings and heating schemes would have a design life of more than 10 years!

10.3 Subsurface heat conduction

The most important mechanism for subsurface heat transfer to a typical vertical closed-loop heating system is conduction. We have already met Fourier's law (which has its hydrogeological analogue in Darcy's law):

$$Q = -\lambda \cdot A \cdot \frac{d\theta}{dL} \qquad (10.1)$$

where Q = flow of heat in Joules per second, or Watts ($J\,s^{-1}$ or W),
λ = thermal conductivity of the material ($W\,m^{-1}\,K^{-1}$),
A = cross-sectional area under consideration (m^2), perpendicular to direction of heat flow,
θ = temperature (K),
L = distance coordinate in the direction of decreasing temperature (note that heat flow

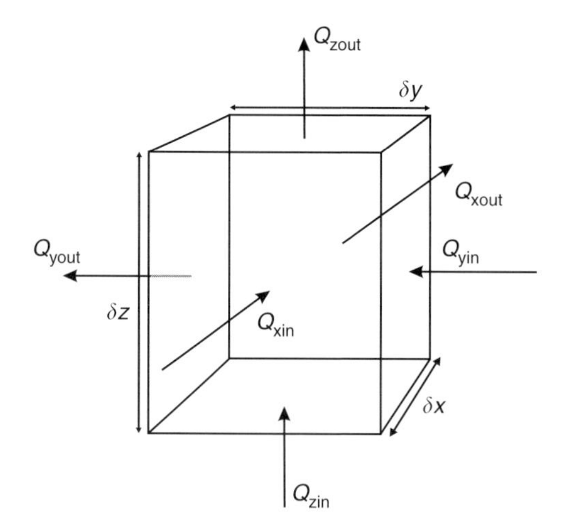

Figure 10.1 Schematic diagram of the conductive heat flow through a representative volume of aestifer, of negligible dimension $\delta x \times \delta y \times \delta z$.

is in the direction of decreasing temperature: hence the negative sign in the equation), $d\theta/dL$ = temperature gradient (K m^{-1}).

Now, we are in a position to consider how 3-dimensional heat conduction in the subsurface causes temperature to change with time. Let us consider a very small volume of aestifer material (V_{aest}) of dimension $\delta x \times \delta y \times \delta z$, isotropic thermal conductivity λ and volumetric heat capacity S_{VC} (Figure 10.1). The heat flow in through the front face of the element (from Fourier's Law) is:

$$Q_{xin} = -\delta y \cdot \delta z \cdot \lambda \cdot \left(\frac{\partial \theta}{\partial x}\right)_x \tag{10.2}$$

and the heat flow out through the rear face is

$$Q_{xout} = -\delta y \cdot \delta z \cdot \lambda \cdot \left(\frac{\partial \theta}{\partial x}\right)_{x+\delta x} \tag{10.3}$$

Thus, the change in heat flow in the x direction is given by

$$Q_{xout} - Q_{xin} = -\delta x \cdot \delta y \cdot \delta z \cdot \lambda \cdot \left(\frac{\partial^2 \theta}{\partial x^2}\right) = -V_{aest} \cdot \lambda \cdot \left(\frac{\partial^2 \theta}{\partial x^2}\right) \tag{10.4}$$

We can perform a similar exercise for the y and z directions. As the dimensions of the volume become negligible, we can construct the following differential equation, based on the assumption that, for a given increment of time, the net heat influx to the volume is equal to the volumetric heat capacity multiplied by the change in temperature:

$$V_{aest} \cdot \lambda \cdot \left(\frac{\partial^2\theta}{\partial x^2}\right) + V_{aest} \cdot \lambda \cdot \left(\frac{\partial^2\theta}{\partial y^2}\right) + V_{aest} \cdot \lambda \cdot \left(\frac{\partial^2\theta}{\partial z^2}\right) = V_{aest} \cdot S_{VC} \cdot \frac{\partial\theta}{\partial t} \quad (10.5)$$

$$\frac{\partial^2\theta}{\partial x^2} + \frac{\partial^2\theta}{\partial y^2} + \frac{\partial^2\theta}{\partial z^2} = \frac{S_{VC}}{\lambda}\frac{\partial\theta}{\partial t} \quad (10.6)$$

This equation essentially states that, for every tiny volume of aestifer, the heat that enters must equal the heat that leaves[1]. If it does not, a progressive temperature change with time results. If more heat enters than leaves, the temperature increases and vice versa. (This is, in fact, the same as the equation given in Box 7.3 (p. 175), minus the term for heat advection with groundwater.) The equation is the foundation of computer-based numerical heat transfer models: the aestifer is mathematically broken down into a finite difference (or finite element) grid of small volumes of rock and Equation 10.6 is solved by the computer simultaneously for each small block of aestifer. Note that Equation 10.6 assumes that thermal conductivity is isotropic – it is the same in the x, y and z directions. In reality, it may not be (Box 10.1), making the equation a little more complex but still tractable.

10.4 Analogy between heat flow and groundwater flow

Surely, if Darcy's law is analogous to Fourier's law and if the heat balance Equation 10.6 has a direct parallel in groundwater flow theory, we should be able to apply analogues of the Theis (Equation 7.8), Cooper–Jacob (Equation 7.11) and Logan (Equation 7.16) equations to the radial flow of heat towards a closed-loop borehole. And we can ... almost! In fact, the 'rules of thumb' of Section 10.1, which relate heat yield to borehole depth, are the thermogeological version of the Logan Approximation.

There is one important difference between hydrogeology and thermogeology, however: this difference lies in the boundary conditions for an aquifer and an aestifer. Let us first consider a typical unconfined aquifer (Figure 10.2). The base of the aquifer is usually some low-permeability aquitard: it can be approximated to a no-flow boundary. The top of the aquifer is a water surface (the water table) with a potentially fluctuating head, but which receives (in the long term) an approximately constant annual supply of rainfall recharge. Conceptually, therefore, it is a constant-flow boundary (or, in the case of a confined aquifer, a no-flow boundary). Theis's and Cooper–Jacob's equations predict that, if we pump a borehole in the aquifer, a cone of drawdown of head develops

[1] The hydrogeological equivalent of this equation states that the amount of water entering each small block of aquifer should equal what goes out: if it does not, it results in a change in groundwater head:

$$\frac{\partial^2 h}{\partial x^2} + \frac{\partial^2 h}{\partial y^2} + \frac{\partial^2 h}{\partial z^2} = \frac{S_S}{K}\frac{\partial h}{\partial t} = \frac{S}{KD}\frac{\partial h}{\partial t} = \frac{S}{T}\frac{\partial h}{\partial t}$$

BOX 10.1 Thermogeological Anisotropy.

In reality, thermal conductivity is not isotropic: it is a tensor and has a magnitude that varies with direction. In particular, horizontal thermal conductivity can be different from vertical conductivity. This may be due to the primary sedimentary or crystalline structure of the rock, it may be due to fracturing or jointing in a lithified rock or it may be due to fine-scale layering or lamination within a sediment or sedimentary rock. Indeed, as early as 1885, M. Jannettaz reported that the thermal conductivity along the planes of foliation in European schists, slates and gneisses was 1.5–1.98 times greater than the conductivity perpendicular to them (Prestwich, 1885).

Consider a volume of sedimentary rock, 10 m thick, comprising 10 000 very thin layers, each of thickness $D \simeq 1$ mm, of alternating sand ($\lambda = 2.4\ \mathrm{W\,m^{-1}\,K^{-1}}$) and clay ($\lambda = 1.6\ \mathrm{W\,m^{-1}\,K^{-1}}$). The (horizontal) thermal transmissivity (T_{the}) of this 10 m thick aestifer can be found by summing the product of conductivity and thickness for each layer:

$$T_{the} = \sum_{n=1}^{10\,000} \lambda \cdot D = (5\,000 \times 0.001\ \mathrm{m} \times 2.4\ \mathrm{W\,m^{-1}\,K^{-1}}) + (5\,000 \times 0.001\ \mathrm{m} \times 1.6\ \mathrm{W\,m^{-1}\,K^{-1}}) = 20\ \mathrm{W\,K^{-1}}$$

The bulk horizontal thermal conductivity λ_{bh} is obtained by dividing this by the total thickness of the aestifer (it is, in fact, the thickness-weighted arithmetic mean of all the layers):

$$\lambda_{bh} = \frac{T_{the}}{10\ \mathrm{m}} = 2.0\ \mathrm{W\,m^{-1}\,K^{-1}}.$$

If we consider the vertical conduction of heat, we can consider the aestifer as a collection of thermal resistances connected in series. The thermal resistance (R) of each layer is given by:

$$R = \frac{D}{\lambda} = 4.2 \times 10^{-4}\ \mathrm{m^2\,K\,W^{-1}} \quad \text{for sand and } 6.3 \times 10^{-4}\ \mathrm{m^2\,K\,W^{-1}} \quad \text{for clay}$$

The total vertical thermal resistance (R_{tot}) of the 10 m thick sequence is the sum of the thermal resistances of 5000 sand and 5000 clay layers, or 5.21 $\mathrm{m^2\,K\,W^{-1}}$. Thus, the bulk vertical thermal conductivity λ_{bv} is given by:

$$\lambda_{bv} = \frac{10\ \mathrm{m}}{R_{tot}} = 1.9\ \mathrm{W\,m^{-1}\,K^{-1}}$$

This is, in fact, the thickness-weighted, harmonic mean of the conductivities of the various layers.

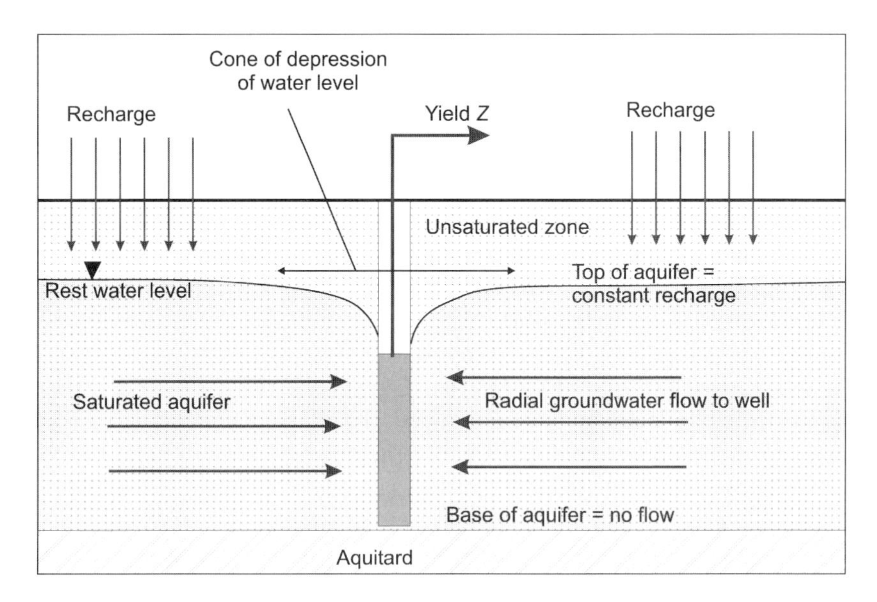

Figure 10.2 Schematic conceptual model of a pumped well in an unconfined aquifer, showing typical boundary conditions.

around the borehole. The cone of drawdown continues to develop spatially (in theory) *ad infinitum* and indefinitely in proportion to the logarithm of time (Figure 10.3)[2].

Let us now consider a closed-loop borehole in an aestifer (Figure 10.4). The aestifer has no physical base, but we can conceptually consider the bottom of our diagram as a constant flow boundary supplying a flux of geothermal energy of several tens of $\mathrm{mW\,m^{-2}}$. The top of the aestifer is not a constant-flow boundary. In fact, the top of the aestifer can be regarded as a constant-temperature boundary in the long term (several years timescale) at the annual average surface temperature. If the ground is hotter than the average surface temperature, hear will be lost via the surface. If we abstract heat from our borehole and cool down the ground, we will induce heat flow from the surface into the ground. The colder the ground, the greater the heat flow induced from then surface. Thus, the boundary conditions for our aestifer are different to our aquifer. The aestifer has a constant temperature boundary, whereas the aquifer (at least in our drawing) has no constant head boundary. Thus, if we pump heat from our closed loop borehole, a zone of depressed temperature develops in the rock around the borehole. Initially, heat stored in the surrounding rock will be conducted radially in towards the borehole (the early phase of Figures 10.5 and 10.6). Later, the ground will cool down to such an extent that heat flow will be increasingly induced from the surface.

[2] In reality, the cone of drawdown will usually stabilise in a steady state condition, but to do this it must either (i) induce recharge from a constant head boundary such as a river, lake or the sea, or (ii) suppress natural discharge such as baseflow to rivers, springs or wetlands. In other words, only head-dependent sources of recharge can stabilise a cone of depression, and not constant-rate sources of recharge (rainfall) – see Theis (1940) and Bredehoeft *et al.* (1982).

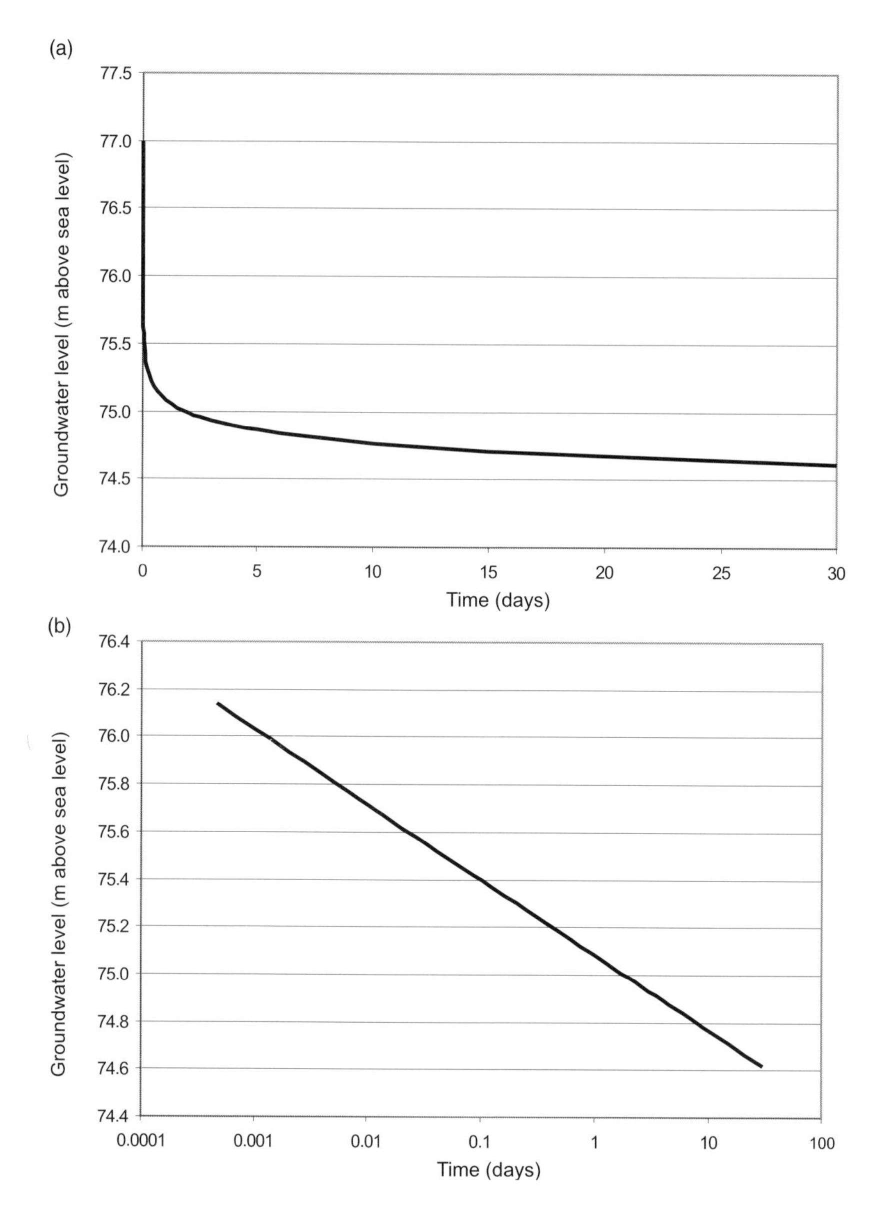

Figure 10.3 The development of groundwater level (a) with time and (b) with $\log_{10}$ (time) in a 200 mm diameter groundwater well, being pumped at a constant rate of $10\ \mathrm{L\,s^{-1}}$. Here, it is assumed that the well is hydraulically efficient and that drawdown is small relative to aquifer thickness. The initial rest water level is 77 m above sea level, the transmissivity $500\ \mathrm{m^2\,d^{-1}}$, storage 10%.

Eventually, the heat flow induced from the surface will balance the heat abstracted and a steady-state condition will develop (the late phase of Figures 10.5 and 10.6).

The other major difference between the aquifer and the aestifer is that, in our aquifer, there is usually little or no vertical head gradient prior to pumping (i.e. head does not

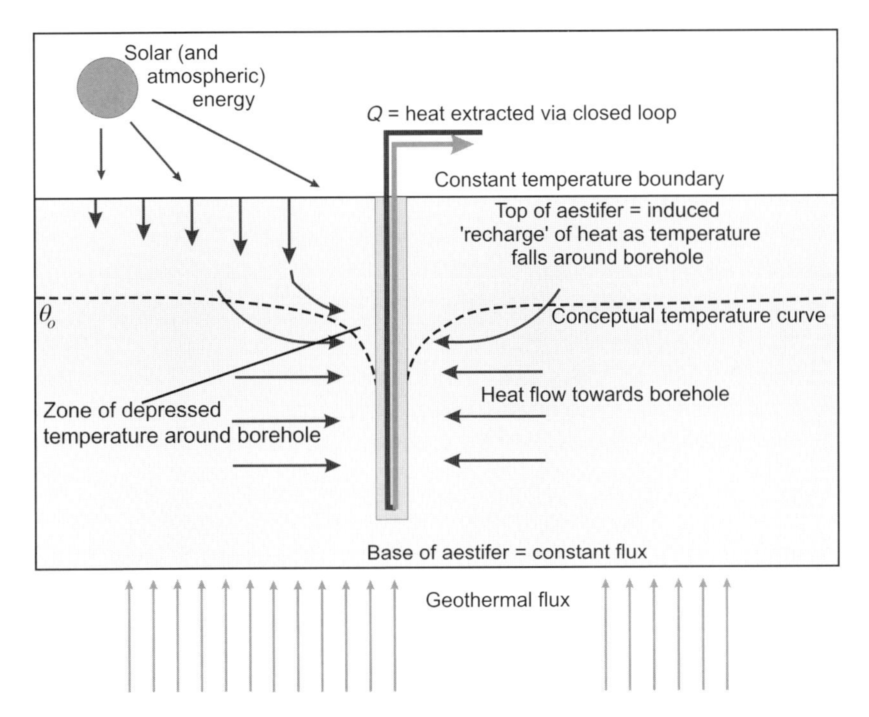

Figure 10.4 Schematic conceptual model of a closed-loop ground source heat scheme, abstracting heat from an aestifer, showing typical boundary conditions. θ_O = initial 'far-field' temperature.

change greatly with depth in many high-permeability aquifers). In our aestifer, we initially have a vertical temperature gradient corresponding to the geothermal gradient. Our borehole has an approximately constant average temperature, however, as we are circulating a carrier fluid down its length. It is the temperature of this carrier fluid that we are interested in simulating for the practical design of our ground source system.

10.5 Claesson and Eskilson's solutions

10.5.1 Early phase of heat abstraction

The fundamental equations governing the radial conduction of heat towards a line 'sink' in the earth (i.e. a borehole) and other subsurface heat exchange geometries were developed in the 1940s, by workers such as Leonard and Alfred Ingersoll, Otto Zobel (Ingersoll *et al.*, 1948) and Guernsey *et al.* (1949). However, the Swedes Johan Claesson and Per Eskilson (1987a,b) have provided a particularly coherent investigation of numerical and analytical solutions to the extraction of heat from a closed-loop borehole in an aestifer, and it is their work that forms the basis for the subsequent discussion. Via numerical modelling, they were able to derive a curve similar to Figure 10.5b, but

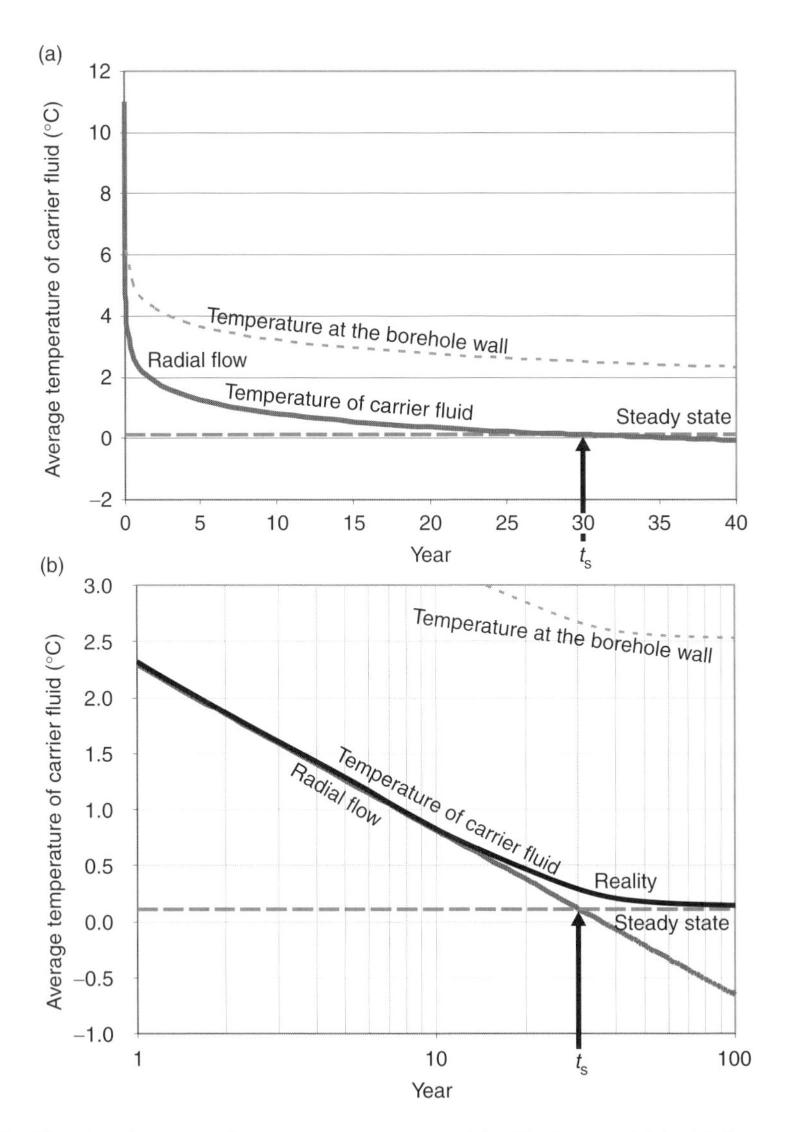

Figure 10.5 The development of average temperature (a) with time and (b) with $\log_{10}$ (time) of the carrier fluid in a closed-loop borehole. Here, the borehole diameter is 126 mm, the constant heat extraction rate is 2 kW, $R_b = 0.12$ K m W^{-1}, ground thermal conductivity 2.48 W m K^{-1}, $S_{VC} =$ 2.4 MJ m^{-3} K^{-1} and the undisturbed ground temperature 11°C over the borehole's 100 m length. The diagram also shows the calculated temperature of the borehole wall (2.4°C hotter than the carrier fluid).

a simple analytical solution proved elusive. They were, however, able to gain traction on the problem by the use of two simplifying assumptions:

Assumption 1: that the geothermal gradient can be neglected and that we can consider the aestifer as initially having a uniform temperature equal to the average temperature

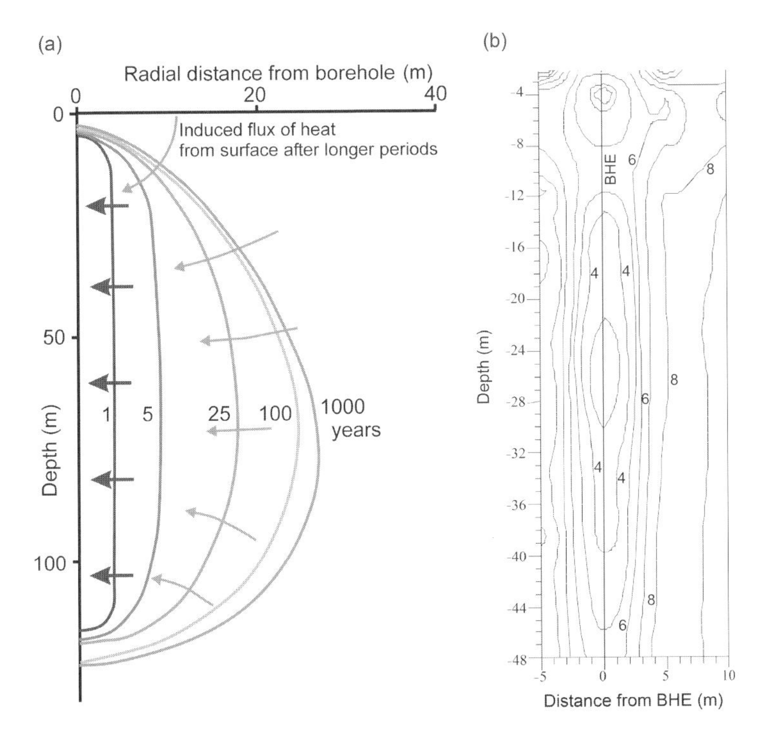

Figure 10.6 The development of the 'thermal front' in the rock surrounding a closed-loop borehole. (a) The contours representing the loci of a 1°C drop in temperature around a closed-loop borehole, extracting heat at a constant rate of 22 W m^{-1} from the aestifer (thermal conductivity 3.5 W m^{-1} K^{-1}, diffusivity 1.62×10^{-6} m^2 s^{-1}), after different periods of time. The black arrows show lines of heat flux after 1 year's operation (radial flux towards borehole), while the grey arrows show lines of flux after 25 years (borehole is beginning to induce heat flow from the surface and to approach a steady state) – *Redrawn from a diagram by Claesson and Eskilson (1987b) and published in Energy, Vol. 13(6), 509–52, © Elsevier (1988)*. (b) Real monitored ground temperature (contours in °C) around a closed-loop borehole (BHE) in Germany, after one heating season (7 months) of operation – *after Sanner (2002) and reproduced by kind permission of Dr. Burkhard Sanner.*

over the borehole length. Thus, if a 100 m borehole (prior to any heat extraction) has a temperature near the surface of 9°C and a temperature at its base of 11°C, we would say that the aestifer has an initial or 'far-field' average temperature θ_o of 10°C. If we imagine that we circulate a carrier fluid around the ground loop, prior to switching on our heat pump, it will tend to equilibrate with this average temperature θ_o.

Assumption 2: that, in the early phase of Figure 10.5, we can neglect induced heat flow from the surface and simply simulate radial heat flow from the aestifer to the borehole. In other words, in the early phase, the closed-loop borehole is simply removing heat from storage in the earth's subsurface reservoir.

We are thus left trying to solve our equation:

$$\frac{\partial^2\theta}{\partial x^2} + \frac{\partial^2\theta}{\partial y^2} + \frac{\partial^2\theta}{\partial z^2} = \frac{S_{VC}}{\lambda}\frac{\partial\theta}{\partial t} \qquad (10.6)$$

in radial coordinates (r), ignoring the geothermal gradient, with the following boundary conditions:

- at $t = 0, \theta = \theta_o$ for all values of r and z.
- as $r \to \infty$, $\theta = \theta_o$ for all values of t.

The solution is not trivial, but the problem is soluble:

$$\theta_o - \theta_b = \frac{q}{4\pi\lambda}E(u) = \frac{q}{4\pi\lambda}\left[-\gamma - \ln(u) - \sum_{n=1}^{\infty}(-1)^n \frac{u^n}{n.n!}\right] \tag{10.7}$$

where $E(u)$ is a Theis-type polynomial expression and $u = (r_b^2 S_{VC})/(4\lambda t)$;
θ_b is the average temperature (°C) of the carrier fluid in the borehole at time t. Here, we assume that this is identical to the temperature at the wall of the borehole – at a radius r_b from the centre of the line heat sink. We are thus assuming that any grout or backfill around the loop is infinitely thermally conductive – an ideally thermally efficient borehole!
$\theta_o - \theta_b$ = temperature 'drawdown' or displacement (K);
q = heat extraction rate per metre of borehole (W m^{-1}); that is, heat extraction rate divided by effective depth, taking account of any upper portion of a borehole that may be thermally insulated from the rock;
λ = thermal conductivity of the rock ($\text{W m}^{-1}\,\text{K}^{-1}$);
r_b = borehole radius (m), and γ = Euler's constant = 0.5772.

This is our thermal analogue of Theis's equation (Equation 7.8). Actually, it is more correct to say that Theis's equation was derived from this radial heat flow equation (Box 7.1, p. 156). As in the case of the Cooper–Jacob approximation (Equation 7.11), we can simplify this expression for low values of u (high values of t):

$$\theta_o - \theta_b \approx \frac{q}{4\pi\lambda}\left[\ln\left(\frac{4\lambda t}{r_b^2 S_{VC}}\right) - 0.5772\right] \tag{10.8}$$

This equation implies that, in the early phase of Figure 10.5, the temperature of the carrier fluid declines with the logarithm of time. This approximation is valid for values of t within the following range:

$$\frac{5r_b^2 S_{VC}}{\lambda} < t < \frac{t_s}{10} \tag{10.9}$$

If the value of t is lower than the lower constraint, the mathematics of the approximation breaks down. If t is too high, the log-linear relationship between time and θ_b begins to diverge, albeit slightly, from the real curve (Figure 10.5) as the system begins to induce heat flow from the ground surface and to slowly approach steady state.

10.5.2 Late (steady state) phase of heat extraction

The time t_s, after which 'steady state' begins to be a better description of the temperature evolution than radial flow, is given by Claesson and Eskilson (1987a) as

$$t_s = \frac{e^{\gamma} D^2 S_{VC}}{18\lambda} \approx \frac{D^2 S_{VC}}{9\lambda} \tag{10.10}$$

where γ = Euler's constant = 0.5772, and
D = borehole depth (m) over which heat extraction takes place.

The steady-state temperature of the carrier fluid, towards which our real temperature curve converges is given by

$$\theta_o - \theta_{s,b} = \frac{q}{2\pi\lambda} \ln\left(\frac{D}{r_b\sqrt{4.5}}\right) \approx \frac{q}{2\pi\lambda} \ln\left(\frac{D}{2r_b}\right), \quad \text{if } D \gg r_b \tag{10.11}$$

where $\theta_{s,b}$ = steady-state temperature of collector fluid in the ideal borehole (K) = steady-state temperature of borehole walls, neglecting thermal resistance of backfill in annulus.

10.5.3 Heat rejection to a closed-loop borehole

For a ground source cooling system, the heat load rejected to the ground (G) is related to the cooling effect (C) and the seasonal performance factor (SPF_C) of the heat pump by:

$$G \approx C\left(1 + \frac{1}{SPF_C}\right) \tag{10.12}$$

Exactly the same Equations 10.7–10.11 can be used as for heat extraction, except that the heat extraction rate will be negative and the temperature θ_b will progressively rise above θ_o.

10.6 Real closed-loop boreholes

10.6.1 Borehole thermal resistance

The equations described in Section 10.5 assume that the closed loop is in ideal thermal contact with the aestifer, with no additional thermal resistance within the borehole itself. In reality, there will be thermal resistances related to (amongst other factors):

- the conductivity of the filling of the borehole annulus (typically grout);
- thermal short-circuiting (heat leakage between the upflow and downflow shanks of the U-tube);
- thermal resistance associated with transfer of heat from the grout, through the U-tube wall, to the carrier fluid.

We must thus take into account this borehole thermal resistance[3] (R_b): it imparts an additional temperature loss between the aestifer and the carrier fluid, over and above that predicted by Equations 10.8 and 10.11. Fortunately, it can be regarded as a linear term with respect to heat abstraction. The equations in question thus become:

$$\theta_o - \theta_b = qR_b + \frac{q}{4\pi\lambda}E(u) \tag{10.13}$$

$$\theta_o - \theta_b \approx qR_b + \frac{q}{4\pi\lambda}\left[\ln\left(\frac{4\lambda t}{r_b^2 S_{VC}}\right) - 0.5772\right] \quad \text{for } \frac{5r_b^2 S_{VC}}{\lambda} < t < \frac{t_s}{10} \tag{10.14}$$

$$\theta_o - \theta_{s,b} = qR_b + \frac{q}{2\pi\lambda}\ln\left(\frac{D}{r_b\sqrt{4.5}}\right) \approx qR_b + \frac{q}{2\pi\lambda}\ln\left(\frac{D}{2r_b}\right), \quad \text{if } D \gg r_b \tag{10.15}$$

The thermal resistance of a borehole is essentially an empirical quantity that should ideally be measured for every borehole by a thermal conductivity test (Chapter 12). However, various authors have suggested formulae for estimating its value: for example, Shonder and Beck's (2000) *gpm* software provides the following very simple formula:

$$R_b = \frac{\ln(r_b/r_U)}{2 \cdot \pi \cdot \lambda_g} \tag{10.16}$$

where r_U and r_b are the radii of the U-tube pipe and borehole, respectively, and λ_g is the thermal conductivity of the grout (or other backfill). Kavanaugh and Rafferty (1997) imply that this is probably too simplistic, contending that:

$$R_b = R_g + R_U \tag{10.17}$$

where R_g is the thermal resistance of the grout or other backfill and R_U is the resistance related to the U-tube. They then argue that, for a situation where the shanks of the U-tube are neither touching each other or the borehole wall:

$$R_g = \frac{(r_b/r_U)^{0.6052}}{17.44\lambda_g} \tag{10.18}$$

and $R_U = 0.043$ K m W^{-1} (for pipe of $SDR^4 = 11$) and 0.054 K m W^{-1} (for $SDR = 9$) for turbulent flow in HDPE U-tube. For transitional flow ($Re = 2300$ to 4000), an additional 0.005 K m W^{-1} can be added to the R_U for turbulent flow; for laminar flow ($Re = 1000$ to 2300) a figure of 0.014 K m W^{-1} is added.

Alternatively, if all this seems far too complicated, analytical software programs such as Earth Energy Designer (EED, Eskilson *et al.*, 2000) can be used to calculate

---

[3] This is the analogue of well loss in groundwater engineering (see Section 7.7.1, p. 160). However, whereas hydraulic well-loss is non-linear relative to abstraction rate, borehole thermal resistance is linear.

[4] SDR is the ratio of a pipe's outer diameter (OD) to its wall thickness

the borehole thermal resistance from given input parameters, including the dimension, configuration and material of the collector loop, its shank spacing, the type and flow rate of the carrier fluid, the borehole diameter and the grout/backfill's thermal properties.

As an example, let us imagine a 125 mm diameter borehole installed with 32 mm OD HDPE U-tube of wall thickness 3 mm (SDR 11). The borehole is backfilled with grout of thermal conductivity 1.5 $\mathrm{W\,m^{-1}\,K^{-1}}$ and the carrier fluid is flowing turbulently. Equation 10.16 yields a value of $R_b = 0.145\ \mathrm{K\,m\,W^{-1}}$. The Equations 10.17 and 10.18 yield values of $R_g = 0.087\ \mathrm{K\,m\,W^{-1}}$ and $R_b = 0.130\ \mathrm{K\,m\,W^{-1}}$. To place these values into context, Mands and Sanner (2001) and Sanner *et al.* (2000) cite values of borehole thermal resistance between 0.06 and 0.50 $\mathrm{K\,m\,W^{-1}}$ derived from empirical thermal response tests in Germany. All except two of the German tests yield values below 0.12 $\mathrm{K\,m\,W^{-1}}$, however, while boreholes filled with thermally enhanced grout yield values of 0.06–0.08 $\mathrm{K\,m\,W^{-1}}$. A 'good' closed-loop borehole thermal resistance would thus be $<0.1\ \mathrm{K\,m\,W^{-1}}$. A poor R_b would generally be considered $>0.16\ \mathrm{K\,m\,W^{-1}}$. In the author's experience from the United Kingdom, borehole thermal resistances of 0.11–0.14 $\mathrm{K\,m\,W^{-1}}$ are typical for polyethene U-tube installations.

10.6.2 Minimizing borehole thermal resistance

Obviously, in designing our closed-loop borehole, it is in our interests to minimise borehole thermal resistance. For example, if a 100 m deep closed-loop borehole, extracting 4.5 kW energy (i.e. 45 $\mathrm{W\,m^{-1}}$), has a borehole thermal resistance of 0.14 $\mathrm{K\,m^{-1}\,W^{-1}}$, then there will potentially be an additional temperature drop of $45\ \mathrm{W\,m^{-1}} \times 0.14\ \mathrm{K\,m^{-1}\,W^{-1}} = 6.3°\mathrm{C}$ between the borehole wall and the carrier fluid. If the borehole was more thermally efficient and if R_b was only 0.1 $\mathrm{K\,m^{-1}\,W^{-1}}$, the temperature drop would be 4.5°C and the heat pump would be operating on a carrier fluid almost 2°C warmer than in the first case.

We can achieve a low value of R_b by ensuring a number of factors are optimised. Figures 10.7 and 10.8 illustrate a number of these. First, we should specify a carrier fluid flow rate adequate to produce turbulent flow in the carrier fluid. Figure 10.8 demonstrates that R_b declines dramatically with the onset of turbulent flow. Second, we should ensure a large (and constant) shank spacing between upflow and downflow tubes to minimise thermal short-circuiting. The effect of shank spacing is shown in Figure 10.8, while Figure 10.9a and b illustrate the difference between a correctly and an incorrectly spaced U-tube. In practice, small plastic or sprung 'spacers' can be utilised to keep the shanks of a U-tube at a constant separation during installation. Shank spacing is especially important if the thermal conductivity of the grout is low.

We should also backfill the borehole around the U-tube with a material that has a high heat transfer coefficient. Pure bentonite and cement-based grouts have rather low thermal conductivities. Moreover, Portland-cement-based grouts can have a tendency to shrink away from the U-tube on setting (unless additives are used to prevent this), both reducing the thermal contact and providing a pathway for contaminated surface water. Furthermore, Portland cement releases considerable heat of hydration on setting,

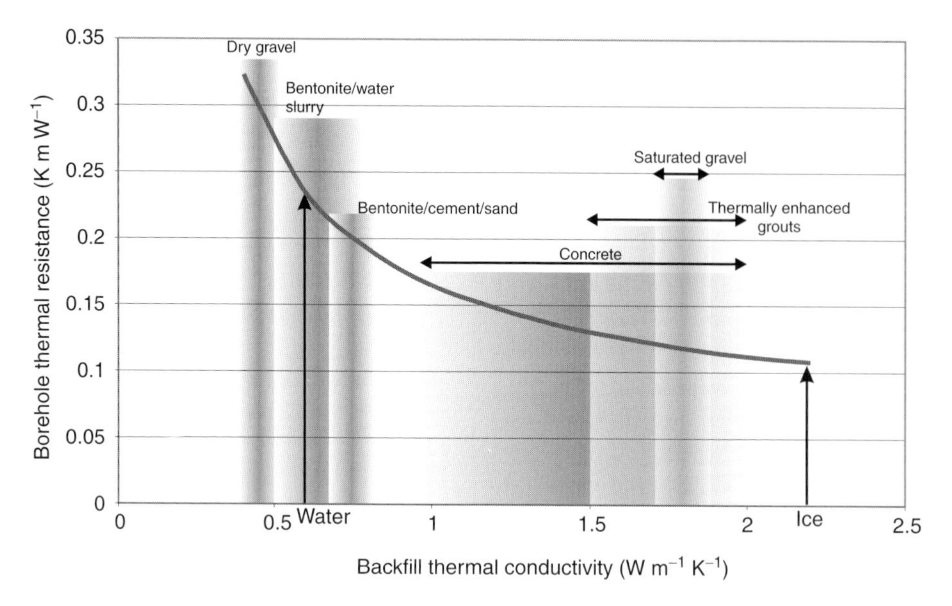

Figure 10.7 The dependence of borehole thermal resistance (R_b) on backfill material, calculated by the programme EED on the following assumptions: 127 mm diameter borehole installed with a single U-tube comprising 32 mm HDPE, of wall thickness 3 mm and shank spacing 58 mm, carrier fluid 25% ethylene glycol circulating at 16.5 L min^{-1}. Typical ranges of thermal conductivities from Eskilson *et al.* (2000) and other sources.

which can damage plastic U-tubes (McCray, 1999). We tend therefore to favour grouts that have a high quartz content (and thus a high thermal conductivity), but which also have a high enough cement or bentonite content to ensure a low hydraulic conductivity and a good hydraulic seal. The so-called 'thermally-enhanced' grouts are typically of two types:

1. A thick slurry of bentonite, fine quartz sand/silt and water, such as 'Thermal Grout 85' (see Glossary). Such thermally enhanced bentonite/sand grouts are reported to have conductivities in excess of 1.5 W m^{-1} K^{-1}, while some manufacturers claim conductivities as high as 2 W m^{-1} K^{-1}. However, Rosén *et al.* (2001) and VDI (2000b) express concern that high-bentonite concoctions may be susceptible to damage upon freezing. VDI (2001a) suggest that mixtures of bentonite, quartz silt or sand, furnace fly ash ± cement will give good mechanical properties and frost resistance down to −15°C, although thermal conductivity is much more modest ('over 0.8 W m^{-1} K^{-1}').
2. Cementitious thermally enhanced grout. In New Jersey, this is defined as (approximately) a 2 to 1 mix by mass of fine silica sand to cement, with small quantities of sodium bentonite and sulphonated naphthalene superplasticiser (GeoExchange, 2003). Allan and Philippacopoulos (2000) found that their 'Mix 111', comprising 2.13 parts sand to 1 part cement, with added superplasticiser, provided a field thermal conductivity as high as 2.19 W m^{-1} K^{-1}. They observed, however, some problems

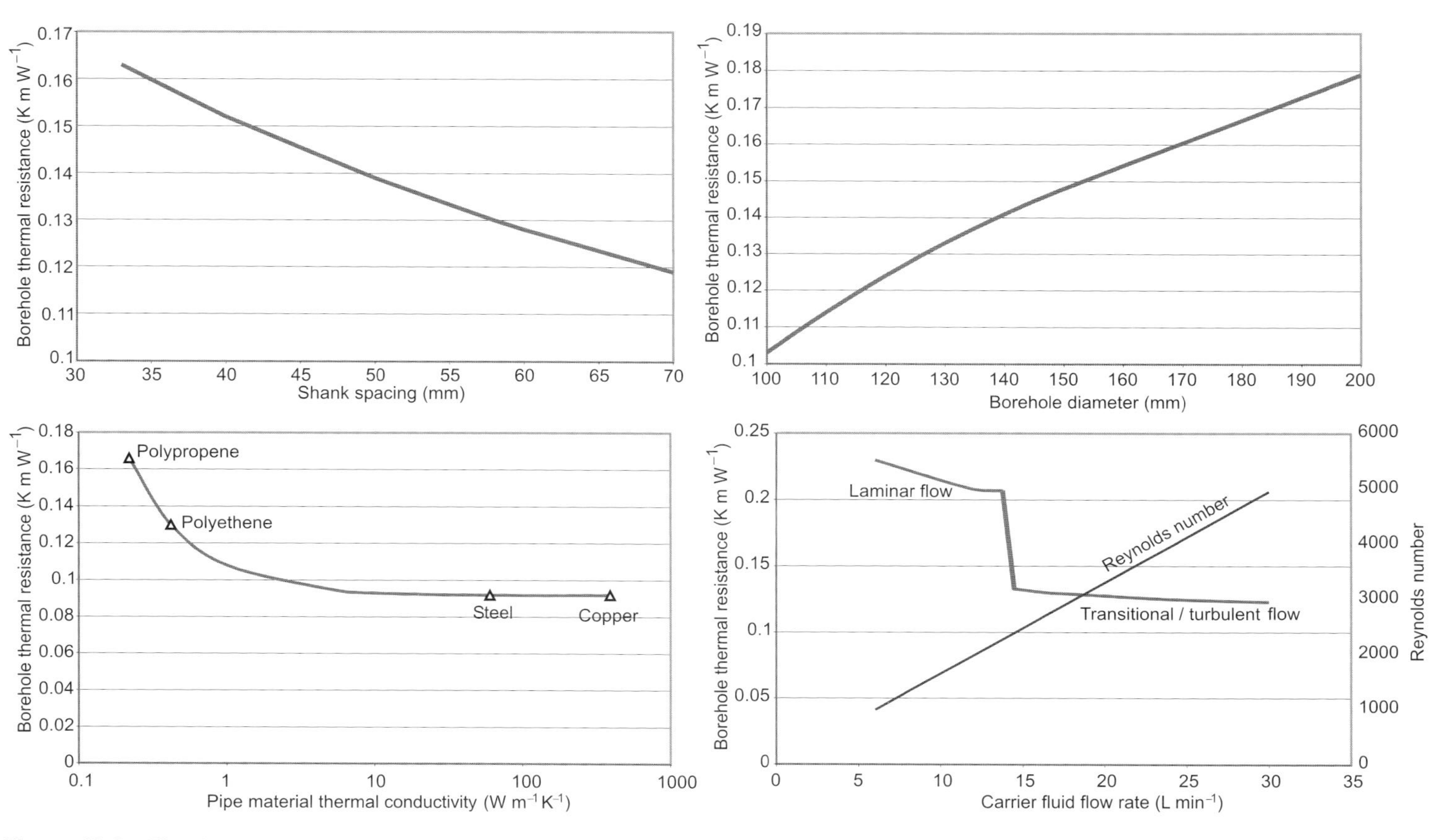

Figure 10.8 The dependence of borehole thermal resistance (R_b) on shank spacing, borehole diameter, U-tube pipe material and fluid flow rate. Note especially, the difference that laminar/turbulent flow transition makes to R_b. Calculations made using the programme EED (Eskilson *et al.*, 2000), which assumes a Reynolds number of around 2300 for the laminar/turbulent transition. The baseline assumptions for each case were: 127 mm diameter borehole installed with a single U tube comprising 32 mm HDPE, of wall thickness 3 mm and shank spacing 58 mm, backfilled with a thermal grout of conductivity 1.5 W m^{-1} K^{-1}, carrier fluid 25% ethylene glycol circulating at 16.5 L min^{-1}.

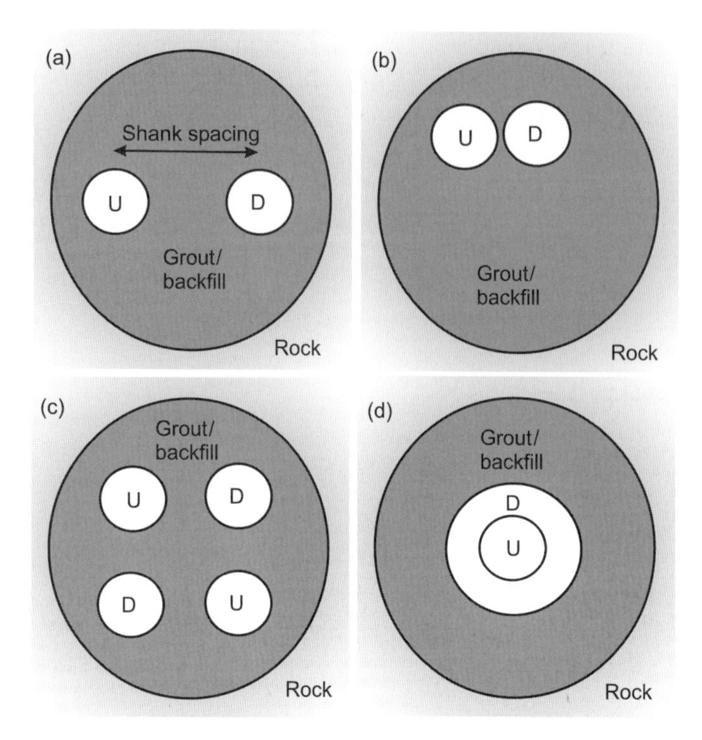

Figure 10.9 Borehole cross-section illustrating different types of ground-loop installations: (a) a correctly installed single U-tube, with large shank spacing; (b) a poorly-installed single U-tube, with small shank spacing; (c) a double U-tube; (d) a closed coaxial collector. 'U' and 'D' show upflow and downflow portions.

> with shrinkage of the grout away from the U-tube, but still claimed a relatively low overall hydraulic conductivity (a little more than 10^{-9} m s^{-1}). They also experimented with 'Mix 114' and 'Mix 115', where some of the cement was exchanged for granulated blast furnace slag and fly ash, respectively, to reduce the heat of hydration.

Any grout should, of course, be emplaced slowly in the borehole from the bottom up, using a tremie pipe, to avoid air pockets and to ensure a good thermal contact between grout and U-tube. The impact of grout thermal conductivity on R_b is shown in Figure 10.7.

Rather than using a grout, the borehole can be backfilled around the ground loop with a quartz sand or gravel, allowing natural groundwater to fill the pore spaces – resulting in a high thermal conductivity (Figure 10.7). A low-conductivity seal is emplaced in the uppermost section of the borehole to prevent ingress of surface water.

Alternatively, the U-tube can be 'dangled' in a borehole filled with natural groundwater (a concept that has been favoured in Scandinavia – Rosén *et al.*, 2001; Skarphagen, 2006). This may not seem a promising solution, given water's low thermal conductivity, but a number of factors can conspire to render overall heat transfer rather effective, including: (i) formation of high-conductivity ice around the loop[5], (ii) forced convection of heat by groundwater flux, (iii) the establishment of free convection cells within in the borehole's column of water, and (iv) a density-driven 'thermosiphon' effect in the aquifer around the borehole (Gehlin *et al.*, 2003). These 'grout-free' solutions pre-suppose a shallow groundwater level and lack of objection from environmental regulators.

One might suppose that the use of HDPE for the U-tube, rather than, say, copper, would result in an unnecessarily high value of R_b. There is some truth in this (Figure 10.8), although HDPE tends to be used in modern systems due to its robustness, cheapness and resistance to corrosion.

10.6.3 Ground-loop configuration

Hitherto, we have considered the use of simple HDPE 'U-tubes' as the means by which we extract heat from the deep subsurface by conduction and convey it to a heat pump (Figure 10.10). There are, of course, other configurations of ground loop that can be used, which will increase the amount of exchange-surface area within the borehole and thus minimise the borehole's thermal resistance. We can, for example, utilise a 'double' (or even a triple) U-tube, with two upflow and two downflow shanks connected either in series or in parallel (Figure 10.9). This does (theoretically) have the capability to significantly decrease the borehole's thermal resistance. However, its disadvantages

Figure 10.10 Heat is conveyed from a closed ground-loop (in a borehole or trench) to the heat pump by a carrier fluid. Where there are two or more carrier fluid circuits in parallel, they will be connected by means of a manifold (with shut-off valves). *Reproduced by kind permission of Kensa Engineering Ltd.*

[5] In Norway, this accumulation of ice around a U-tube is called an 'ispølse' or ice sausage!

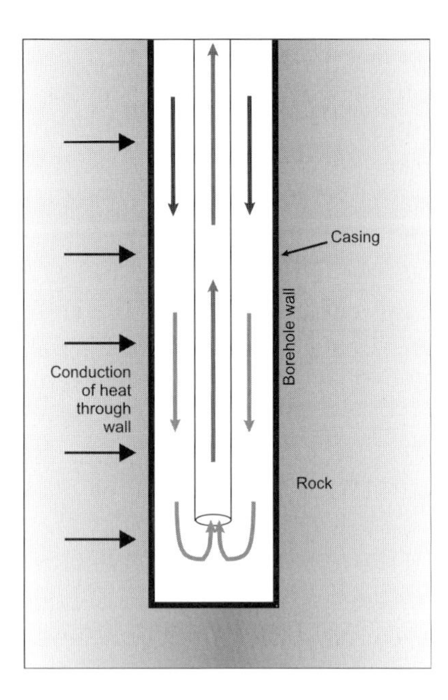

Figure 10.11 An 'open' coaxial collector, where the drilled borehole itself forms the outer flow tube. In low-permeability, stable, self-supporting formations, the borehole wall may not be lined with casing (if this is the case, antifreeze or other potentially polluting substances will not be able to be used).

are (i) that it is more difficult to install and grout than a single U-tube and (ii) it may involve a decreased hydraulic efficiency (higher hydraulic head losses and/or higher carrier fluid flows).

Other types of ground loop are available, including a coaxial tube configuration (Figure 10.9). Canadian researchers (Rosén *et al.*, 2001) have even utilised a spiral ground loop – a form of vertical slinky – in large diameter boreholes. A final possibility, in some circumstances, would be to employ an 'open' coaxial configuration where the borehole wall forms the outer flow conduit. The borehole wall may consist of bare rock, or of steel/plastic casing. The circulating carrier fluid (which may be natural groundwater) thus absorbs heat directly from the borehole wall (Figure 10.11). Clearly, environmental regulators are likely to raise objections to the use of antifreeze or other additives in such systems if the borehole wall is unlined, due to the possibility of their release to the wider groundwater environment.

10.7 Application of theory – an example

10.7.1 Design constraints

Equations 10.13–10.15 can be used as the basis for the design of a closed-loop heat extraction system. Our fundamental design constraint will relate to the temperature of the ground and/or the carrier fluid during heat extraction. The carrier fluid temperature should not drop too low during operation of a heat pump system, for several reasons:

i. *Economic:* the heat pump COP_H may decrease to a point where its efficiency is unacceptably low or where any efficiency advantage over, say, an air-sourced heat pump, is lost.
ii. *Geotechnical:* although there is no fundamental operational reason why some degree of ground freezing cannot be accepted (and can, indeed, be advantageous, given the high thermal conductivity of ice); extensive ground freezing may be regarded as undesirable for geotechnical reasons, especially if buildings are located above or in close proximity to the boreholes. Furthermore, the mechanical properties of some clays will be damaged by repetitive freezing and thawing.
iii. *The bottom line:* temperatures should not drop so low that the carrier fluid freezes (25% ethylene glycol has freezing point of around −14°C).

The following design criteria are therefore suggested, based on accepted best practice in several countries (although they should not be regarded as absolutes – temperature constraints will ultimately be defined by geotechnical and efficiency considerations and operational constraints):

i. The mean temperature of the carrier fluid under average 'base-load' conditions should not drop significantly below 0°C over the design life of the system. Operational carrier fluid temperatures in the range −4 to 0°C are typical.
ii. If constraint (i) is too onerous, the average temperature of the carrier fluid under average 'base-load' conditions should not drop below a temperature corresponding to $-qR_b$°C over the design life of the system, where q is the specific heat absorption rate. This means that, for a typical average heat absorption base load of $20\ \mathrm{W\,m^{-1}}$ and a borehole thermal resistance of $\sim 0.1\ \mathrm{K\,m\,W^{-1}}$, we calculate that the average temperature drop between the borehole wall and carrier fluid will be 2°C. Thus, we *can* operate with an average (base-load) carrier fluid temperature as low as −2°C without the risk of freezing the surrounding rock or sediment (although the temperature of the grout backfill may drop below 0°C).
iii. The minimum temperature of the carrier fluid should not approach its freezing point during 'peak-load' conditions. Remember that the down-flow temperature of the fluid from the heat pump may be 3–5°C cooler than the upflow temperature and thus around 2°C lower than the average carrier fluid temperature θ_b. Bear in mind the possibility that crystal formation may start to occur on the evaporator well before the bulk carrier fluid temperature approaches it freezing point. Thus, using 25% ethylene glycol with a freezing point of −14°C, the average carrier fluid temperature should not drop below, say, about −9°C under peak-load conditions.

VDI (2001a) corroborate this interpretation by suggesting that, under German conditions, the return temperature of the carrier fluid should not be more than 11°C lower than undisturbed ground temperature under base-load conditions (weekly average) and not more than 17°C lower than undisturbed ground temperature under peak-load conditions.

In designing a closed-loop system, our task is therefore to ensure that, over the system's design life and under prevalent operational conditions, the ground-loop carrier fluid temperature (θ_b) remains at acceptable levels.

10.7.2 An example

Equations 10.14 and 10.15 can readily be programmed into a computer-based spreadsheet, or the calculations may be performed manually. Let us consider a domestic heat pump system, with a COP_H of 4 and a peak heat demand on the coldest winter days of 6 kW. This implies that the 'peak-load' heat extraction rate from the ground is 4.5 kW (the remaining 1.5 kW comes from the heat pump compressor). Let us assume that, averaged out over the year, the heat load is around 33% of this figure (and maybe less). Let us assume that this long-term average 'base-load' extracted from the ground is 1.5 kW (13.2 MWh yr^{-1}).

Let us further assume that the ground source system utilises a 126 mm diameter, 100 m borehole drilled into granite with a thermal conductivity (λ) of 2.48 $\text{W m}^{-1}\text{ K}^{-1}$ and a volumetric heat capacity of 2.4 $\text{MJ m}^{-3}\text{ K}^{-1}$. Finally, let us assume a borehole thermal resistance (R_b) of 0.12 K m W^{-1} and an initial average ground temperature (θ_o) of 11°C. All of these figures are considered typical for a UK installation. We can thus use Equation 10.14 with the following input parameters:

$$\theta_o = 11°\text{C}$$

$$S_{VC} = 2\,400\,000\ \text{J m}^{-3}\text{ K}^{-1}$$

$$\lambda = 2.48\ \text{W m}^{-1}\text{ K}^{-1}$$

$$R_b = 0.12\ \text{K m W}^{-1}$$

$$r_b = 0.063\ \text{m}$$

First, let us simulate the annual average long-term base load (1.5 kW) over a 25 year design life of the heating system ($q = 15\ \text{W m}^{-1}$, $t = 25$ years $= 789 \times 10^6$ s). Equation 10.14 predicts that the average carrier fluid temperature will evolve as shown in Figure 10.12), reaching a value of 2.92°C after 25 years. Remember that we can calculate the temperature drop within the borehole itself as $qR_b = 0.12\ \text{K m W}^{-1} \times 15\ \text{W m}^{-1} = 1.80°\text{C}$. This implies a temperature at the borehole–rock interface of 4.72°C. So far, so good.

We now need to consider the peak heating condition, that on the coldest day of winter, heat is being extracted at the maximum 'peak-load' rate of $q = 45\ \text{W m}^{-1}$ for a duration of maybe 24 h ($t = 86400$ s). Here the Equation 10.14 predicts a temperature displacement of 11.06°C, 5.66°C of which takes place in the aestifer and 5.40°C of which takes place across the borehole backfill (qR_b). Thus, after 24 h of peak load, our

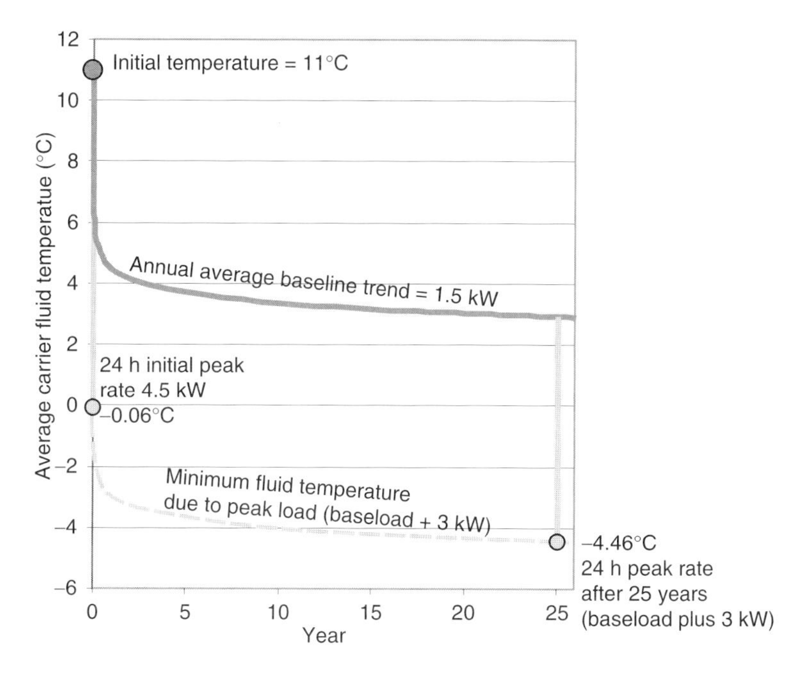

Figure 10.12 The evolution of average carrier fluid temperature over 25 years for a 100 m deep closed-loop ground source heat borehole, delivering a peak heat load of 6 kW with a COP_H of 4. The solid curve shows the annual average *trend* of carrier fluid temperature (corresponding to a continuous heat extraction of 1.5 kW), while the lower, dashed curve shows the approximate minimum fluid temperature under short-term (24 h) peak heating load.

average carrier fluid temperature may have dropped to −0.06°C. This still seems quite acceptable.

However, we must assume the worst: that this peak load takes place on top of the base-load condition at the end of our 25 year design life. In other words, we are imparting an additional 3 kW (30 W m^{-1}) of heat extraction on the coldest day of the 25th year, in addition to our 1.5 kW base load. The additional 30 W m^{-1} of peak load for 24 h gives a temperature displacement of 7.38°C. Thus, the minimum peak-load temperature of the carrier fluid after 25 years is predicted to be 2.92°C − 7.38°C = −4.46°C.

In summary, we can see that our borehole is adequately designed; the baseload temperature does not drop below 0°C during its design life, and the peak-load minimum temperature remains well above the −14°C freezing point of our 25% ethylene glycol carrier fluid. Indeed, the system may even be slightly overdesigned – we could take a few metres off the borehole length if we really wanted to.

If we wanted to make a more sophisticated calculation, rather than simply considering a 25 year 'average' and a 24 h 'peak pulse', we could also consider the effect of a 'winter heating season' pulse of duration 5 months, or a 'diurnal' signal of duration corresponding to the typical daily operation of the heat pump. It is by consideration of several such superimposed heat 'pulses' that many computerised analytical design models function.

From this simple exercise, we can see whence our '100 m borehole for a peak heat demand of 6–10 kW' rule of thumb comes from. But we can also see that a brief design calculation takes a matter of minutes and we do not really need to rely on 'rules of thumb'. Indeed, we should have become aware of the importance of the assumptions we make about:

- the duration of peak load (12 h, 24 h, 1 week?);
- the percentage of peak load representing average base load (this will depend heavily on the occupancy patterns of the house);
- what we assume to be the minimum acceptable carrier fluid temperatures.

These assumptions will significantly affect the outcome of our design.

10.7.3 *Mathematical checks*

We have previously stated that Equation 10.14 is valid for:

$$\frac{5r_b^2 S_{VC}}{\lambda} < t < \frac{t_s}{10}$$

In our case, this means 5 h $< t <$ 3 years. Below 5 h, the mathematical assumption underlying the approximation breaks down. After 3 years, the equation begins to very slightly overestimate the real temperature drop (i.e. underestimate the carrier fluid temperature) as the system begins to induce heat flow from the surface. Thus, a degree of conservatism is built into our approach. Nevertheless, we can conclude that our use of the equation is broadly mathematically valid.

10.7.4 *Steady state*

At time t_s the steady-state Equation 10.15 becomes a better approximation of reality than the radial flow Equation 10.14. In our example above, Equation 10.10 predicts that this is around 30 years. After this time, the temperature evolution flattens out and begins to tend towards the predicted steady-state average fluid temperature $\theta_{s,b}$ corresponding to the 1.5 kW 'base' heat extraction rate. In our example, from Equation 10.15, we calculate that this is 2.83°C. The corresponding minimum temperature during a 24 h event imparting an additional 3 kW peak load at times >30 years would be 2.83°C − 7.38°C = −4.55°C.

We should remember, however, even after time t_s, that our borehole may still be some way from true equilibrium! Claesson and Eskilson (1987a,b) provide a rather complex expression to estimate the proportion of heat derived from the surface (i.e. heat replenishment from the soil and atmosphere) at any given time. They indicate that, for a typical closed-loop borehole after 25 years heat extraction, only some 32% of the extracted heat is derived via the ground surface. The remaining 68% is still derived from thermal storage in the rocks.

10.8 Multiple borehole arrays

10.8.1 Simulations using analytical computer models

Such simple calculations and spreadsheet-based solutions run into difficulties when we are dealing with large, complex ground source heating and cooling schemes (e.g. Box 10.2). In such situations:

- There may be a very complicated pattern of heating and cooling throughout the year. In some months heat may be both abstracted from the ground (in the early morning, for example) *and* dumped to the ground (in the afternoon).
- There will be an array comprising a number of boreholes, which will thermally interfere with each other.

A number of analytical computer programs are available to simulate how ground-loop fluid temperatures evolve in such complex scenarios. Two of the most widely used are

- EED (Earth Energy Designer), from Sweden (Eskilson *et al.*, 2000);
- GLHEPro (Ground Loop Heat Exchanger Professional) from Oklahoma State University, USA (Spitler, 2000). This is mathematically analogous to EED but with slightly more flexible input options.

In both of these programs, multiple borehole arrays are dealt with by factoring the calculations using a so-called 'g-function' – a mathematical function dependent on the geometry and shape of the array. One can simulate lines of boreholes, rectangular blocks of boreholes, 'open rectangles', where boreholes are spaced around the perimeter of an area, and even L-shaped arrays.

The complex heating and cooling patterns are simulated using successive 'step functions'. The type of calculation performed in Section 10.7.2 involved a type of 'step function' where we superimposed a 24 h 'step' of peak loading on top of a long-term base load. In analytical simulation models, the base load is typically specified as the heating and cooling load *per month* of a typical year, and is simulated as the combination of sequential monthly steps. Moreover, for each month, the magnitude and duration of the peak heating and cooling loads is specified for the coldest and hottest days. A load input file for EED or GLHEPro may therefore look something like Table 10.1: in this example, the scheme has a peak output of 95 kW. During the coldest months, however, 20 MWh month^{-1} of heat is supplied (corresponding to only 28 kW on average, implying, say, up to 8 h operation per day). If the seasonal performance factor of the scheme (SPF_H) is 3.5, then 71% of the heat supplied comes from the ground (i.e.

BOX 10.2 Case Study: Dunston Innovation Centre, Chesterfield.

The Dunston Innovation Centre, Chesterfield. Photo by D Banks.

The Dunston Innovation Centre is a 3300 m^2 (gross floor area) complex of office and conference space, located near a former colliery site on the northern outskirts of Chesterfield, UK. The site was developed by Chesterfield Borough Council to a high standard of thermal efficiency and is rented to a number of tenants, all of whom will have their own heating or cooling requirements. A distributed ground source scheme was thus selected (Section 6.6.4, p. 144), using around 97 small reversible console-type water-to-air heat pumps (see Figure 4.8b, p. 77), producing a flow of warm or cool air. The console units are individually controlled by and metered to the tenants (Climate Master, 2004). The heat pumps are mounted on a loop through which carrier fluid is circulated to the ground array. The central conference facility, in a 'Rotunda', is heated and cooled by a somewhat different philosophy, using a 22 kW reversible water-to-water heat pump unit. Here, heat is drawn from (or rejected to) the ground loop, transferred to a small secondary fluid circuit and distributed to the conference area by fan coil units.

The ground loop itself comprises 32 boreholes of 60 m depth, drilled through the Carboniferous Coal Measures strata below the site and laid out in a grid beneath a landscaped area to the rear of the building. The scheme supports peak cooling and heating loads of 242 and 130 kW respectively and was commissioned in 2001/02 (Climate Master, 2004; Earth Energy, 2005).

Chesterfield Borough Council anticipated a significant annual saving in costs, compared with traditional air-conditioning and heating systems. Sometimes, it is quite difficult to demonstrate that these cost savings are real, but the Council fortunately has another similarly-sized complex, in nearby Tapton, which uses conventional gas-fired heating. The first year's operation revealed that Dunston's total energy bill was around £5500 cheaper than Tapton's, despite the fact that Tapton had no air-conditioning. If one takes the latter factor into account, together with maintenance costs, one would calculate a realistic payback time on the initial

BOX 10.2 *(Continued)*

investment. The ultimate test of the scheme is, of course, client satisfaction: both users of the complex and Chesterfield Borough Council appear highly satisfied with the outcome, to the extent that a closed-loop ground sourced approach has also been utilised at the town's new Tourist Information Centre and the new Venture House office complex at Dunston (J. Vaughan, Chesterfield BC, pers. comm.).

Table 10.1 An example of a heat-load profile for a hypothetical building that has been used as the basis for examples in Section 10.8. No peak loads are entered for the summer months as they have negligible impact on the outcome of modelling simulations.

Month	Monthly heat load (MWh)	Peak heat load (kW)	Duration (h)
J	20	95	8
F	20	95	5
M	15	65	4
A	14	65	3
M	5	20	3
J	0.5	0	0
J	0.4	0	0
A	0.2	0	0
S	3	15	3
O	16	65	3
N	18	70	5
D	20	95	8
Total	**132.1**		

68 kW of the peak demand and 14.3 MWh of the monthly heat load for the coldest month), and 29% from electrical energy input.

Given a peak heating load of 95 kW, we might make a first guess that $95\,000\ \mathrm{W}/60\ \mathrm{W\,m^{-1}} \approx 1\,500$ m of drilled borehole would be required, or 15×100 m boreholes. Let us use EED (Eskilson *et al.*, 2000) to simulate the evolution of carrier fluid temperatures given that:

- the aestifer is an arkosic sandstone with thermal conductivity $2.9\ \mathrm{W\,m^{-1}\,K^{-1}}$ and specific heat capacity $2.0\ \mathrm{MJ\,m^{-3}\,K^{-1}}$.
- the boreholes are 133 mm in diameter and the available drilling rig cannot drill deeper than 100 m
- the U-tube is of 32 mm OD HDPE with a wall thickness of 3 mm. Shank spacing is 58 mm.

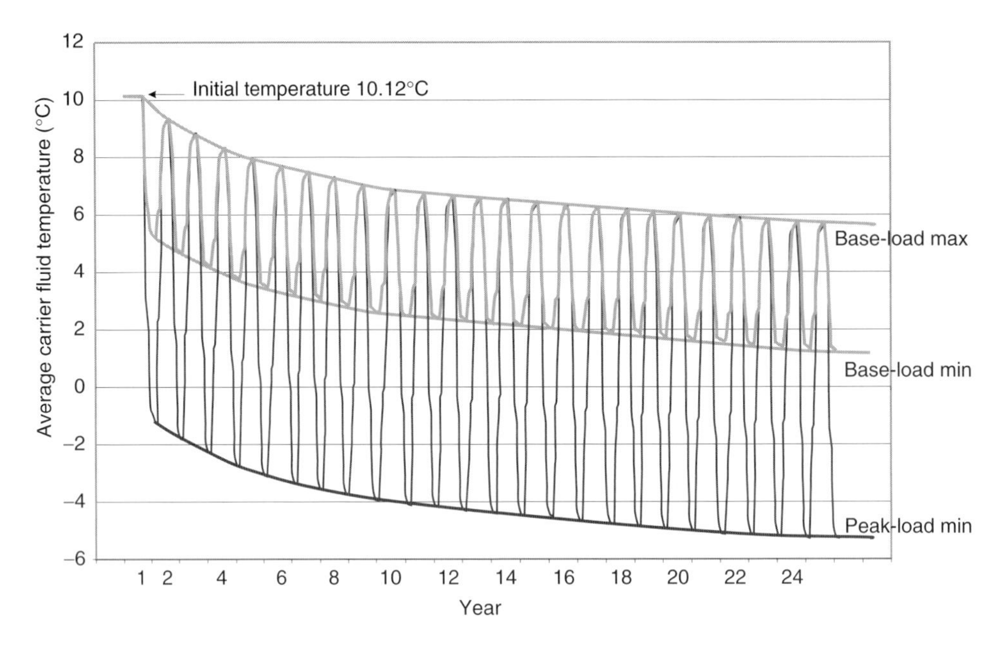

Figure 10.13 The evolution of carrier fluid temperatures over a 25 year period in a 15-borehole closed-loop array, with a peak heat output of 95 kW. See text for further details.

- the boreholes will be backfilled with a thermal grout of conductivity 1.8 $\mathrm{W\,m^{-1}\,K^{-1}}$
- the carrier fluid is 25% ethylene glycol, flowing at a rate of 20.3 $\mathrm{L\,min^{-1}} = 0.00034\ \mathrm{m^3\,s^{-1}}$ per borehole
- the average ground surface temperature is 9°C, with a geothermal heat flux of 65 mW $\mathrm{m^{-2}}$ (this allows the initial average temperature of the ground θ_o to be calculated)

The program calculates the borehole thermal resistance to be 0.119 $\mathrm{K\,m\,W^{-1}}$ and the average initial ground temperature θ_o (for the first 100 m depth) to be 10.12°C. If we assume that the 15 boreholes are arranged in a 3×5 grid, with a spacing of 10 m, EED predicts the carrier fluid temperature evolution depicted in Figure 10.13. There are two sinusoidal curves plotted: the upper, thicker curve shows the evolution of fluid temperature under the monthly 'base load'; the lower, thinner curve represents the minimum temperature every month under peak loading conditions. The envelopes 'base-load min' and 'base-load max' bracket the base-load curve; the 'peak-load min' curve represents the limit of minimum temperatures under peak loading in the coldest month. Of course, even these curves are only temperature *trends*; the real fluid temperatures will exhibit a complex pattern of diurnal fluctuations representing the daily switching on and off the heat pump. The 'base-load' curves show the general trend of fluid temperatures; the 'peak-load' curve represents the minimum fluid temperatures under 'worst case' conditions. We can see that, at the end of the 25 year period, typical 'base-load' carrier fluid temperatures vary between 1.26°C in February and 5.71°C in August. In January, under peak-loading conditions, average carrier fluid temperatures

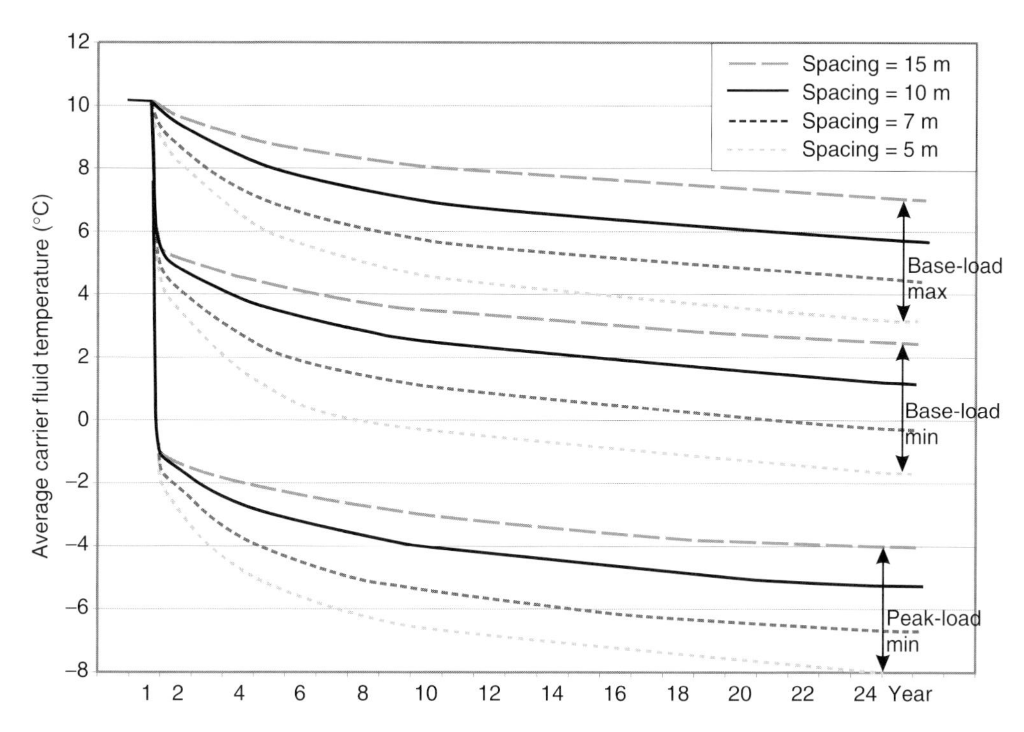

Figure 10.14 The influence of borehole spacing on performance. The fluid temperature evolution in the 15 borehole array from Figure 10.13 is shown here, with borehole spacings ranging from 15 m to 5 m.

may drop as low −5.21°C. Remember, this figure is an average of the downflow and upflow temperatures: the downflow temperature may be 2–2.5°C colder than this. Nevertheless, it seems that our system will perform acceptably: the peak-load temperature is well above the freezing point of the antifreeze. The baseload temperature does not drop below 0°C. Indeed, given an R_b of 0.119 $\mathrm{K\,m\,W^{-1}}$ and a specific heat absorption rate of 13.1 $\mathrm{W\,m^{-1}}$ in the coldest months, the temperature at the borehole wall would be 0.119 $\mathrm{K\,m\,W^{-1}} \times 13.1\ \mathrm{W\,m^{-1}} = 1.56$°C higher than the baseload temperature. In other words, the ground temperature would not drop below 2.82°C for long periods and there is no risk of extensive ground freezing.

Indeed, we could argue that our system is over-performing. We might be able to save capital costs by fewer or shallower boreholes. Our next step might be to simulate the system with 14 boreholes. In fact, EED has an inverse modelling option to predict the number of drilled metres required to meet certain fluid temperature criteria.

10.8.2 Effect of borehole array geometry

In the example above, we considered an array of 15 boreholes, in a 3 × 5 grid, with a spacing of 10 m. We can also ask our software to calculate the result if we alter the spacing to, say, 15, 7 or 5 m. The results are shown in Figure 10.14. We can see

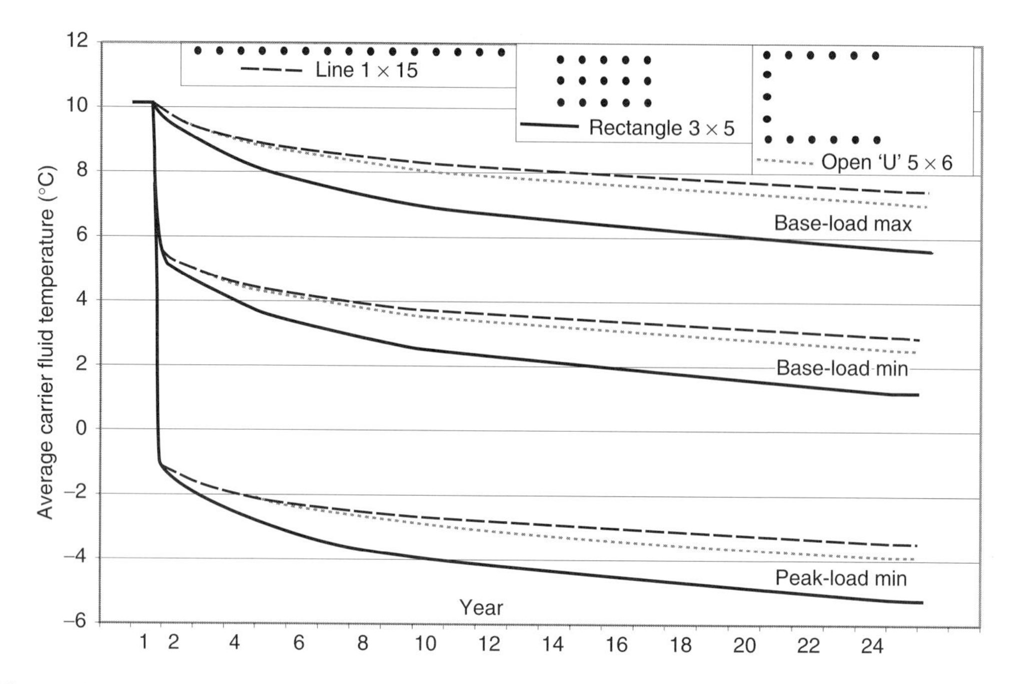

Figure 10.15 The influence of borehole array geometry on performance. The fluid temperature evolution in the 15-borehole, 3 × 5 array from Figure 10.13 is shown here, compared with linear and 'open U' borehole arrays, all with borehole spacings of 10 m.

that decreasing the spacing to 7 m results in a general drop in fluid temperatures of around 1.5°C, while the decline in spacing to 5 m results in a further similar decline. Conversely, increasing the spacing to 15 m results in an increase in fluid temperatures of 1.5°C. Thus, whether we choose a spacing of 10 or 5 m can have a major impact on the efficiency of our scheme, to the tune of 3°C in our carrier fluid temperatures. If we have enough space, we should maximise the distance between our closed-loop boreholes. Unfortunately, in today's urban areas, land prices are very expensive and we may have a limited area within which to work. We may be forced to live with a spacing of <10 m and we may find that we have to consider drilling deeper boreholes to deliver a system capable maintaining a given output.

The shape of the system also has an impact. Figure 10.15 compares our original 3 × 5 borehole array, with the performance of a single line (1 × 15) of boreholes and an open 'U' shape, with six boreholes along two sides and five along the base of the 'U'. In all cases, the spacing is 10 m, but we see that the open arrays, especially the line array, perform significantly better than the tightly packed array.

If we really wanted to optimise a linear array, we could drill angled boreholes (Figure 6.15, p. 130) with alternate boreholes drilled in opposite directions. Here we are minimising the thermal interaction between adjacent boreholes. Despite the small surface footprint of such a borehole array, it accesses a very large subsurface volume of rock (Figure 10.16a).

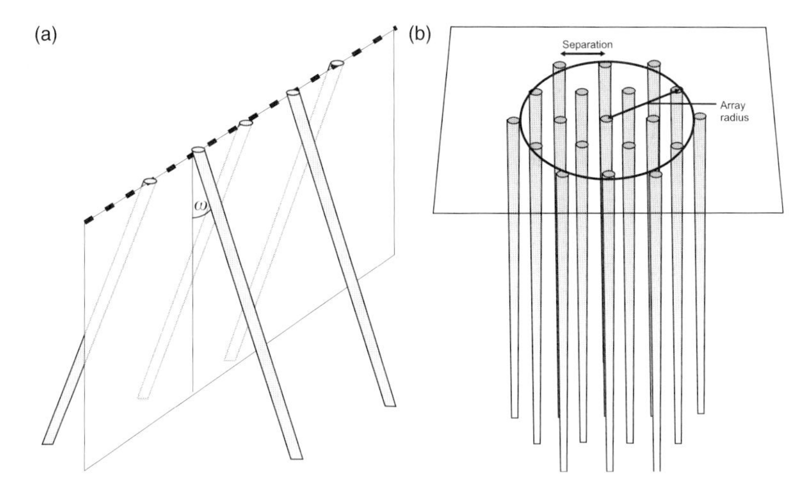

Figure 10.16 (a) An 'open' linear borehole array optimised for continuous heat extraction or rejection. Boreholes drilled at an angle ω to the vertical, with alternate boreholes in opposing directions. This minimises thermal interference and allows a large volume of rock to be accessed from a narrow surface footprint; (b) a 'closed' hexagonal borehole array designed for UTES: subsurface storage of energy rejected in summer, such that it can be extracted in winter.

10.8.3 Simulating cooling loads

Thus far, we have considered situations where heat is only extracted from a closed-loop array. We can also use the techniques described above to simulate the delivery of a cooling effect by 'dumping' waste heat into an aestifer. We can simply substitute a 'negative' value of q into Equations 10.13–10.15. In this case, of course, the temperature of the ground and carrier fluid rises with time. Eventually, we would hope that a steady state would be reached, where the heat loss via the borehole array is balanced by a heat flux from the ground *to* the surface. Our job is to design a borehole array that allows the heat flux to be dumped without excessive temperature rises in the ground or the carrier fluid.

What do we mean by 'excessive'? In the case of heating systems, there are obvious lower temperature limits imposed by the desire not to cause widespread ground freezing or freezing of the carrier fluid. In cooling mode, there are no such obvious upper limits. True, excessive ground heating can cause a small amount of ground movement by thermal expansion (see Chapter 13), but in practice, our upper acceptable temperature limit is more likely to be determined by efficiency factors: the hotter the carrier fluid, the less efficiently our heat pump will operate. Indeed, we may suspect that if we are operating with carrier fluid temperatures consistently in the upper twenties or thirties degrees Centigrade, any competitive advantage that a ground 'sink' heat pump has over an air 'sink' heat pump (i.e. conventional air-conditioning) begins to be eroded (see Chapter 4). Kavanaugh and Rafferty (1997) suggest that carrier fluid temperatures should typically be between 5°C and 11°C below the ground temperature in heating mode and 11–17°C above it in cooling mode. In a UK context, where the ground temperature may be around

10°C, this implies typical minimum average ground-loop temperatures of around −1°C in the heating mode, and maximum temperatures of 27°C in the cooling mode. VDI (2001a) suggest that, under German conditions, the return (uphole) temperature of the carrier fluid should not deviate by more than be more than 11°C from the undisturbed ground temperature under base-load conditions (weekly average) and not more than 17°C under peak-load conditions.

10.8.4 Simulation time

We can see from the examples of Sections 10.8.1–10.8.3 that the majority of evolution in temperature occurs in the first 5 years (or even less) of the closed-loop system's operation. Nevertheless, we have also seen (Section 10.7.4) that it can take more than 30 years before we achieve anything approaching a steady state condition. Analytical computer codes such as EED offer us the possibility of specifying the time period that we wish to simulate. Architects design our buildings to have a lifetime of at least several decades (and hopefully more), while a heat pump may have a useful life of 20–25 years. It seems reasonable then to run our simulations for a period of at least 25–30 years, at which point we might hope that some form of steady-state condition begins to 'kick in'.

10.9 Balanced UTES (Underground Thermal Energy Storage) systems

In many schemes, we may have a cooling demand in the summer and a heating demand in the winter. We thus end up 'dumping' waste heat to the ground in the summer and abstracting heat in the winter. Here, however, we are not really 'dumping' the heat: we are using the geological environment to store it, such that we can re-abstract in the winter. Of course, if we are depositing more heat in the summer than we re-abstract in the winter, the ground temperatures will steadily creep up with time (and this can be simulated with the analytical tools discussed in Section 10.8.1 or via numerical heat flow models). If, however, we consider a 'balanced' heating and cooling load, where the amount of injected heat approximately balances that abstracted on a yearly cycle, we would expect some kind of 'steady state' to be achieved: a sustainable ground source heating and cooling scheme utilising Underground Thermal Energy Storage (UTES – Sanner and Nordell, 1998).

As an example, let us reconsider the heating loads in Table 10.1, which have already formed the basis for our examples in Section 10.8. Let us, however, add the cooling demand specified in Table 10.2 and assume a seasonal performance factor (SPF_C) of 2.5 in cooling mode. We now have a situation where 132.1 MWh of heat is delivered annually, corresponding to a heat extraction from the ground of 132.1 MWh $\times$ $(1 - 1/3.5) = 94.4$ MWh. 60 MWh of heat is removed from the building annually, corresponding to a heat transfer rate *to* the ground 60 MWh $\times$ $(1 + 1/2.5) = 84$ MWh. We

Table 10.2 An example of a cooling-load profile for a hypothetical building that has been used, in addition to the heating loads in Table 10.1, as the basis for the example in Section 10.9.

Month	Monthly cooling load (MWh)	Peak cooling load (kW)	Duration (h)
J	0	0	0
F	0	0	0
M	0	0	0
A	3	50	2
M	5	50	2
J	10	70	6
J	15	70	8
A	15	70	6
S	10	50	3
O	2	0	0
N	0	0	0
D	0	0	0
Total	60		

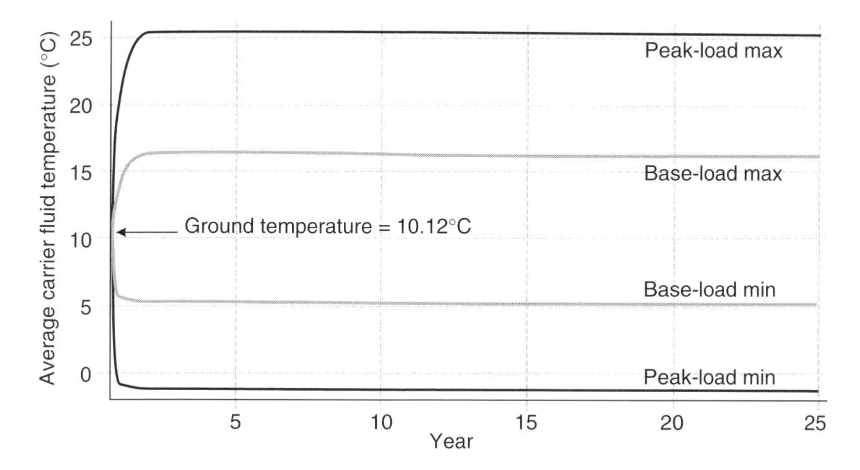

Figure 10.17 The evolution of carrier fluid temperatures over a 25 year period in a 15-borehole closed-loop array, with approximately balanced summer heat rejection and winter heat abstraction (based on data shown in Tables 10.1 and 10.2). See text for further details.

thus have a net heat extraction rate of only around 10 MWh yr^{-1}. The temperature evolution in our 15 boreholes, spaced at 10 m in a 3×5 grid, is shown in Figure 10.17. We seem to achieve a steady temperature trend within only 1–2 years, with typical 'base-load' (monthly average) temperatures of 16°C in the summer and 5°C in the winter, with peak-load temperatures of around 25°C (summer) and −1°C (winter). In fact, the temperatures are not quite steady – there is a slight decline with time due to the small net heat abstraction. We should note two important features of such approximately balanced schemes:

i. Provided we could accept higher or lower peak temperatures, we could operate the scheme with significantly fewer boreholes and still maintain the 'base-load' (monthly average) temperatures within acceptable limits, without any highly detrimental long-term trend.
ii. Because we are extracting and replenishing heat on an annual cycle, rather than simply extracting heat over a period of decades, we need only concern ourselves with the migration of a heat front in the ground over a time scale of 1 year, rather than 30 years or more. Thus, the issue of thermal interference between boreholes is less critical and we can space our boreholes closer together. In fact, reducing the borehole spacing to 5 m has very little effect on the fluid temperature evolution in Figure 10.17.

If the heating and cooling demands of a building are not approximately 'balanced', we *may* find it advantageous to incorporate conventional heating (e.g. gas) or cooling (cooling towers) into our building design such that they can absorb any excess heating or cooling load. Such hybrid systems can be manipulated to result in approximately balanced heat fluxes to and from the ground, thus permitting a robust UTES scheme.

10.9.1 Design principles of balanced UTES schemes

We have seen that, for schemes where heat is being continuously extracted (in heating schemes) *or* dumped (in cooling schemes) to an aestifer:

- Temperatures evolve over a period of decades until a quasi-steady-state begins to be established due to heat exchange via the ground surface.
- Schemes with a large surface area will encourage heat exchange via the surface.
- Thermal interference between boreholes becomes important and spacing of boreholes is critical (at least 10 m is recommended, if possible).
- A linear or 'open' array performs better than a tightly packed, closed array of boreholes, as it encourages exchange of heat with the broader aestifer external to the array.
- Any groundwater flow through the closed-loop borehole array will be beneficial, tending to replenish heat to a heat extraction array of boreholes and removing heat from a heat rejection array, by advection.

If, however, our objective is to *store* heat in the aestifer, by injecting it in the summer and re-abstracting it in the winter, many of these considerations do not apply. Indeed, as VDI (2001b) points out:

- Temperatures in the system reach a dynamic steady state relatively quickly, although they may vary over an annual cycle.
- Thermal interference between boreholes is of lesser importance and closer borehole spacings are possible.

- To avoid leakage of heat outside the array, a more 'closed' array shape will be preferable to an open or linear array. Indeed to minimise the array's overall surface area-to-volume ratio, an equidimensional cylindrical or hexagonal array may be preferable (Figure 10.16).
- If we wish to minimise heat loss from our subsurface heat store, large throughflows of groundwater will be disadvantageous. Thus, either the hydraulic conductivity should be low to modest *or* there should be a low groundwater hydraulic gradient in the area.
- More extreme peak temperatures may be possible or even desirable.

Figure 10.16 compares a linear 'open' array suited to continuous heat extraction, with a more 'closed', hexagonal array, suited to heat storage. The total thermal storage (H) of a cylindrical array of boreholes of depth D and effective array radius r_{array} is given by:

$$H = S_{VC} \cdot \pi \cdot r_{\text{array}}^2 \cdot D \cdot \Delta\theta \tag{10.19}$$

where S_{VC} = volumetric heat capacity of the aestifer material and $\Delta\theta$ is the average change in ground temperature of the ground enclosed by the borehole thermal array. This thermal storage is the amount of heat rejected to the ground in summer and abstracted in winter.

One example of a major UTES scheme, based on closed-loop boreholes, is that at Richard Stockton College in New Jersey, where a volume of 1.2 million m^3 of sediments is accessed by an array of 400 boreholes of 135 m depth. The cooling load (over 5000 kW) is reportedly larger than the heating load, so that the ground will likely have a tendency to warm up over time, although this may be mitigated by heat loss due to groundwater advection (Stiles, 1998).

Clearly, the amount of heat that can be stored is maximised if $\Delta\theta$ is maximised. In northern countries, where storage of surplus heat in the summer is encouraged for subsequent abstraction in the winter, carrier fluid temperatures in excess of 40°C are commonly considered viable. Usage of heat pumps to reject heat at such temperatures may not be especially energetically efficient: but if the waste heat is already available at a high temperature (for example, waste heat from industrial processes) and can be stored underground, it will permit highly efficient extraction of heat in winter. Indeed, one of the earliest UTES schemes, using 121 boreholes to store waste industrial heat in a rock volume of some 120 000 m^3, in Luleå, Sweden, operated at temperatures of at least 65°C (Sanner and Knoblich, 1998) and reportedly up to 82°C in heat rejection mode (G. Hellström, *pers. comm.*). VDI (2001b) and Gabrielsson *et al.* (2000) even discuss temperatures as high as 90°C for UTES schemes. Temperatures as high as this will require careful consideration of the materials employed in borehole construction: polyethene becomes unviable as a ground-loop material above 60–70°C (VDI 2001b) and alternatives such as steel or polypropene may need to be considered. Simulation tools and models should also be carefully assessed for applicability. Bear in mind the temperature dependence of thermal properties of materials and also that vapour movement and fluid convection may become significant at highly elevated temperatures, in addition to conduction.

We must remember that waste heat from space-cooling heat pumps need not be our only source of surplus heat in summer. If we have a high heating demand in winter that must be satisfied, it becomes increasingly attractive (VDI, 2001b) to indulge in *supplementary* recharge of heat, via the ground loop, to the subsurface by:

- installing solar thermal panels to collect summer heat (Gabrielsson, 1997; Sanner and Hellström, 1998; Nordell and Hellström, 2000);
- harvesting surplus summer heat from engines or combined heat-and-power installations;
- collecting heat from loops installed under black surfaces, such as roads or car parks.

Here, we are explicitly manipulating the subsurface as a massive storage heater (see Chapter 3). Such solutions are most attractive in northern climes, where the major challenge is to provide winter space heating to buildings. The closer to the tropics one moves, the more prominent the issue of providing space cooling becomes. Nevertheless, with a little imagination, an understanding of heat fluxes and a concept of the ground as a huge thermal battery, the possibilities for manipulating heat on a large scale are manifold.

11

Standing Column Wells

The older view of the nature of heat was that it is a substance, very fine and imponderable indeed, but indestructible, and unchangeable in quantity, which is an essential fundamental property of all matter ...We must rather conclude from this that heat itself is a motion, an internal invisible motion of the smallest elementary particles of bodies.

Hermann von Helmholtz

11.1 'Standing column' systems

We have, by now, become familiar with 'open-loop' ground source heat systems, where heat is transported from the subsurface to a heat exchanger by forced convection (or advection) with groundwater flow. We have also dealt with closed-loop systems, where the dominant heat transport mechanism in the rock or sediment is usually conduction.[1] This section deals with systems that are neither open nor closed loop, but halfway between the two. These are called standing column wells. But before we examine these in earnest, let us briefly recap a type of closed-loop configuration we discussed in Chapter 10.

[1] Although, in closed-loop schemes, the dominant heat transport process in the rock or sediment surrounding the closed loop will usually be conduction, convective processes in vapour or groundwater may also be important in some circumstances. Heat will pass from the aestifer through any borehole grout and through the walls of a polyethene closed-loop pipe by conduction. However, the process by which heat is transferred from the pipe walls to the internal carrier fluid will involve convection. Furthermore, the carrier fluid conveys heat to the heat pump by forced convection. We can thus see that heat transfer from the aestifer to the heat pump in a closed-loop system is by a complex combination of processes.

11.1.1 'Open coaxial' closed-loop boreholes

The title of this section suggests that we are reaching the limits of available terminology! Chapter 10 (Section 10.6.3, p. 233) briefly dealt with coaxial closed-loop heat exchangers. Figure 10.11 showed us that an 'open coaxial' closed-loop system could be established by circulating water down the bore of a cased borehole (or, in stable low-permeability formations, an unlined borehole) and then pumping it back up a rising main along the centre of the borehole. Although the arrangement may sound attractive because it can minimise the thermal resistance of the borehole itself, this arrangement is seldom used because

i. Heat transfer within the aestifer to the borehole wall is dominantly by conduction (especially if the borehole is lined)[1].
ii. In permeable formations, unless the entire borehole is lined with plain casing, we are unable to use antifreeze in the carrier fluid (it would be a potential pollutant). This constrains the minimum temperature permissible in the fluid (water) to a few degrees above 0°C.
iii. If we propose to use antifreeze solution as the carrier fluid, the entire borehole length should be lined with plain casing, which must prevent any leakage of antifreeze. This results in a rather expensive borehole.

In other words, the gains in heat transfer efficiency over a grouted U-tube arrangement are modest, but the cost of the borehole (due to the need for casing) may be significantly greater. This 'open coaxial' closed-loop configuration would thus be most applicable in stable, self-supporting, low-permeability formations, where open holes can be constructed without large lengths of casing.

11.1.2 Standing column wells

However, if we could induce heat transfer from the geological formation to the borehole by advection in groundwater flow as well as by conduction, the arrangement outlined above may start to look more attractive. In this case, groundwater needs to be able to enter the well, so we require either an 'open' (unlined) borehole or a water well screen (see Section 6.3.1, p. 112). Let us therefore consider the situation in Figure 11.1a, where we re-circulate a carrier fluid down the well bore and up a rising main but where we bleed a certain proportion (B) of that flow to waste (following passage through the heat pump). This will have the effect of lowering heads within the borehole and inducing the influx of a corresponding amount of new groundwater to the borehole. This groundwater will also transfer a new 'load' of heat to the system by advection. We might consistently bleed a certain percentage of the total borehole circulation during operation (although this would not be the normal mode of operation). More likely, bleed would be automatically invoked at times of peak heating (or cooling) demand: if the temperature falls below a certain level in the circulating fluid (in heating mode),

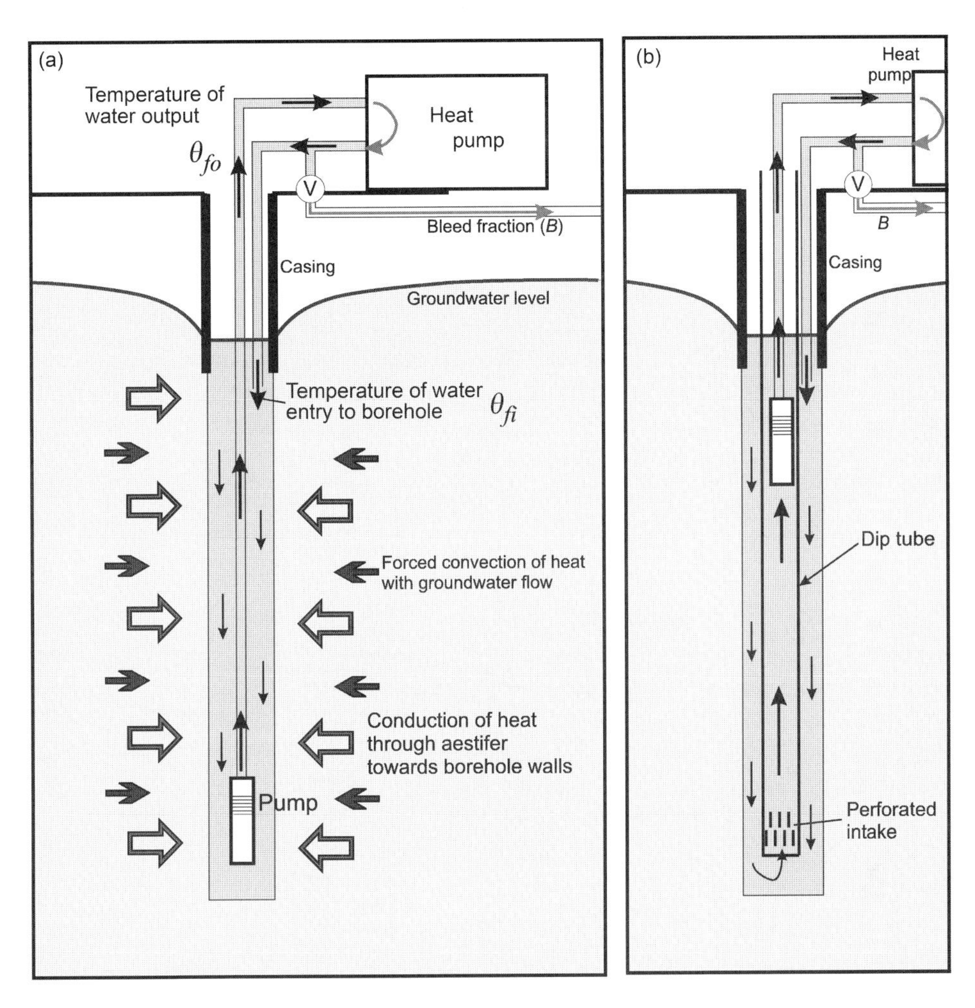

Figure 11.1 (a) Schematic diagram showing a standing column well and (b) a standing column well suited to larger capacity installation where the length of rising main is minimised by using a pump 'sheath' or 'dip tube'. In both cases, the wells are operating in 'bleed' mode. In 'non-bleed' mode, there would be little or no drawdown, but some circulation of groundwater through the formation, due to temperature and pressure gradients along the borehole.

the bleed valve is opened to induce the flow of fresh (warmer) groundwater into the borehole. In the cooling mode, bleed will be invoked when the temperature rises above a certain level, inducing inflow of cooler groundwater to the bore. This ensures that fluid temperatures entering the heat pump remain within an acceptable range and thus avoids poor values of COP.

If the bleed percentage is 0, we essentially have a 'coaxial' closed-loop-type system of the type described in Section 11.1.1, where heat transfer from the aestifer to the borehole wall is dominated by conduction. If the bleed percentage is 100, none of the pumped water is re-circulated to the borehole and we have an open-loop system, where heat transfer from the aestifer is by advection with groundwater. The intermediate arrangement outlined above (where bleed percentage may typically be in the region of 5–30) is termed a 'hybrid' or 'standing column well' (SCW) system. Around 1000 SCW systems are believed to be operational in the USA (O'Neill *et al.*, 2006). Such wells are typically 6–8 in. (150–200 mm) in diameter and in the USA are usually >100 m deep. The shallower (typically smaller, residential) schemes will have an electrical submersible pump placed at the base of the well (Figure 11.1a), while re-injected water is returned just below the surface of the standing water column (under all operating conditions). In deeper wells, it is common for the submersible pump to be placed within a ≈100 mm diameter pipe – termed a dip tube. The dip tube extends to the base of the well, where it has a perforated intake section. The pump itself is placed a little way beneath the water level of the standing column of water, allowing lengths of rising main and cable to be minimised (Figure 11.1b). The flow and return mains are sized for flow velocities of ≈1.5 m s^{-1} (Orio *et al.*, 2005). Depending on water quality, the pumped groundwater can flow through the evaporator of the heat exchanger or through a prophylactic heat exchanger (Section 6.3.2, p. 115).

As an example of an SCW system, Deng *et al.* (2005) documented a system at a public library in Massachusetts, drilled in Cambro-Ordovician metasediments, where a 10% bleed rate, of 30 min duration, automatically kicks in when the well temperature drops below 4.4°C.

The SCW system has not been widely researched, but the papers by Mikler (1993), Deng (2004), Deng *et al.* (2005) and O'Neill *et al.* (2006) provide a good introduction to the performance of such systems. These authors discovered that the most important design variables for such systems are

- borehole depth;
- rock thermal conductivity;
- rock hydraulic conductivity;
- bleed rate.

The mechanisms of heat transfer within the SCW system include

- heat conduction through the surrounding rock to the borehole wall (and thence by convection into the carrier fluid);
- convective transfer at the borehole walls and surfaces of pipework;
- forced convection/advection with groundwater flow during bleed periods;
- a small component of forced and buoyancy-driven convection within the formation itself, which will be related to the head gradients and temperature gradients along the borehole.

With these many variables and heat transfer mechanisms, we should not be surprised that there are no simple 'rules of the thumb' for designing these systems. Deng *et al.* (2005) propose a 'simplified model' for simulating the performance of the standing column. Even this simplified model requires considerable mathematical insight and computing skills, but it is currently being incorporated into some design software packages.

11.2 The maths

The simplified model of Deng (2004) and Deng *et al.* (2005) is a one-dimensional (radial) model, with the assumptions of homogeneous and isotropic aquifer conditions and a negligible natural geothermal temperature gradient. It solves heat transport within the aquifer by the one-dimensional advection/conduction model (which is a version of the mathematics encountered in Box 7.3, p. 175, in radial coordinates):

$$S_{VCaq}\frac{\partial \theta}{\partial t} + S_{VCwat}\nu_D\frac{\partial \theta}{\partial r} = \lambda_{eff}\left(\frac{\partial^2 \theta}{\partial r^2} + \frac{1}{r}\frac{\partial \theta}{\partial r}\right) \tag{11.1}$$

$$\nu_D = \frac{FB}{2\pi r D} \tag{11.2}$$

where S_{VCaq} and S_{VCwat} are the volumetric heat capacities of the saturated aquifer and of water, respectively, λ_{eff} is the effective (enhanced) thermal conductivity, B is the bleed rate as a fraction of the total pumping rate F, θ is the temperature, t is the time and r is the radial distance from the borehole. v_D is the Darcy velocity (flow rate per unit aquifer area – see Glossary), D is the borehole depth within the aquifer or the effective thickness of the portion of the aquifer under consideration.

Here, a so-called enhanced thermal conductivity that includes both genuine thermal conduction and the modifying effects of convective processes related to the pumping borehole is used. The fluid temperature within the borehole is calculated by invoking the heat transport calculated from this Equation 11.1 and a separate and explicit term representing the inflow from 'bleed'. It should be obvious that the average temperature of the fluid in the SCW (θ_f) is given by

$$\theta_f = \frac{(1-B)\theta_{fi} + B\theta_b + \theta_{fo}}{2} \tag{11.3}$$

This allows us to construct a heat balance for the borehole including both the 'bleed' term and 'quasi-conductive' transfer of heat from the borehole wall. Without reconstructing the whole of Deng O'Neill's argument, this takes the form:

$$V_{SCW}S_{VCwat}\frac{d\theta_f}{dt} = 2FS_{VCwat}\left[(1-B)\,\theta_{fi} + B\theta_b - \theta_f\right] + \frac{(\theta_b - \theta_f)}{R_b}D \tag{11.4}$$

where V_{SCW} is the volume of fluid in the standing column; θ_b is the fluid temperature at the borehole wall (which is assumed to be the temperature of the groundwater bled into the well) derived from Equation 11.1; R_b is a borehole resistance term; and θ_{fi} and θ_{fo} are the temperatures of the recirculating water's entry to and exit from the well.

This simplified model makes a number of very 'dodgy' assumptions: it implicitly assumes that the aquifer approximates to a homogeneous porous medium and also that there is near-instantaneous thermal equilibrium between the rock–sediment matrix and groundwater (see Section 7.11, p. 174), to the extent that a single temperature can describe both phases. These assumptions will, unfortunately, be least valid in exactly those aquifers where SCWs are most commonly used – fractured, well-lithified aquifers. Nevertheless, the model has been validated against a rather complex two-dimensional, numerical model and found to be broadly satisfactory. However, as one might expect, discrepancies between the simplified model and the complex model do arise in fractured rock aquifer systems, and Deng *et al.* (2005) suggest that these might be of the order of 2°C over a 1 year simulation. Discrepancies also arise in shallow boreholes because of end effects, and in situations with high natural geothermal temperature gradients.

11.3 The cost of SCWs

O'Neill *et al.* (2006) compared the requirements for drilled borehole length for three types of ground array, based on the heating and cooling demands of a real building in Boston, Massachusetts. The site was underlain by fractured igneous and metamorphic bedrock of assumed thermal conductivity 3.0 W m^{-1} K^{-1} and specific heat capacity 2.6 MJ m^{-3} K^{-1}. The undisturbed earth temperature was considered to be 12.2°C. The three ground arrays considered were as follows:

- A 'conventional' grouted U-tube, closed-loop array comprising a line of 8 boreholes, drilled to 82 m (total drilled depth 653 m). The borehole thermal resistance was assumed to be 0.14 K m W^{-1}.
- A 152 mm diameter standing column well, with no bleed.
- A 152 mm diameter standing column well with 10% bleed. For the SCW systems, an 'enhanced' thermal conductivity of 3.5 W m^{-1} K^{-1} was applied.

The study found that, to satisfy the same heating and cooling demand, the SCW with no bleed needed to be 391 m deep, while the SCW with 10% bleed needed to be 263 m deep, compared with a total drilled depth of 653 m for the conventional closed-loop boreholes. O'Neill *et al.* (2006) found, furthermore, that the borehole thermal resistance of the SCWs was only 0.0011 K m W^{-1}. It was found that the operating costs of the three types of scheme over a 20 year life cycle were very similar. The 'cheapest' system was thus largely determined by capital cost. In the selected geological terrain, where the SCW wells could be constructed simply as open holes, the capital costs were strongly related to drilled depth and the SCW well with bleed had the cheapest capital (and thus overall) cost.

It should be noted that, if the geology had required specialist construction of water wells, with a lot of casing, well-screen and/or gravel pack, the costs of the SCW would probably have been dramatically higher than the conventional closed-loop well. Given that the SCW only gives just over double the heat yield per drilled metre, the analysis would likely have shown that the conventional closed-loop approach would have been cheaper under those circumstances. One could argue (S. Rees, *pers. comm.*), however, that polyethene U-tubes become increasingly difficult to install in very deep boreholes (Orio *et al.*, 2005, consider depths of up to 450 m). Thus, if available land area is very constrained and if very deep boreholes are required, SCWs may be the favoured option from a constructional viewpoint in certain lithologies.

Moreover, O'Neill *et al.* (2006) assumed a shallow water table in their analysis. They went on to demonstrate that, if the water table was deep (around 30 m in their example), the additional pumping costs of the SCW system with high rates of bleed began to make it uncompetitive with closed-loop or low-bleed options. We can thus summarise that SCW wells (with bleed) are best applied in a rather limited range of geological and logistical circumstances:

- well-lithified or crystalline rock aquifers that are self-supporting and do not require long casing lengths or well screens;
- situations where available land area is tightly constrained;
- aquifers with relatively modest available yields and buildings with modest heat demands (if yields are potentially large, open-loop systems may well be preferable);
- aquifers with relatively shallow water tables (otherwise pumping costs of lifting the bleed water may render the economics unattractive);
- situations where the well can be used to provide a domestic water supply as well as heating/cooling.

It is thus not coincidental that most of the SCWs found in the USA are in the crystalline rock terrains of the north-east of the country or the Pacific north-west (Orio *et al.*, 2005).

11.4 SCW systems in practice

Orio *et al.* (2005) conducted a study of around 20 commercial and residential SCW schemes in north-eastern USA, ranging in size (installed output) from 17 to 700 kW, typically dominated by heating loads. Depths of boreholes ranged from 73 to 457 m, and some of the larger schemes comprised several SCW wells (up to six wells in one instance). In such multiple-well schemes, automated variable-speed pumps were deemed to be advantageous.

SCWs were typically drilled at a diameter of 6 in. (152 mm), while the static groundwater level was within 37 m of the surface in all cases and typically <20 m from the surface. Orio *et al.* (2005) regard bleed rates of between 2% and 25% as typical. The number of metres of standing column to support 1 kW load ranged from 2.6 to 7.4 m kW^{-1}, with upper and lower quartiles of 3.4 and 5.2 m kW^{-1} and a median of

4.8 m kW^{-1}. The submersible pumps' power rating was typically in the range 1–3% (i.e., 10–30 W per kW load) of the total output of the heat pump (although it reached as high as 7% in one instance).

Collins *et al.* (2002) state that a 1500 ft (457 m) SCW should be adequate to supply a heating or cooling demand of 420 000–480 000 Btu hr^{-1} (123–141 kW) – equivalent to 3.2–3.7 m kW^{-1}. They also recommend a spacing of 15–23 m between individual SCW boreholes.

In over 90% of the residential SCW systems surveyed by Orio *et al.* (2005), the SCW systems bled water to domestic water supply systems. This highlights the final major advantage of SCW systems: their ability to produce domestic heat and domestic potable water supply simultaneously.

11.5 A brief case study: Grindon camping barn

At Grindon, on a popular walkers' route along Hadrian's Wall in Northumberland, UK, a former telephone repeater station was converted to a hostel to provide accommodation and food for ramblers. The owner of this remote building expressed interest in a borehole to provide a domestic water supply and a base load of ground source heat to the building (to be supplemented by wood fires in the coldest weather). A groundwater sourced system was thus constructed, supplied by a 104 m deep borehole (drilled using Down-the-Hole Hammer methods – Figure 11.2) through overlying Carboniferous mudstones and sandstones, the dolerite Whin Sill (along whose outcrop Hadrian's Wall runs) and into the underlying saturated limestones and sandstones. It was initially intended that the system would run as an open loop, with chilled water being discharged to two shallow soakaway boreholes. It soon became clear, however, that the soakaways were unable to accept the entire pumped volume, so the system was reconfigured as an SCW. Groundwater was pumped from the well at a design rate of around 34 L min^{-1}. Part of this flow was used for domestic water supply, while the remainder was circulated directly through the evaporator of a 17 kW heat pump. Around 80% of the water was re-circulated to the abstraction well: it was not returned directly to the well's interior, however, but to the gravel pack installed between the PVC well screen and the borehole wall. The waste bleed water (c. 20%) was disposed to the two shallow soakaway boreholes.

Unfortunately, after around 2 years of operation, some problems emerged that appeared to be related to the groundwater's chemistry, which seemed to be somewhat reducing in nature (containing H_2S gas and around 0.5 mg L^{-1} dissolved iron). The coaxial heat-exchange element of the heat pump evaporator began to develop symptoms of iron incrustation and had to be cleaned of ochreous precipitates by pressure rinsing. Furthermore, the well's gravel pack has also become increasingly clogged, quite possibly with iron oxyhydroxide deposits, and has become progressively less able to accept the re-circulated water. The system has thus reverted to open-loop mode, although the site owner has had to establish supplementary disposal routes for 'waste' water.

Figure 11.2 A Down-the-Hole (DTH) hammer rig in operation, drilling the standing column well at Grindon Camping Barn. *Photo by Jonathan Steven and reproduced by kind permission of Geowarmth Ltd.*

This experience underlines the extent to which an understanding of groundwater chemistry is important for open-loop and SCW systems. It also emphasises the fact that while such systems utilising natural ground waters often require fewer drilled metres, they generally have far greater maintenance requirements than closed-loop systems, whose fluid composition can be controlled.

11.6 A final twist – the Jacob Doublet Well

A final means of exploiting hydrogeological stratification to increase the sustainability of an abstraction/recharge operation in a single well grows out of the idea of the 'Doublet Well', proposed by the hydrogeologist C.E. Jacob in the 1960s. In Jacob's original concept, the Doublet Well consists of an upper portion of the well bore (used for abstraction) and a lower portion (used for circulation and recharge), separated from each other by a system of packers. Jacob proposed the use of this well in situations where a less dense fluid was 'floating' over a denser fluid in an aquifer. For example, in coastal aquifers, fresh groundwater could be abstracted from the upper chamber of the well, while the lower chamber is used for the recirculation and reinjection of the underlying, denser saline groundwater. Jacob also noted that the Doublet Well could be used to

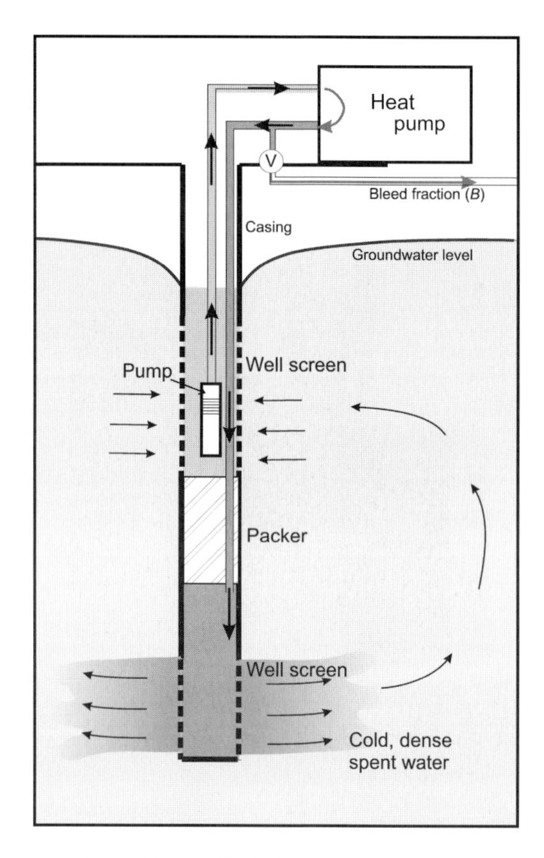

Figure 11.3 A simple vertical doublet well. In fact, C.E. Jacob's original 'doublet well' concept was more complicated than this, utilising three separate chambers for abstraction, recirculation and reinjection – see Wickersham, 1977).

maximise the abstraction of an oil phase, floating on top of denser formation water in an oil reservoir.

Wickersham (1977) noted that the Doublet Well could also be used in the context of heat pumps. Groundwater could be abstracted from the upper chamber of the borehole and passed through a heat pump to provide space heating (Figure 11.3). All or some of the cold, thermally spent water could then be re-injected into the well's lower chamber. This arrangement is not, strictly speaking, an SCW: we are forcing the water to travel through the surrounding aquifer formation, increasing its contact time with geological materials and the available heat exchange area. The extent to which the spent water migrates laterally into the aquifer would be related to the aquifer's anisotropy: a high ratio of horizontal to vertical hydraulic conductivity would increase the length of the aquifer flow path from the lower chamber back to the upper chamber. In a sense, we could argue that the Doublet Well is a type of vertically oriented open-loop doublet system, where the abstraction and recharge wells are combined into a single structure, making use of the aquifer's hydraulic stratification.

12

Thermal Response Testing

Logarithmic plots are a device of the devil.

Attributed to Charles Francis Richter

12.1 Sources of thermogeological data

We have seen in Chapter 10 that when we are planning a borehole-based closed-loop heating or cooling system, we require values of the ground's thermal conductivity and its volumetric heat capacity as input parameters to the design process. We can acquire these by three routes:

- Generic tables of values (such as Tables 3.1, p. 35, or 3.3, p. 53). These can be found in many publications; alternatively, databases may be bundled with design software such as Earth Energy Designer (Eskilson *et al.*, 2000). In some countries, geological surveys may be able to provide a tailored report of the likely thermogeological parameters at a specified site. In the UK, the British Geological Survey can produce such a 'GeoReport' for a fee.
- Laboratory testing. A core or sample of geological material can be returned to a laboratory and the thermal conductivity determined by means of measuring the heat flux when the material is subject to a known temperature gradient.
- *In situ* field tests.

Clearly, a laboratory test is of some value. Indeed, many of the data reported in literature (including those that form the basis of reports produced – at least at the time of writing – by the British Geological Survey) are largely based on laboratory determinations. However, such determinations are subject to considerable limitations. One can query whether they are representative: they determine the properties of a sample of only a few centimetres in dimension. In an operational ground source heat scheme, we try to induce heat flow through many hundreds or thousands of cubic metres of rock. If major fractures or discontinuities are present in the aestifer, they are likely to reduce the bulk thermal conductivity of the rock. Such features are unlikely to appear in a core recovered from a borehole (fracturing makes core recovery difficult). Furthermore, ambient groundwater flow in an aestifer may enhance heat transport by advection: laboratory tests reveal nothing of this phenomenon, but it can be identified in field tests (Sanner *et al.*, 2000).

12.2 The thermal response test

The typical field test used to estimate thermogeological parameters is the thermal response test (TRT). We have already seen (Equation 10.14) that the early evolution of average fluid temperature (θ_b) in a closed-loop borehole with time (t) is governed by the equation:

$$\theta_o - \theta_b \approx qR_b + \frac{q}{4\pi\lambda}\left[\ln\left(\frac{4\lambda t}{r_b^2 S_{VC}}\right) - 0.5772\right] \text{ for } \frac{5r_b^2 S_{VC}}{\lambda} < t < \frac{t_s}{10} \qquad (10.14)$$

where θ_o = average 'far-field' (initial) temperature of the ground over the length of the borehole (K); $\theta_o - \theta_b$ = temperature 'drawdown' or displacement (K); q = heat extraction rate per metre of borehole (W m^{-1}); λ = thermal conductivity of the aestifer (subsoil or rock) in W m^{-1} K^{-1}; r_b = borehole radius (m); S_{VC} = volumetric heat capacity of aestifer (J m^{-3} K^{-1}); t_s = time at which steady state begins to be a better representation of fluid temperature than radial flow; and R_b = borehole thermal resistance (K m W^{-1}).

Note that the logarithmic term in Equation 10.14 is written as a natural logarithm (ln or log$_e$). If we now consider heat *injection* to the borehole at rate q, and if we convert the natural logarithm to log$_{10}$ by multiplying by 2.303, this equation can be rearranged as follows:

$$\theta_b - \theta_o \approx qR_b + \frac{q}{4\pi\lambda}\left[2.303\log_{10}\left(\frac{4\lambda}{r_b^2 S_{VC}}\right) - 0.5772\right] + \frac{2.303q}{4\pi\lambda}\left[\log_{10} t\right] \qquad (12.1)$$

Thus, if we inject heat to a closed-loop borehole at a constant rate q, the temperature will evolve in proportion to $\log_{10}t$. From the rate at which the temperature evolves, we should be able to deduce the ground's thermal conductivity λ. Figure 12.1 shows just such a TRT: heat is being injected to the ground via the closed loop at a rate of around 5 kW. The carrier fluid enters the ground loop at a temperature some 5–6°C higher than the return temperature. The mean fluid temperature (θ_b) is calculated simply as the average of the uphole and downhole temperatures.

We can also see that Equation 12.1 should correspond to a straight line if we plot temperature displacement ($\theta_b - \theta_o$) against the logarithm (log$_{10}$) of time (t). The gradient

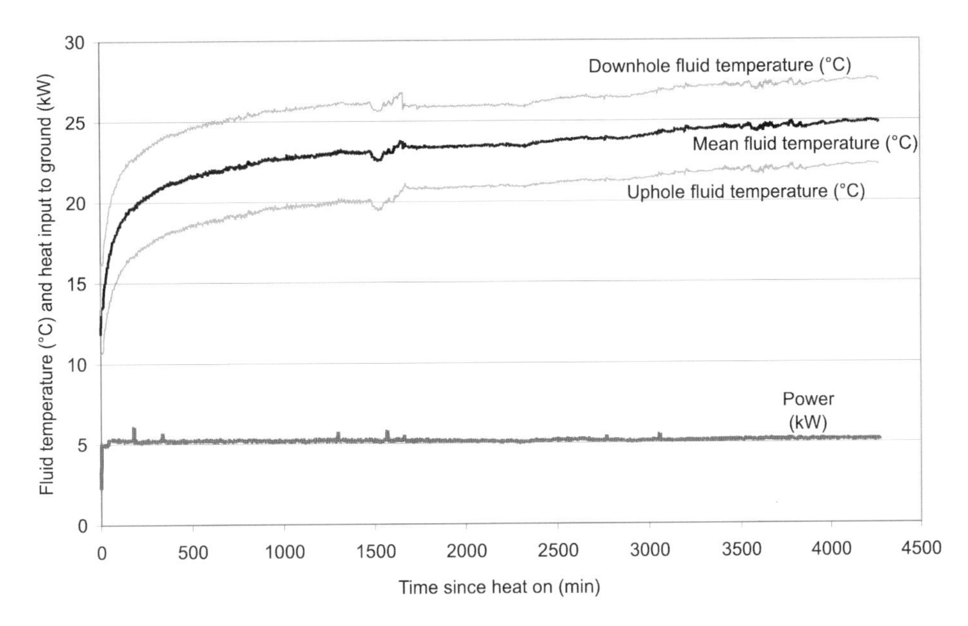

Figure 12.1 Example of output from a TRT. *Based on data provided by GeoWarmth Ltd. of Hexham, UK.*

of the straight line is

$$\text{Gradient} = 2.303q/4\pi\lambda \tag{12.2}$$

Because we know q, this enables us to find the thermal conductivity λ. Furthermore, the intercept of the straight line on the temperature axis at $\log_{10}t = 0$ is given by the expression:

$$\text{Intercept} = qR_b + \frac{q}{4\pi\lambda}\left[2.303\log_{10}\left(\frac{4\lambda}{r_b^2 S_{VC}}\right) - 0.5772\right] \tag{12.3}$$

With this equation, we can deduce the value of the borehole thermal resistance (R_b), provided we have a good estimate of S_{VC} (or vice versa).

So much for the theory – let us try it out. If we plot the curve in Figure 12.1 on a logarithmic ($\log_{10}$) scale, we can see that the data form an approximately (but not exactly) straight line. We will discuss later the reasons why the line may not be precisely straight. For the sake of argument, let us draw a straight line through the later part of the data, as shown in Figure 12.2. This line has a gradient of around 3.9 K per $\log_{10}$ cycle. If the heat input during this later part of the test is, on average, 5200 W and the borehole is 110 m deep, then the heat input rate (q) is 47.3 W m^{-1}. We can use Equation 12.2 to calculate the thermal conductivity as ~2.2 W m^{-1} K^{-1}.

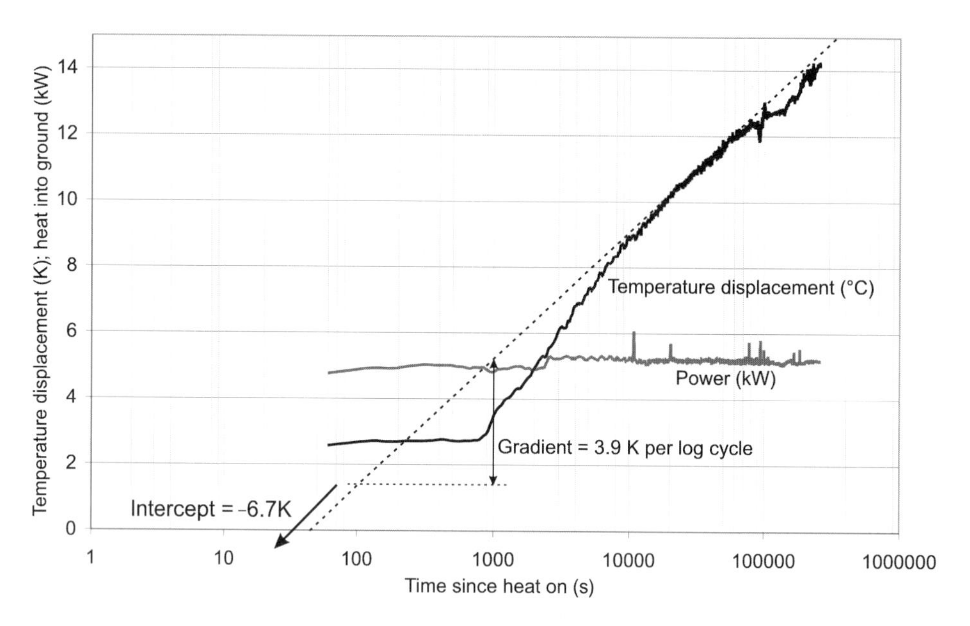

Figure 12.2 Example of output from a TRT, plotted on a log (time) scale, based on the data from Figure 12.1, provided by *GeoWarmth Ltd. of Hexham, UK*. The temperature displacement is the difference between the average fluid temperature θ_b in the ground loop and the initial average ground rest temperature θ_0.

Furthermore, we can estimate that the intercept at $\log_{10}(t) = 0$ (or $t = 1$ s) is -6.7 K. By rearranging Equation 12.3:

$$R_b = \frac{\text{Intercept}}{q} - \frac{1}{4\pi\lambda}\left[2.303\log_{10}\left(\frac{4\lambda}{r_b^2 S_{VC}}\right) - 0.5772\right] \tag{12.4}$$

Given that the borehole radius is 0.07 m, and assuming a volumetric heat capacity for the rock of 2 000 000 J m^{-3} K^{-1}, which is typical for a sandstone, we can estimate that the borehole thermal resistance is around 0.13 K m W^{-1}.

We can then re-substitute these values of λ and R_b back into Equation 12.1 and plot the temperature evolution curve to check that it provides a good fit to the observed data.

For those interested in further details of the theory and practice of TRTs, excellent reviews and assessments have been produced by Gehlin (2002), Brekke (2002) and Signorelli *et al.* (2007).

12.3 Sources of uncertainty

It is important not only to derive values of λ and R_b from the TRT but also to be able to give the client some idea of the level of confidence in these values. Signorelli *et al.* (2007) believe that the accumulated error in the results from a 'line gradient' analysis, of the type detailed in Section 12.2, can easily reach 10%. There are three main sources of uncertainty in a TRT: if we can identify and quantify these, we can begin to place a quantitative value on the uncertainty in our result:

- uncertainty in measured data: temperatures and power input;
- uncertainty in assumed input values, especially of S_{VC};
- uncertainty in the 'straight line' fit to the log-transformed data.

As regards the first factor, the acceptable levels of uncertainty and the necessary precision of measurement are defined by international guidelines (see Section 12.6). As regards the second factor – we have three unknowns (S_{VC}, λ and R_b) in our Equation 12.1, but we can only solve it to yield values of two of them. The confidence that we have in the assumed value of S_{VC} (which does not vary greatly between geological media, but which can vary by, say, 10–20%) will be reflected in the confidence that we have in our result for R_b.

It would seem (from Figure 12.2) that a major source of error in our result is the way in which we draw a straight line through the data on the plot of temperature versus log time. Clearly, in Figure 12.2, we could have given more emphasis to the early data (1000–7000 s), resulting in a steeper gradient and a lower value of λ. In particular, the value of R_b, which depends on the intercept, will be sensitive to the gradient of the chosen straight line. There are, perhaps, three reasons why we should be a little cautious of placing too much emphasis on the early data from the test:

- In the very earliest data, the 'slug' of unheated water initially present in the ground loop is still circulating in the system and has not been homogenised with the remainder of the flow (up to around 900 s or 15 min in Figure 12.2).
- Initially, the ground loop extracts heat predominantly from the grout (or other backfill) surrounding the loop, rather than from the rock itself. The zone of depressed temperature is still expanding through the borehole annulus. The slope of the temperature curve may thus reflect the properties of the backfill/grout rather than the properties of the rock itself. Only at later time will the heat flow through the grout approach an approximately stable value: the term qR_b in Equation 12.1.
- Equation 12.1 is only a mathematical approximation of a more complex expression, and is only strictly valid for times greater than $t > 5r_b^2 S_{VC}/\lambda$. In the example of Figure 12.2, this critical time is 22 000 s (or 370 min).

We can thus see why it may have been sensible to use the later data in Figure 12.2 as the basis for a straight line, although the final decision clearly has a degree of subjectivity involved. We can also see why we probably need to collect several days', rather than several hours', worth of data if we are to use graphical techniques to derive values of thermogeological parameters.

As if this were not complicated enough, external factors can also affect the measured gradient of the curve of temperature evolution. Equation 12.1 assumes, for example, that all heat transport from the borehole is by radial conduction. In reality, and especially in aquifer strata, groundwater flow may also contribute to removing heat from the borehole by advection. This effect is usually relatively minor, but it may lead to a shallower gradient on the plot of temperature versus log time, and a higher effective thermal conductivity than would otherwise have been the case (especially for groundwater flow velocities of greater than 0.1 m day^{-1}; Signorelli *et al.*, 2007). Advection with groundwater flow often manifests itself as a progressive shallowing of the gradient of the temperature versus log time curve with time, producing a characteristic upwardly convex shape (Sanner *et al.*, 2000). There are possibly signs of such a decreasing gradient with time (even after we have considered the various factors affecting early data – see above) in Figure 12.2. Could this be due to the effects of groundwater flow? We really need more test data to be sure.

Finally, in Figure 12.2 one can see a downward deviation in temperature between around 90 000 and 200 000 s into the test. This 'blip' suggests that the ground loop is losing heat at a higher rate than would have been expected. When such a feature is observed, one should examine whether this could be due to heat losses from the ground loop at the surface. Unless very well insulated, the ground-loop pipe can lose a significant amount of heat, especially on cold days, on its short journey from the top of the borehole to the test rig, and back. Conversely, on hot summer days, the loop fluid may actually heat up if exposed at the surface (and especially if the pipe is black polyethene). In this case, reflective pipe insulation is recommended.

12.4 Non-uniform geology

Our test analysis assumes that the borehole penetrates a single rock type. In reality, the TRT delivers an integrated value of thermal conductivity over the length of the borehole – effectively, a thermal transmissivity. Thermal transmissivity is defined as the product of thermal conductivity and thickness for each individual horizontal stratum (units W K^{-1}). If various strata are present throughout the borehole length, the transmissivity of each unit is added to yield a total thermal transmissivity.

For example, if our borehole penetrates 20 m of clay ($\lambda = 1.6$ W m^{-1} K^{-1}) and then 90 m of sandstone ($\lambda = 2.3$ W m^{-1} K^{-1}), the thermal transmissivity integrated along the borehole length would be $(2.3 \times 90) + (1.6 \times 20)$ W K^{-1} = 239 W K^{-1}. The mean thermal conductivity (which is the result that the test gives us) is then given by the thermal transmissivity divided by the borehole depth = 239 W K^{-1}/110 m = 2.2 W m^{-1} K^{-1}.

As the TRT only provides a value of mean conductivity, we can only deduce the thermal conductivity of the sandstone if we have information (or make an assumption) about the conductivity of the clay.

12.5 The practicalities: the test rig

To carry out a TRT, we need, in essence, the following:

- a representative, completed closed-loop borehole;
- a circulation pump for the carrier fluid;
- a source of constant heat input – this can be a large electrical resistance element, powered by a generator, or a reliable, calibrated gas burner;
- temperature sensors mounted at the well head on the upflow and downflow shanks of the ground loop;
- a means of recording data, such as a data logger or computer.

These elements are shown schematically in Figure 12.3. Figure 12.4 shows an actual test rig (using gas burners as a heat source).

The mean fluid temperature (θ_b) is calculated as the average of the upflow (θ_u) and downflow (θ_d) temperatures:

$$\theta_b = \frac{(\theta_u + \theta_d)}{2} \tag{12.5}$$

The rate of heat rejection to the ground can be calculated by

$$q = F \cdot S_{VCcar} \cdot (\theta_d - \theta_u)/D \tag{12.6}$$

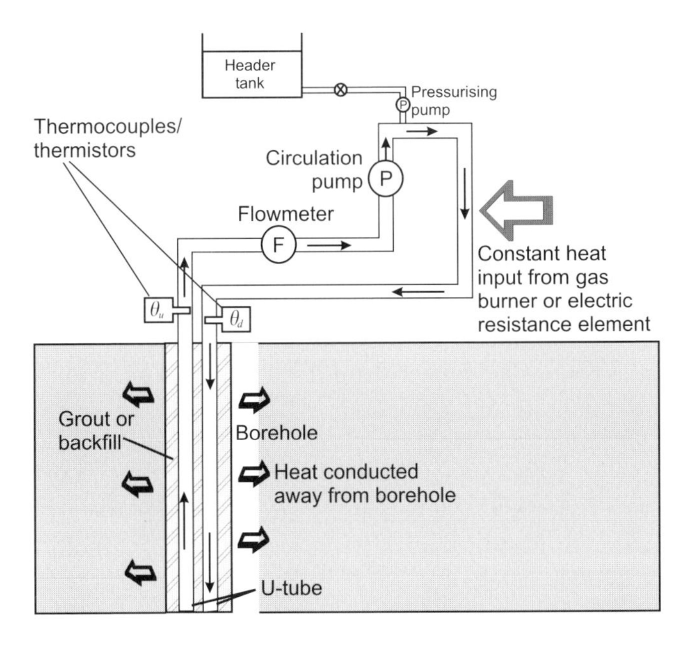

Figure 12.3 Schematic figure of the operational principles behind a TRT rig. A real rig will also include bleed valves and purging circuits to remove air and solid debris from the ground loop.

Figure 12.4 (a) The head of a closed-loop borehole installation in Sheffield entering the TRT rig (*Photo by David Banks*). (b) The ground loop running from a test borehole at the Annfield Plain site (Box 8.2, p. 200) to the test rig (*Reproduced by kind permission of Pablo Fernández Alonso*). (c) The interior of the jointly owned Kensa Engineering/GeoWarmth Ltd. test rig, which is powered by liquid petroleum gas (*Photo by David Banks*).

where F is the flow rate of the carrier fluid and S_{VCcar} is its volumetric heat capacity. D is the effective depth of the borehole containing the ground loop. Remember: if we wish to obtain a value of R_b that is representative of the operational conditions, we must ensure that the flow rate F is adequate to give turbulent flow. It should also be selected to simulate the temperature differential $(\theta_d - \theta_u)$ that would be typical for a heat pump system ($\approx$5°C). The carrier fluid itself can be a solution of antifreeze that will be utilised under operational conditions. Alternatively, as we are heating (and not cooling) the fluid, it can simply be water. Water has the advantage of a low viscosity, which decreases further with increasing temperature. It is thus easy to achieve turbulent flow conditions. Remember, when analysing the data, that the volumetric heat capacity of the carrier fluid will vary a little with temperature and may need to be corrected to give a reliable value of q.

In practice, using water as a carrier fluid in a 26 mm ID ground loop, a flow rate of around 3.2 L min^{-1} kW^{-1} is able to deliver both turbulent flow (see Box 8.1, p. 185) and a reasonable temperature differential with a heat input of, say, around 5 kW.

We should note, in passing, that we can also perform a TRT by extracting heat from the loop (i.e., chilling the carrier fluid) using a heat pump. The analysis procedures are the same as for heat addition, except that we plot $(\theta_o - \theta_b)$ against $\log_{10}t$, rather than $(\theta_b - \theta_o)$. Signorelli *et al.* (2007) believe that a 'cooling' test produces slightly more representative results, but it is rather more logistically difficult to perform. One advantage of the 'cooling' test might be that it avoids the drying out of the sediment/rock due to vapour migration during heat injection – this could affect thermal conductivity.

12.6 Test procedure

When performing a test, we need to (i) obtain a good estimate of the initial 'rest' temperature of the ground (θ_o, averaged over the borehole length) prior to the test; (ii) ensure accurate measurement of fluid temperature and flow rate (turbulent); (iii) ensure a stable power input; and (iv) continue the test for a sufficient time so that we obtain a representative temperature evolution curve that reflects conditions in the aestifer and not the borehole backfill.

There are, in fact, at least two sets of international guidelines on the performance of TRTs: those of the American Society of Heating, Refrigerating and Air-Conditioning Engineers (ASHRAE, *2002*) and those published by a working group of the Implementing Agreement on Energy Conservation through Energy Storage of the International Energy Agency (Sanner *et al.*, 2005). The salient points of these guidelines can be summarised as follows:

- There should be a delay after drilling, loop installation and loop filling, and before a closed-loop borehole is tested, such that the borehole can attain a thermal equilibrium with the ground. ASHRAE (*2002*) suggest a minimum of 3–5 days (longest delay in low-conductivity formations). Remember also that cement has an exothermic setting reaction (it releases 'heat of hydration' during curing), so that considerably longer delays may be necessary if cementitious grouts have been used.
- Any surface portions of the ground-loop circuit should be short and thermally insulated (Sanner *et al.*, 2005). Moreover, the current author suggests that they should be covered in reflective material to minimise absorption of solar radiation. The ambient (air) temperature should also be recorded during the test, so that possible interference from heat 'leakage' at the surface can be identified.
- Before testing, the initial average ground temperature should be measured either (a) by lowering a temperature probe down the fluid-filled U-tube, measuring the temperature at regular depth intervals (evenly spaced over borehole depth) and averaging these, or (b) by circulating the carrier fluid without heat input for 10–20 min, and averaging the temperature readings as the first volume of carrier fluid (corresponding to the volume of the ground loop) emerges from the upflow shank.

- The system should be purged of air before testing commences (Sanner *et al.*, 2005). It will usually also be pressurised to simulate operational conditions and to prevent pump cavitation.
- The test should be at least 36–48 hr long (ASHRAE, *2002*). Sanner *et al.* (2005) recommend at least 50 hr. In practice, the test should be long enough to yield an interpretable straight-line response (Figure 12.2) that is representative of the aestifer's thermal properties.
- The fluid temperature should be measurable to an accuracy of <0.3°C, the power input (heater plus circulation pump) to <2% and the fluid flow rate to <5% accuracy (ASHRAE, *2002*).
- A flow rate should be selected that results in a temperature difference of 3–7°C between the upflow and downflow fluid fluxes (ASHRAE, *2002*). The flow should also be turbulent (Sanner *et al.*, 2005).
- The selected heat input rate should correspond to around 50–80 W m^{-1} of drilled depth (ASHRAE, *2002*) or 30–80 W m^{-1} (Sanner *et al.*, 2005). The lower rates correspond to lower-conductivity formations.
- During testing, the standard deviation in heat power input should be less than 1.5% (and spikes should be less than 10%), according to ASHRAE (*2002*).
- If the borehole needs to be re-tested, at least 10–14 days need to elapse following the cessation of the first test, to allow the borehole to thermally 'recover' fully (ASHRAE, *2002*).

12.6.1 Prior to testing

Prior to testing, we have seen that we need to measure the initial 'rest' temperature of the ground, averaged over the borehole length. If the borehole has equilibrated with the ground, this is effectively reflected in the initial temperature of the carrier fluid. For example, in the UK, the initial temperature of the carrier fluid in a 100 m deep borehole may increase from 10°C at the surface to 12°C at the base of the U-tube, corresponding to the geothermal gradient (at least, under ideal, undisturbed conditions). θ_0 would then be $\approx$11°C.

We can determine θ_0 by running a weighted temperature sensor, attached to a measuring tape, down the U-tube (or simply down the groundwater-filled borehole annulus itself, if the borehole has not been backfilled). We might take measurements every 2 m. We would thus end up with a fluid temperature log of the U-tube and could simply take an average of these measurement (50 measurements in a 100 m borehole) to obtain a good estimate of θ_0.

A quicker, though possibly less accurate, method is to circulate fluid through the U-tube and the test rig, without any heat input. A 100 m deep, 26 mm ID U-tube contains $\approx$106 L of carrier fluid. If the carrier fluid is circulating at 16 L min^{-1}, it will take $\approx$7 min to flush the carrier fluid once around the ground loop. Thus, if we switch on the circulation pump, with no heat input, and monitor the upflow fluid temperature for $\approx$10 min, the average reading can be taken as representative of θ_0. However, take care not to continue measurements for too long – the circulation pump may itself contribute a few hundred Watts of heat to the carrier fluid, causing its temperature to increase slowly with time, even if the main heaters are not switched on.

Before we start the test, the carrier loop should be purged of any air or solid debris and the circuit pressurised. Most test rigs will be constructed with a side-circuit and appropriate valve work to purge and bleed air from the system. Moreover, the rig will usually incorporate some form of filter to ensure that solid debris does not enter the heater/burner assembly.

12.6.2 After the test

After our test is complete, our instinct will probably be to turn everything off and return home. Before we do this, we should make sure our data are safely saved in a logger or computer (and backed-up). We should check that power input throughout the test has been constant and that we have achieved an adequate temperature response.

If we are in any doubt about the quality of our data, we can perform a *thermal recovery test*. Here, we turn the heat input off, but continue circulating the fluid. The temperature curve slowly returns towards θ_0 in a shape that is the inverse of Figure 12.1 and which can be analysed in exactly the same way. The main drawbacks of the recovery test are, first, the additional time required to perform it (and time is money in our society!) and, second, the complicating factor that the circulation pump continues to add a small amount of heat to the carrier fluid during the recovery. The latter effect can, however, be factored into the recovery test analysis, provided we can estimate the magnitude of heat input from the pump.

12.7 Non-constant power input

If we have been unable to maintain a constant power input over the course of the test (e.g. because the efficiency of heat transfer from a gas burner to the carrier fluid decreases slightly as the carrier fluid's temperature increases), do not despair. Provided we know how the power input has varied with time (and we can calculate this from Equation 12.6), we can still solve the maths. We need to do this by inverse modelling techniques, however, and for this we must use a computer. Fortunately, Shonder and Beck (2000) of the USA's Oak Ridge National Laboratory have developed a neat piece of software (named 'gpm' – Geothermal Properties Measurements) to fit parameters to a data set by inverse modelling, producing output in terms of ground thermal conductivity, θ_0 and borehole thermal resistance, all with associated estimates of uncertainty.

12.8 Analogies with hydrogeology

The hydrogeologists among you will have spotted that the TRT is the thermogeological analogue of the water well pumping test. In the former, we stress the aestifer by imposing a constant heat input (q), and measure its response in terms of temperature change (θ_b). In the latter, we stress the aquifer by extracting a constant rate of groundwater flow (Z), and measure its response in terms of head change. We analyse aquifer pumping tests by a method exactly analogous to that described in Section 12.2: the Cooper and Jacob (1946) method, which is an approximation of the Theis (1935) equation.

Borehole thermal resistance also has a hydrogeological analogue – well loss or well inefficiency (Bierschenk, 1963; Hantush, 1964). Whereas the thermal inefficiency of a borehole is expressed in Equation 12.1 as the term qR_b (i.e. a linear relationship with q), hydrogeological well loss is non-linear and commonly expressed as a power law such as CZ^2 or CZ^n. Here, C and n are simply constants (see Section 7.7.1, p. 160).

13

Environmental Impact, Regulation and Subsidy

The ground remembers major events in its surface temperature history ...
Lachenbruch and Marshall (1986), also cited by Chapman (2001)

13.1 Introduction

Thus far, this book has argued that the heat stored in the subsurface environment is a resource that can be utilised for space heating and cooling. Moreover, we have seen that it is a resource that can be utilised with a rather low 'carbon footprint' and minimum visual impact. Finally, it is a resource that can be, in many cases, economically competitive with conventional space heating and cooling solutions.

Surely, this technology should be welcomed with open arms and our policy-makers should be removing all barriers to its uptake? If only life were so simple! In practice, the installer of any ground source heat pump (GSHP) system will need to overcome certain bureaucratic and regulatory hurdles in many countries. Only a very generic snapshot of the current regulatory status within the European Union and elsewhere will be presented in this chapter. The uptake of ground source heat technology proceeds rapidly, whereas regulations and policy vary between nations and are changing fast.

In many areas of ground source heat technology, there is currently a 'regulatory vacuum' – laws simply do not exist in many countries to govern the use and abuse of the subsurface heat resource. In the absence of binding legislation, regulators and professional trade organisations are seeking to develop codes of best practice. As examples of these, take a look at the recent draft policy document from the Environment Agency of England and Wales (EA, 2007), the guidelines developed for the US Industry by the

NGWA (McCray, 1998, 1999) and the codes of good practice developed by the International Ground Source Heat Pump Association (IGSHPA). Even in Russia, despite the relatively low uptake of ground source heat systems, there is an architectural guideline (Vasil'ev *et al.*, 2001; Vasil'ev, 2003) whose title roughly translates as 'Management of heat pump applications utilising secondary energy resources and non-traditional renewable energy resources'.

In this chapter, we will try to identify the primary impacts that regulators and developers of 'best practice' will seek to address. These typically take three forms:

- Regulations regarding buildings' thermal efficiency and the performance of heat pump systems,
- Concerns over groundwater contamination and the hydrogeological impacts of ground source heat schemes,
- Concerns over the impact of changing the temperature of the subsurface by extraction or rejection of heat ('thermogeological impact').

13.1.1 Constraints on building standards

This book is primarily focussed on the geological aspects of ground source heat systems. However, any thermogeologist needs to be aware that, as a result of international treaties, such as the Kyoto Protocol, many nations are seeking to improve energy efficiency and to restrict CO_2 emissions to meet specified future targets. For example, within the European Union, Directive 2002/91/EC on the Energy Performance of Buildings (EU, 2002) seeks to enhance the thermal performance of buildings and their energy (and thus, CO_2) efficiency. The Directive emphasises the importance of insulation in the winter and thermal performance in the summer, with an emphasis on passive rather than active cooling. The Directive (like all EU Directives) is implemented by translating its fundamental principles into national legislation in the EU member states. In the United Kingdom, for example, it has been implemented in the 'Buildings Regulations, Part L', which describe in detail how to calculate U-values, insulation standards, heat losses and the cost and CO_2 performance of different space heating solutions (via SAP, 2005). This sounds daunting but, in fact, works in favour of ground source heat technology. While conventional space heating technologies may struggle to meet the demands of national buildings regulations under the influence of Directive 2002/91/EC, ground source heat provides one tool for developers to satisfy these regulations.

In addition to satisfying any national buildings laws and regulations, a developer may need to gain planning permission from a local or regional authority. Planning authorities may impose their own (sometimes apparently rather arbitrary) constraints to ensure that new developments have a minimised CO_2 footprint. For example, at the time of writing, several local authorities in the United Kingdom, including London, operate a policy of demanding a certain percentage of renewable energy (10% in London) to be incorporated into certain new developments. Again, this plays to the strengths of ground source heating and cooling solutions and can be seen as a major reason for the recent stimulation of the GSHP market in Britain.

If we focus on heat pumps as a means of achieving such policy objectives, however, we need to be sure that they are performing efficiently. In Canada, energy efficiency regulations state that all ground- or water-sourced heat pumps of less than 35 kW must have a COP_H greater than 3.0 (Bouma, 2002). In the UK, in order to claim tax relief, a water-to-water heat pump must have been shown under test conditions (standard EN14511) to have a $COP_H > 3.7$ and a $COP_C > 3.0$, while for water-to-air heat pumps, the criteria are 3.4 and 3.3, respectively (ECA, 2006a,b).

13.1.2 Concerns over hydrogeological impact

Environmental regulators often have concerns over the potential hydrogeological impact of borehole drilling and ground source heat technology on aquifers. For closed-loop systems, as there is usually no actual abstraction or discharge of groundwater, or any actual discharge of pollutants (only the future possibility of a leakage), the legal tools may not exist for environmental regulators to control such systems. Regulators will thus be concerned that uncontrolled drilling of closed-loop systems, possibly on contaminated sites, by inexperienced drillers, could result in the pollution of aquifers by migration of surface contaminants down poorly constructed boreholes, the connection of separate aquifer horizons and the unwitting penetration of artesian aquifers (Figure 13.1; Section 13.3). Furthermore, regulators may be concerned about the pollution potential from leakage of refrigerants (and even antifreeze solutions) into the ground from closed-loop systems. Be aware, however, that in some countries, closed-loop systems may fall under specific laws relating to geothermal energy (even if these laws were designed with high-enthalpy systems in mind). Elsewhere, closed-loop systems may be regarded as 'wells' and be regulated under well-drilling regulations. In Germany, the fact that closed-loop systems affect the temperature of groundwater means that they can be regulated under the water law (Rybach, 2003b).

As regards open-loop systems, regulators will typically have some existing legal framework to exercise control: typically water resources, control of pollution, mining or even specific geothermal resources legislation. Within the European Union and in many other countries, substantial abstractions from, or discharges to, groundwater bodies will require some form of license or permit. The environmental regulator will usually seek to limit detrimental effects on aquifers, and to protect the interests of the environment and other users. Thus, in the case of open-loop systems, a regulator may seek to

- prevent wastage of groundwater reserves by placing limits on the net abstraction rate permissible (or insisting on 100% re-injection of used water back to the original aquifer);
- prevent detrimental changes to the temperature or chemical quality of surface water recipients receiving discharges from open-loop systems;

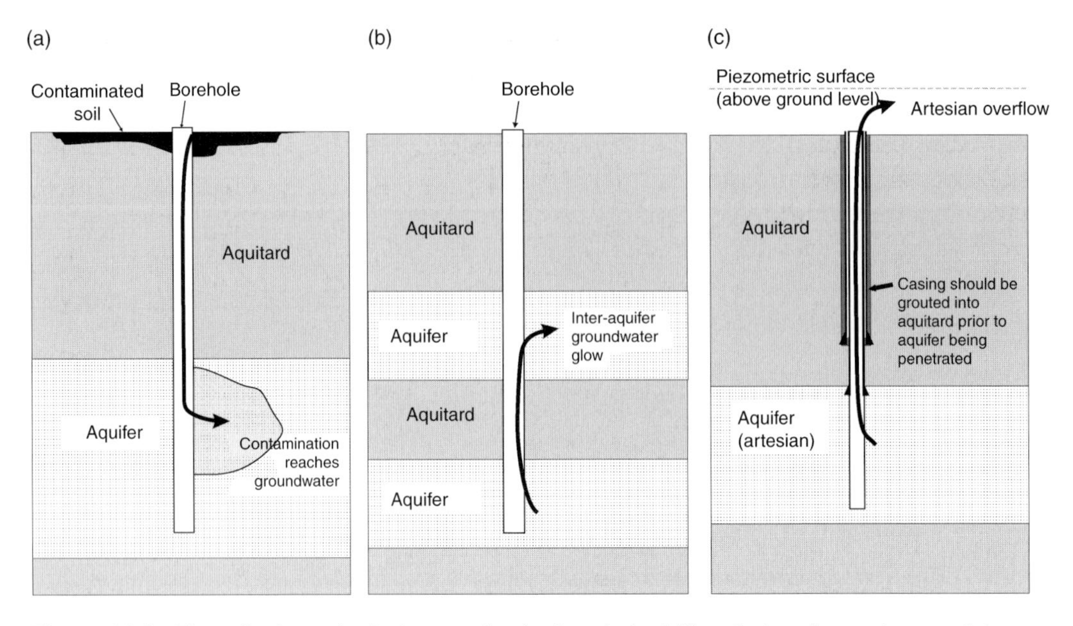

Figure 13.1 Three hydrogeological scenarios for borehole drilling that make environmental regulators nervous: (a) drilling in contaminated land can provide a conduit for contaminants to enter an aquifer; (b) drilling through two aquifers can provide a hydraulic connection and leakage of water from one aquifer to another (unless the borehole is either cased through the upper aquifer, or securely grouted/sealed in the section through the intervening aquitard); (c) inadvertent penetration of an artesian aquifer. If casing has been grouted into the overlying aquitard prior to penetration, any artesian flow can be controlled.

- prevent or limit widespread (or sometimes even localised) changes in aquifer water levels as a result of abstraction or re-injection;
- prevent large-scale changes in aquifer groundwater temperatures beyond the immediate location of the ground source heating or cooling schemes. Thus, the regulator may stipulate operational constraints on 'doublet' systems and may limit (a) the net amount of heat that can be abstracted or re-injected or (b) the maximum or minimum acceptable temperature of re-injection of wastewater.

13.1.3 Concerns over 'thermogeological impact'

Clearly, even outside aquifer units, the abstraction or injection of heat via closed-loop systems has potential for localised changes in ground temperature that could conceivably have detrimental effects. These include

- Extraction of heat causing ground freezing and frost heave. Freezing of ground causing damage to built structures, buried services or plant roots.
- Warming of ground causing thermal expansion with geotechnical consequences, or desiccation of soils due to vapour migration.

- Thermal interference between neighbouring closed-loop schemes, decreasing the efficiency of the schemes for their respective users.

In many (but not all) countries, the concept of heat in the shallow subsurface as a resource is too new for there to exist many legal tools to regulate the thermogeological impact of closed-loop ground source heat systems at the present time.

13.2 Heat as a pollutant

We have been dealing with subsurface heat as a resource, but it is widely recognised that heat *can* also be a pollutant. For example, the European Water Directive defines pollution as 'the direct or indirect introduction, as a result of human activity, of substances or heat into the air, water or land which may be harmful to human health or the quality of aquatic ecosystems or terrestrial ecosystems directly depending on aquatic ecosystems, which result in damage to material property, or which impair or interfere with amenities and other legitimate uses of the environment' (EU, 2000). Interestingly, this definition does not allow for the cooling down of a water environment (by extraction of heat) to be defined as pollution, although this activity could potentially damage ecosystems, damage material property (through frost heave) or reduce the utility of the environment.

Even if we accept that heat is a pollutant, difficulties soon arise because the EU Water Framework Directive, its daughter Groundwater Directive and its incarnations in national legislation tend not to regard heat as a 'substance'. The Directive and subsidiary legislation are typically concerned with controlling the discharge of chemical 'substances' to water and monitoring the 'chemical status' of water bodies. Thus, although heat is recognised as the potential cause of pollution, the legislation is far from unambiguous in giving regulatory bodies powers to control it as a polluting 'substance'. Thus, in many cases, the management and regulation that environmental authorities can exercise over ground source heat schemes may be limited to:

i. Regulation of groundwater abstractions and spent water discharges. Temperature (or other) conditions may be attached to permits to abstract groundwater or to discharge water to surface waters or to an aquifer.
ii. The planning process, where environmental authorities may be statutory consultees to new developments. Inappropriate ground source heat schemes may thus be identified and regulated via planning consents.

Inevitably, even within the European Union, there is considerable variation in how ground source heat is regulated (Rybach, 2003b; EGEC, 2006), for several reasons:

- Different EU member states may interpret the Water Framework Directive differently.

- They may even have specific national legislation over and above the requirements of the Directive, enabling them to regulate the use of subsurface heat in greater detail.
- They may have mining or geothermal resources laws under whose jurisdiction ground source heat is judged to fall.

In general, however, there does appear to be something of a regulatory vacuum within the European Union around some types of ground source heat scheme at the current time (especially closed-loop systems). In the coming years, regulators will need to decide whether stricter regulation is necessary: if they do, they may risk smothering a new 'green energy' technology in the cradle.

Informally, some regulators within the United Kingdom have expressed the opinion that smaller closed-loop schemes probably do not require a strict degree of state regulation (although self-regulation by professional trade bodies should be encouraged), while larger closed-loop schemes can be controlled via the planning process. Regulators appear most concerned with mitigating the possible hydrogeological and heat pollution impacts of the larger open-loop schemes, and these typically fall fairly and squarely under existing water resources and control of pollution legislation.

13.2.1 Thermogeological 'background' conditions

It is axiomatic that, in order to be able to identify the effects of anthropogenic pollution and environmental impact, we need first to understand the 'background conditions' (Banks, 2004). For example, if we detect high concentrations of arsenic in an aquifer, we might be tempted to jump to the conclusion that it is due to 'pollution' from a local wood impregnating works. However, we must first establish that the groundwater does not contain naturally high background concentrations of arsenic before we can begin to blame anyone. The same applies to thermal pollution. If we spot an area of apparently elevated groundwater temperatures in a city, we should not automatically assume that it is heat pollution from an open-loop cooling scheme. We first need to identify the natural distribution of groundwater temperatures in the aquifer beneath the city (and it is a three-dimensional distribution).

Furthermore, we need to recognise any temporal trends in ground or groundwater temperature. Although ground temperatures are generally fairly stable, they are not completely static and they reflect a dynamic global and local climate. We have already seen that temperature profiles in boreholes still bear the thermal signatures of the past ice ages and of isostatic crustal rebound (Šafanda and Rajver, 2001). Ground temperatures are also slightly affected by recent global climate change: they would be expected to reflect steady increases in annual average air temperature (Lachenbruch and Marshall, 1986; Beltrami and Harris, 2001; Bodri *et al.*, 2001; Chapman, 2001; Majorowicz *et al.*, 2006). They might also be affected by changes in the depth and duration of snow cover and the timing and amount of recharge (Box 3.7, p. 47).

Finally, we need to recognise the possibility of other sources of 'diffuse' thermal pollution, in addition to localised sources such as ground source heating and cooling schemes. The 'urban heat island' effect, where the air temperatures in cities may exceed rural temperatures by several degrees (Henson, 2006), due to changes in the albedo, radiative and storage properties of the urban environment, may be reflected in the temperature of the subsurface. Moreover, surprisingly large changes in ground temperature may be ascribable to all the heat lost by downward conductive heat flow from warm house basements and floors, buried services or even by leakage of warm effluent from sewers.

13.2.2 Diffuse thermal pollution: urban temperature change

While we have been busy monitoring groundwaters for minute concentrations of trace organic contaminants in city environments, it seems that hydrogeologists may have overlooked one of the most obvious forms of ground pollution to affect urban areas – diffuse thermal pollution. In a recent, but remarkable, piece of detective work, Ferguson and Woodbury (2004) have mapped groundwater temperatures beneath the city of Winnipeg, Canada. The regional aquifer temperature outside the city is typically below 6°C. However, below the urban area, temperatures were found to be some 2–3°C higher than this background, and in some cases 5°C higher (Figure 13.2). Ferguson and Woodbury considered whether this could be due to global or local atmospheric warming: they found, however, that Winnipeg's annual average air temperature had only risen by around 1°C over the past century (from +1–2°C to +2–3°C). They also considered whether the increase in aquifer temperature could be due to 'thermal plumes' from active ground source cooling schemes (where warm wastewater is injected to the aquifer). They found that, while this could have important local effects, it could not account for the regional anomaly beneath the entire city area (Figure 13.2). They eventually concluded that the effect was largely due to simple conductive heat leakage from the floors and basements of buildings into the ground. Unlikely though this seems, Ferguson and Woodbury marshalled evidence from borehole temperature logs to demonstrate that, beneath old built-up parts of the city, the geothermal gradient was reversed (i.e. temperatures decrease with depth) down to depths of up to 100 m. This demonstrates that there has been downward conduction of heat into the subsurface from the surface during the past century or so. The downward speed of this temperature 'signal' is governed by the thermal diffusivity of the rock (which in most cases is of the order 10^{-6} m^2 s^{-1}). Modelling of downward heat conduction or 'leakage' from buildings demonstrated that this effect is able to account for the observed gradient reversals. Does this constitute 'thermal pollution'? In one sense it does, as it is a significant anthropogenic disturbance of natural conditions. In another sense, it may be beneficial – the elevated ground temperature allows ground source heating schemes to perform more efficiently. The downside, of course, is that ground source cooling schemes would perform less efficiently.

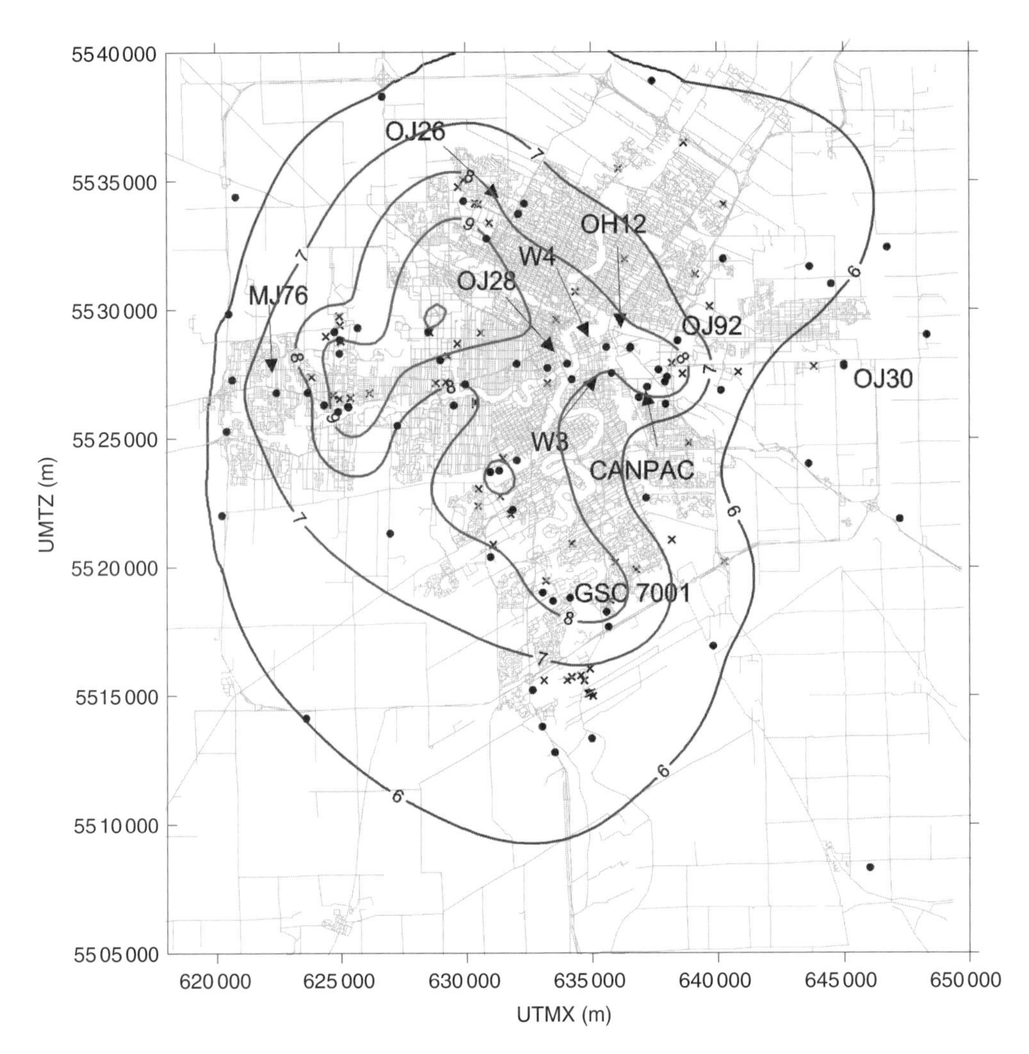

Figure 13.2 Contour map of the Winnipeg area showing temperatures in °C (nominally at ~ 20 m depth) in the Upper Carbonate aquifer, and locations of temperature measurements (dots). The regional background aquifer temperature is believed to be <6°C. The road network is shown in grey. Locations of injection wells are shown as crosses (after Ferguson and Woodbury, 2004). *Reproduced by kind permission of © American Geophysical Union.*

The Winnipeg story is far from the only example of a 'thermogeological urban heat island': similar results are being reported from Sweden (G. Hellström, *pers. comm.*), Osaka, Japan (Taniguchi. and Uemura, 2005), Ireland (Allen *et al.*, 2003; Goodman *et al.*, 2004) and Gateshead, UK (Banks *et al.*, in press). The author suspects that reported groundwater temperatures of >14°C from the Chalk aquifer below London *may* indicate a similar effect.

13.3 Environmental impact of closed-loop systems

Closed-loop ground source heat schemes are, in many countries, pretty much unregulated. The larger of such schemes, especially those involving redevelopment of 'brownfield'[1] sites, will often form part of an application for planning permission and may thus be subject to some form of control by environmental regulators. Smaller schemes may, however, go beneath the regulators' 'radar'. There is a fear that some borehole schemes may be constructed by contractors who are more accustomed to drilling shallower site investigation boreholes, rather than by more expensive specialist well drillers. Regulators fear that such installers may not necessarily be fully aware of the various 'searches' that need to be made before commencing drilling a deep borehole (Box 13.1). Moreover, they may not be aware of the various hydrogeological risks that may be encountered during uninformed drilling.

BOX 13.1 Before Drilling a Borehole

There are various precautions that should be undertaken before drilling a borehole. Some of these may be prescribed by national legislation, others are simply good practice and common sense. If you get it wrong and drill through an artesian aquifer, an underground railway or a government nuclear bunker, you can expect a call from an angry environmental regulator/insurance assessor/secret service agent (delete as appropriate!).

The following check list is not exhaustive and the possible risks associated with drilling are described in more detail by Misstear *et al.* (2006).

- Obtain permission from the landowner/site occupier.
- Check that the rig has access to the site (e.g. bearing capacity of the access road).
- Assess any geotechnical risks from drilling close to buildings.
- Is the site likely to be contaminated (if yes, develop a specific drilling plan in liaison with local and environmental authorities).
- Is there an issue of noise or vibration? What working hours will be permitted?
- Develop a health and safety plan (see Misstear *et al.* 2006).
- Do you need a permit from the environmental regulator, a water authority or geological survey to commence drilling? Even if you do not need permission to drill, it is often a very good idea to contact the environmental authority informally to let them know your plans and to obtain advice.

continued

[1] A 'brownfield' site is a former industrial site, typically in an urban environment, that is being redeveloped. There is a risk of contaminated soil or water at such sites.

BOX 13.1 (*Continued*)

- Are there likely to be any buried services beneath the site: gas mains, electricity lines, telephone or communications lines, water mains, drains or sewers, service tunnels, basements, bunkers? Contact the relevant utilities companies to obtain maps of buried services and to obtain permission to excavate/drill if necessary.
- Are there any overhead power lines at the site? What is the minimum safe distance to erect a drilling mast from such lines? Is the rig earthed?
- Are there any transport tunnels (e.g. underground railways) beneath the site?
- Is there any likelihood of drilling through mine workings (beware of contaminated mine water or methane gas)? In the United Kingdom, for example, if you are drilling through coal seams or mined strata, you will need permission from the Coal Authority.
- Are there any water wells or ground source heat schemes in the vicinity that might be affected by your drilling or your ground source heat scheme?
- Are you planning to drill within the inner protection area(s) of a potable groundwater well (where certain activities and installations may be prohibited)?
- How will you dispose of drilling cuttings (especially if contaminated)?
- What is the hydrogeology beneath the site? Do you know which strata you will encounter? Is there any risk of connecting independent aquifer horizons?
- Is there any risk of encountering artesian groundwater conditions? Do you have a contingency plan if this occurs?
- On completion of drilling, send a drilling log and site location plan to the national repository of geological information (often, the Geological Survey). This may be required by law (in the United Kingdom, it is obligatory for any borehole deeper than 15 m to be reported to the British Geological Survey). In any case, it is good practice: if all drillers report data diligently, an excellent database can be acquired that is of benefit to all.

Even in the case of horizontal 'trenched' closed-loop systems, many of the above points apply, especially those regarding contaminated land and buried services.

13.3.1 Hydrogeological risks

There are probably three main types of hydrogeological risk from unregulated drilling that concern environmental authorities. The first of these relates to drilling at a contaminated site. The borehole may constitute a pathway whereby contaminants can migrate from the surface to an aquifer, during or after drilling (Figure 13.1a). In order to minimise this risk:

- If you are drilling on a suspected contaminated site, develop a drilling plan in conjunction with the environmental authority.

- Always install a string of steel surface casing (see Figure 6.16, p. 133, as an example of steel casing around a closed-loop borehole) at least a few metres in length, preferably grouted into 'bedrock' beneath any loose surficial materials or made ground, or at least grouted into natural strata beneath any zone of possible surficial contamination. The annulus behind the casing should be securely grouted with a low permeability grout (or bentonite slurry) to ensure that the space behind the casing does not constitute a pathway for flow of surficial, contaminated water.
- If the borehole top is secured with such a length of casing then, strictly speaking, there is no overwhelming need to backfill the borehole around the closed-loop with a grout or bentonite/sand slurry. Indeed, in Scandinavia, it has been regarded as acceptable practice to simply 'dangle' a U-tube in a groundwater-filled borehole (Rosén *et al.*, 2001). Backfilling with a low-permeability, high-thermal-conductivity grout does, however, provide an additional element of protection against contaminant transport and may be required by some national regulations (e.g. in some states of the US).

The second type of hydrogeological risk relates to the unwitting interconnection of two separate aquifer horizons by the borehole. Such a connection may allow flow of groundwater from one horizon to the other along the borehole (Figure 13.1b). This may play havoc with the management of groundwater resources and raise issues of geochemical compatibility (e.g. if a saline aquifer flows into a freshwater one). Indeed, aquifer interconnection may constitute a case of illegal abstraction (from one aquifer) and discharge (to the other) under water resources or control of pollution legislation. The message is clear: know your geology before you start drilling. If there appear to be multiple aquifers below your site, obtain advice from the authority responsible for groundwater management, and develop a drilling plan to ensure that either (a) the upper aquifer horizon is sealed off by a string of casing (which may be temporary, if the hole is subsequently grouted) or (b) a low-permeability grout or bentonite seal is emplaced in the borehole between the aquifer horizons.

The third hydrogeological risk is that of drilling unwittingly through an artesian aquifer horizon. If this occurs, and adequate casing has not been installed, it can be a nightmare trying to control an artesian flow (Figure 13.1c). Uncontrolled overflow can significantly deplete groundwater resources and *may* be regarded as an illegal abstraction under water resources legislation. If there is a risk of artesian conditions, ensure that your driller has a contingency plan: he should ensure that a string of casing is securely grouted into the confining aquitard before the artesian aquifer is penetrated. Even if this is done, you should carefully consider how you can possibly install a ground loop in a strongly artesian borehole (see Figure 13.3).

13.3.2 Thermogeological risks

When we abstract heat from a closed-loop borehole, we deplete the temperature in a zone around the borehole. This zone does not extend very far: the radius of influence typically does not reach much more than 20 m (Figure 10.6, p. 225; Gabrielsson *et al.*, 2000), and most of the long-term temperature drop occurs within 10 m of the

Figure 13.3 An artesian borehole at High Wycombe, UK. Even when steel casing has been securely installed in an artesian borehole, stemming the flow can be a nightmare! *Photo by David Banks.*

borehole. It is possible to conceive of a situation, however, where two independent closed-loop borehole schemes are drilled in close proximity to each other such that thermal interference occurs. In such a case, the owner of the first system could claim that the second scheme has infringed upon his or her existing right to extract heat. As far as this author is aware, such cases have not yet been documented, and concepts of 'ownership' of the subterranean heat resource vary widely from country to country. In the United Kingdom, by analogy with hydraulic interference between adjacent groundwater wells, the first owner may be able to make a case for 'Nuisance' or negligence under Common Law, but the legal situation is far from clear (and will remain so until a test case is brought).

Some closed-loop schemes (which may operate at sub-zero temperatures) may create a 'halo' or 'sausage' of frozen ground around the ground loop (Section 8.6, p. 198). This may have benefits in terms of the enhanced thermal conductivity of frozen ground, but may also create geotechnical impacts in terms of ground disturbance or frost heave. This is most likely to be a concern in the immediate vicinity of surface or subsurface structures. If the ground loop is close to any sensitive structures, the operator may be advised to design and operate the scheme so as to avoid extensive ground freezing. Furthermore, VDI (2001a) advise that the cold parts of a closed-loop system should be at least 70 cm from any water supply pipe or buried sewer/drain.

For closed-loop systems operating in cooling mode, one should be aware of the potential that exists for vapour migration and progressive drying (and even shrinkage) of soils (Section 8.6, p. 198). Significant heating of the ground can also lead to consolidation and creep settlement in clayey, unconsolidated soils (Gabrielsson *et al.*, 2000) or to thermal expansion of rocks (Section 10.9.1, p. 248)! According to Skarphagen (2006), a terrain 'heave' of 12–17 mm was measured in the early 1980s above a closed-loop thermal storage array in Luleå, Sweden, that was operating at a temperature around, or in excess of 60°C. Linear thermal expansion coefficients of rocks measured in 'normal' temperature ranges include: 3 to 7×10^{-6} K^{-1} (granite and sandstone; Park *et al.*, 2004) and 5.2 to 7.4×10^{-6} K^{-1} (mean values for granites; Janio de Castro Lima and Braga Paraguassú, 2004).

As regards impacts on vegetation, the author is unaware of such effects being documented for responsibly designed systems operating at relatively low temperatures. Ball *et al.* (1993) and Skarphagen (2006) comment that a correctly dimensioned horizontal closed-loop system in a trench should not delay the winter thaw and the growing season of a lawn above it by more than around 2 weeks. However, Skarphagen also recommends avoiding constructing closed-loop systems within the radius of the crown of a tree (which will often approximately correspond to the radius of the root bole). It is unclear if this recommendation is simply to avoid damage to roots during excavation or drilling, or to avoid damage due to temperature changes during operation. For systems rejecting heat at high temperatures, VDI (2001b) suggest that changes in microbiological communities in soils may occur and that a risk assessment should be carried out to identify risks to macroflora. At temperatures in excess of 60°C, soil sterilisation may begin to occur.

13.3.3 Chemical risks

We have already mentioned the risks of drilling through contaminated land (Section 13.3.1), but let us also remember that we, in closed-loop systems, are often circulating potentially contaminating chemicals through the ground. In direct circulation (direct expansion) closed-loop systems, we are using a refrigerant in the ground loop. If it is based on fluorinated hydrocarbons, it may be deemed to fall under the category of halogenated hydrocarbons, which figure on EU nations' 'List 1' of priority pollutants – that is, substances that should be prevented for entering groundwater. It can be argued that, as long as the closed-loop is intact, there is no leakage and no offence being committed, but it certainly has the possibility of making some environmental regulators nervous.

For this reason, indirect closed-loop systems may be preferable to some regulators. Indirect systems typically use solutions of antifreeze, based on salts, glycols or alcohols, as a carrier fluid. These are typically relatively environmentally benign and (in the case of alcohols and glycols) readily biodegradable in the event of a leakage. From a typical, well-constructed and well-operated closed-loop stream, significant leakage from a polyethene ground loop is unlikely and the quantity of antifreeze that would be released in such an event would be modest.

Ethylene glycol is often used in European indirect closed-loop systems: it has better physical properties than most alternative carrier fluids, but is somewhat toxic. Indeed, the IGSHPA have tended to prefer potable water, potassium acetate or propylene glycol solutions due to their lower toxicity (Den Braven, 1998), although there may be penalties to pay in terms of increased corrosivity or viscosity. This preference is reflected in the regulations governing the use of carrier fluids in some US states (see Section 13.6.5). Particularly for closed loops installed in sensitive aquatic environments, such as lakes, surface waters or intensively utilised aquifers, in the proximity of drinking water wells, consideration should be given to the toxicity of the antifreeze used.

13.4 Environmental impact of groundwater-based open-loop systems

Many of the comments noted above in Section 13.3 and Box 13.1 are relevant also to groundwater-based open-loop systems. Additionally, we must remember that we are physically abstracting groundwater from an aquifer and discharging it to another recipient or back into the aquifer. In EU member states (and many other nations), these activities require a permit or license, at least if they involve water quantities greater than a certain threshold value.

Before commencing an open-loop abstraction well, you may be required to apply for a drilling consent from the water, mining or environmental regulator (in England and Wales, for example, this is a 'Consent to Investigate a Groundwater Source' from the Environment Agency). This may come with conditions attached; you may be required:

- to carry out a survey or risk assessment of other nearby wells and environmental features (springs, wetlands) that rely on the aquifer;
- to perform a program of test pumping;
- to monitor sensitive aquatic features during test pumping.

You may then have to apply for a permit or license to abstract groundwater. Whether you obtain this depends on whether your program of test pumping has demonstrated that the abstraction will not have a significant deleterious effect on other aquifer users or the environment, and on whether there are perceived to be available groundwater resources in the aquifer. In intensively utilised aquifers, all the available resources (as defined in regulators' catchment management plans) may already be committed. In this case, you will be refused a permit for net abstraction of groundwater. However, if you propose a well doublet scheme, where the thermally 'spent' groundwater is re-injected back to the original aquifer, the regulatory agency may be more inclined to consent your abstraction.

To take a few examples: the Hebburn Eco-Centre's open-loop scheme (Box 4.2, p. 63) abstracts water from a saline aquifer, which has little water resource value and no great environmental significance. It was regarded as unproblematic to permit this abstraction of water, with simple discharge of spent water to the River Tyne's estuary.

The Chalk of London has, for some time, been regarded as an under-abstracted aquifer. Indeed, due to the closure, in the mid-to-late twentieth century, of much industry that formerly abstracted Chalk groundwater, there was concern over the rate at which groundwater levels were recovering towards their pre-industrial condition (Banks, 1992). For a period, the regulatory authorities tried to encourage increased abstraction in an attempt to stabilise the rise in groundwater levels (and thus protect underground structures from flooding). Indeed, Ampofo *et al.* (2004) document a number of open-loop groundwater cooling schemes in London, all of them with spent water being run to sewers or to the River Thames, rather than being injected back into the Chalk. More recently, however, the rate of abstraction has increased to a level where the Environment Agency is once again beginning to limit abstraction licenses in several areas of the city (EA 2005, 2006). Those proposing open-loop groundwater heating or cooling schemes are now being encouraged to re-inject spent water back to the Chalk.

The London Chalk is a re-enactment of the situation that took place in Brooklyn and Long Island in the United States in the 1920s and 1930s (Kazmann and Whitehead, 1980). Here, there was reportedly a rapid uptake of the use of groundwater for providing cooling for buildings. The spent groundwater was simply run to waste, until the local authorities became concerned that groundwater levels in the aquifers were being depleted. Users of groundwater were then encouraged to return spent water to the aquifers. Problem solved? Not wholly! Kazmann and Whitehead report that concern then steadily grew over the possibility of regionally rising water temperatures within the aquifer system.

13.4.1 Thermal pollution of groundwater

If we are operating an open-loop groundwater-based heating or cooling scheme *without* re-injection, we will typically be rejecting spent cold or hot water to an estuary, a surface water body (river, stream or lake), a sewer or even another, different aquifer unit.

If we reject the water to a sewer, we may not need a permit from an environmental regulator, but we will need permission from the sewer owner/operator (often a utility company) and will typically need to pay a charge.

If we reject water to a natural surface or groundwater body, we will often need a permit (a 'discharge consent' in British parlance) from the environmental regulator. The temperature at which we can discharge water will usually be controlled (EGEC, 2006). To obtain such a permit, we will need to assure the regulator that the water we are discharging will not

- cause thermal pollution of the receiving water: that is, the resulting temperature difference in the water body must not be so great that it has significant negative impacts on the environment, on ecosystems or on downstream users of that water.
- cause unacceptable chemical change in the receiving water. Remember that groundwater, even if uncontaminated, can have a very different chemistry to surface water.

It may be more saline, more reducing, lower in dissolved oxygen or even contain natural substances (such as hydrogen sulphide) that are toxic.
- contain chemical substances whose discharge to the environment is prohibited.
- cause unacceptable changes in water levels or flooding risk in the receiving surface water or aquifer.

In a well 'doublet' system, spent water from the heating or cooling system is re-injected back to the original aquifer via an injection well some distance away from the abstraction well. Here, there is no net abstraction of water. The chemical quality of the water being returned should be the same as that being abstracted. However, the issue of heat migration (thermal pollution) remains.

There is both an internal and an external risk of thermal pollution. The internal risk of warm re-injected water from a cooling scheme, or cold water from a heating scheme, flowing back to the abstraction well ('thermal feedback or breakthrough') has been dealt with in Chapter 7 (Figure 7.10, p. 179). This is essentially a risk for the owner of the scheme rather than for the environmental regulator. If it is excessive, then the system will ultimately lose efficiency and fail.

We have also seen that, unless the injection well is situated directly up the groundwater gradient from the abstraction well (Figure 7.8, p. 171), there will usually be some 'leakage' of heat (or 'coolth' from a heating scheme) from the doublet. This will form a thermal 'plume' of warm or cold water migrating down the hydraulic gradient (Figures 7.6, p. 167, and 7.7, p. 170). The plume will be attenuated by dispersion and dilution with ambient groundwater as it moves down-gradient, and some heat may be lost by conduction to overlying or underlying strata. However, there is a risk that the plume will eventually impact upon another groundwater user (a potable water supply well that will suffer an increase in temperature, or another ground source cooling scheme whose efficiency might be impaired) or an environmental recipient (a spring or wetland, whose dependent ecosystems may be temperature-sensitive). The job of the environmental regulator should be to assess whether such potential downstream 'risk recipients' exist, and whether the magnitude of any thermal impact on them is likely to be acceptable or not.

Ferguson and Woodbury (2005) demonstrated that individual open-loop 'doublet' ground source cooling schemes in the dolomite/limestone aquifer beneath Winnipeg, Canada, were susceptible to the internal risk of 'thermal breakthrough', rendering them unsustainable in the long term (Figure 7.10, p. 179). The same authors (2006) also considered four groundwater cooling schemes operating in one region of Winnipeg and observed that the thermal plumes related to the schemes were overlapping with each other to create a larger coherent region of elevated groundwater temperature. In other words, the four schemes were thermally and hydraulically interfering with each other. They concluded that there is a limit to the number of groundwater-based doublet cooling schemes that a given aquifer unit can support. Clearly this will depend on the properties of the aquifer and on the water and heat fluxes from each scheme, but in the Winnipeg case, Ferguson and Woodbury (2006) recommended that schemes should be spaced no closer than 500 m from each other (i.e. four systems per square kilometre). Gringarten (1978) provides a mathematical analysis of the optimum patterns and spacings for adjacent well doublets in geothermal reservoirs.

Of course, we could conceive of a situation where a thermal plume of warm water from one groundwater cooling scheme encountered a ground source heating scheme down-gradient. In this case, the efficiency of the heating scheme would actually be enhanced by the warm groundwater: not thermal pollution at all!

When deciding whether to permit a well doublet scheme, a regulator may opt to limit the permissible temperature differential between natural groundwater and injected water. In Manitoba, Canada, regulators limited groundwater cooling schemes to a 5°C differential (Ferguson and Woodbury, 2005). In England, there is no formal limit (at the time of writing), but regulators in London are beginning to tend towards a differential of <10°C being acceptable. In Lombardy, Italy, legislation is also currently unclear, but regulators talk of differentials no higher than 3–5°C being permissible (U. Puppini, pers. comm. 2006). In Holland, permitted re-injection temperatures vary from province to province but are typically no cooler than 5°C and no warmer than 25°C (Vos, 2007). The logic behind these differentials is hard to fathom and it may be that limits on quantities of heat (in Joules) being abstracted or rejected are more meaningful than temperatures.

Moreover, given the inherent limits on the sustainability of unbalanced, non-reversible open-loop doublet schemes, some regulators (EA, 2007; Vos, 2007) are beginning to express a strong preference for an approximate balance between heat re-injected to and heat abstracted from groundwater in well doublet schemes. In other words, they are promoting aquifer thermal energy storage (ATES) rather than simple unidirectional heating or cooling schemes.

13.4.2 Impact on aquifer permeability and porosity

We should also remember that the subsurface is an environment in which groundwater reacts with minerals, and that these reactions may be temperature-dependent. Younger (2006) has speculated that ground source heat schemes that result in cooling of a limestone aquifer may conceivably lead to increased rates of limestone dissolution, because the solubility of carbon dioxide (the dissolving agent or 'acid') increases as temperature decreases:

$$CaCO_3 + CO_2 + H_2O = Ca^{++} + 2HCO_3^- \tag{13.1}$$

We must also wonder what impacts might be expected from ground source cooling schemes where warm water is rejected back to the aquifer. Injection of warm water might result in clogging of aquifer porosity with exsolving gas bubbles (most gases become less soluble on heating), or with precipitating minerals (the reverse of Equation 13.1). Excess warmth *may* also stimulate bacterial growth, resulting in accelerated formation of biofilms on well screens or within the aquifer itself (Cullimore and McCann, 1977). Much of the above is speculative, however, and empirical research is needed on the impacts of ground source heating and cooling on the fabric of aquifer materials.

13.5 Decommissioning of boreholes

A 100-m hole in the ground represents a potential pathway for contaminants from the surface to deep groundwater. Thus, when a borehole is redundant and no longer actively managed, it should be responsibly decommissioned, not just abandoned. If the borehole or well is open, it is probably worth enquiring whether the regional agency responsible for groundwater management can convert it to an 'observation well' for monitoring groundwater quality and levels. If not, we may be forced to backfill the well. A concise guide to well decommissioning is provided by SEPA (2004), which broadly coincides with advice given by NGWA (Mc Cray, 1998, 1999). The overriding objectives of decommissioning are (i) to remove any hazard to the public from an open void in the ground, (ii) to remove any carrier fluid from a ground loop, (iii) to remove any headworks and pump, (iv) to prevent any pollution from entering the subsurface, (v) to prevent the borehole acting as a conduit for groundwater flow, or for gases such as methane or carbon dioxide (both of which can be a health hazard) to the surface, and (vi) to return the ground to a condition near to its natural state.

In order to achieve these objectives, we should restrict access to the well site until the well is safely decommissioned, and then remove the pump, rising main and any headworks. Then, having obtained a copy of the original drilling log and construction records, we should commission a specialist contractor to backfill the well. The well can be backfilled along its length by non-polluting, low-permeability, non-shrinking grout or bentonite. Alternatively, if the geology is known, the borehole can be backfilled with clean permeable backfill, sand or gravel adjacent to aquifer horizons, but with low-permeability bentonite or grout adjacent to aquitard horizons. This preserves the integrity of the hydrostratigraphic structure of the aquifer and prevents the well from acting as a conduit for groundwater flow between aquifer horizons. The uppermost section of the well (>2 m) should always be backfilled with a low permeability concrete, cement or bentonite grout medium. At the surface, a concrete cap should extend >0.5 m beyond the well's perimeter (Figure 13.4).

In the case of a closed-loop borehole, the situation is a little trickier. If the ground loop is simply installed in an open, water-filled borehole, the loop can be removed and the borehole backfilled as described above. If the loop has been grouted into the borehole, however, the only solution is to remove any carrier fluid from the loop and then to grout the U-tube with a low-viscosity grout mix, injected under pressure.

After decommissioning, notification of backfilling should be sent to the national or regional custodian of geological records (often, the Geological Survey).

13.6 A whistle-stop tour of regulatory environments

This section provides a snapshot of regulatory status on ground source heat in a few selected nations. Legislation and guidance is changing fast and this section has the

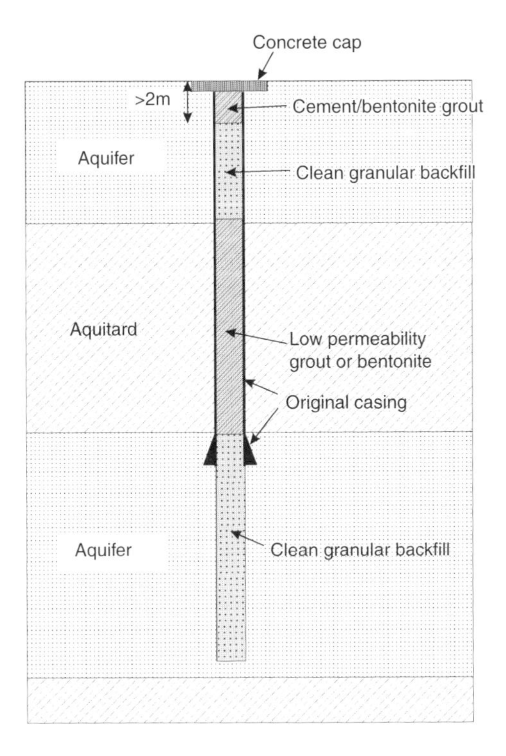

Figure 13.4 One method of backfilling a decommissioned borehole. The backfill type matches the hydrostratigraphy (high-permeability backfill in the aquifer sections, low permeability grout in the aquitard sections). Alternatively, the entire borehole could have been backfilled with low-permeability, non-shrinking grout.

potential to become rapidly outdated. Any GSHP installer or driller is advised to seek specific advice on the relevant legislation applicable in specific nations or states at the current time.

13.6.1 England and Wales

In England and Wales, anyone drilling to explore for or abstract groundwater is required to obtain a consent from the Environment Agency before commencing. This consent may specify the need for a pumping test and some form of environmental impact assessment. Following drilling and test pumping, a license to abstract water may be granted. Open-loop systems fall within this consenting and licensing system, unless they are very small (abstraction of $<20\ m^3\ day^{-1}$ is exempt from licensing). An abstraction license will often be time-limited, which can pose problems for operators of GSHP schemes, as they need to guarantee a prolonged period of operation in order to achieve the long payback time that capital expenditure on GSHP systems often entails. Monitoring of the scheme may also be required.

The operator of an open-loop scheme will also need to apply for a discharge consent in order to return 'spent' water to a natural recipient (a river or back to an aquifer). This could be subject to constraints on re-injection temperature and quality.

Currently, borehole-based closed-loop systems are deemed to fall outside the consenting and licensing system, as they are not specifically drilled to explore for groundwater. The Environment Agency is usually a statutory consultee on planning applications for major land redevelopments and may seek to exercise some control over large closed-loop schemes via the planning process. The agency has also released a draft policy statement (EA, 2007) as part of their new Groundwater Protection Policy. It provides little additional guidance on closed-loop systems, other than strongly recommending 'that systems do not use hazardous substances'. It warns that the use of List 1 contaminants in closed loops may result in the EA issuing an enforcement notice to prevent this. Also, one statement in the policy document reads 'We expect GSHP systems to be operated sustainably. In most cases this means there should be a balance between heating and cooling demand across a year. This will avoid unacceptable heating or cooling of the ground and groundwater'.

Any well or borehole penetrating strata that contain coal deposits or mine workings requires consent from the UK Coal Authority. Furthermore, all boreholes > 50 ft (~ 15 m) deep must be reported, with location, construction details and drilling log, to the British Geological Survey.

13.6.2 Germany

In Germany (Rybach, 2003; EGEC, 2006), geothermal resources fall both under mining *and* water resources legislation. Such heat energy does not belong to a private landowner but to the Federal state, and its use needs to be licensed by mining authorities, unless it is utilised wholly on the site where it is abstracted. There is also an exemption from geothermal/mining legislation for boreholes less than 100 m deep (thus leading to a profusion of 99 m ground source heat boreholes!).

Even small ground source heat schemes, less than 100 m deep, may be subject to water resources legislation at the state (*Länder*) level. Abstraction of groundwater for open-loop schemes requires a license. Even closed-loop schemes may require licensing as they are deemed to 'use' groundwater by changing its temperature. Furthermore, a series of guidelines for construction and operation of ground source heat schemes has been produced by the German Engineers' Association (*Verein Deutscher Ingenieure* – VDI 2000, 2001a, b, 2004).

13.6.3 France

France has had a long tradition of using open-loop well doublet schemes in connection with the exploitation of geothermal energy. Mining law specifically requires permits to be obtained for the exploration and operation of geothermal schemes. This legislation was presumably developed with 'hot' geothermal energy in mind, but there is no exemption for very-low-temperature ground source heat schemes (at least at the time of the Rybach 2003 and EGEC 2006 surveys). There is, however, an exemption for schemes that are <100 m deep and those that abstract less than a certain threshold (<200 thermie hr^{-1}) of heat. Even such relatively small, shallow schemes may still be regulated by environmental protection and water resources law.

13.6.4 Switzerland

According to Rybach (2003), the regulation of geothermal energy is not well defined in federal or cantonal legislation. In practice, the use of ground source heat in open-loop systems is largely regulated at canton level via water resources laws.

Closed-loop boreholes are potentially regulated under both water and environmental protection legislation. This is related to the fear that closed-loop boreholes can lead both to groundwater contamination and to hydraulic effects (aquifer interconnection). Of particular importance is the prohibition of borehole-based closed-loop systems within source protection zones around potable groundwater abstraction wells and in certain high-value aquifer units. Indeed, Rybach (2003) reports that maps of exclusion zones have been published by several cantons.

13.6.5 United States

In the United States, open-loop systems are typically regulated under normal water resources law pertaining to water wells and discharges. The situation as regards closed-loop systems is highly variable, with regulation varying from state to state. An excellent review of the regulatory framework for ground source heat is provided by Den Braven (1998), who notes that in approximately half of the United States' 50 states there is no or minimal regulation of closed-loop GSHP systems. In the other half, the degree of regulation varies from relatively light to excessively onerous (to the extent that Den Braven expresses concern that overzealous legislation could stifle the technology).

Of the 26 states that regulate the type of carrier fluid that may be used in closed-loop systems, the overwhelming preference is for water, potassium acetate solution and propylene glycol solution, with ethylene glycol being permitted in only one of the 26 states. For direct circulation systems, R-22 is often cited as the refrigerant of choice but, at present, the Environmental Protection Agency is trying to phase out its use in favour of other, more atmospherically benign, refrigerants. Similarly, around half of the 50 states offer no legislation on construction and grouting of closed-loop systems. Of those that do, there is an emphasis on achieving leak-proof ground loops and a tendency to prefer grouting of boreholes with low-permeability materials.

The Geothermal Heat Pump Consortium (Geoexchange, 2003) maintains a website summarising the legislation regarding ground source heat in each US state[2]. New Jersey

[2] http://www.geoexchange.org

is one of the states that have a fairly tightly regulated regime. Here, horizontal closed-loop systems are largely unregulated, although vertical closed-loop systems are categorised as wells, must be consented before drilling commences and must be constructed by a licensed well driller. Moreover, the diameter of the borehole must be at least the size of the sum of the ground loop's internal diameter, plus its outer diameter, plus an additional 4 in. (100 mm). In practice, this implies a borehole diameter of 150 mm or more. In New Jersey, all closed-loop boreholes must be grouted with a low permeability grout, which must be one of the following (see Chapter 10):

- High-grade bentonite grout (bentonite plus water), which has a low hydraulic conductivity, but also a rather low thermal conductivity.
- Thermal grout 85, which comprises a mixture of bentonite, fine silica sand and water.
- Cementitious thermally enhanced grout, which is a mixture of silica sand, sodium bentonite, cement, sulphonated naphthalene plasticizer and water. This last grout is the only type to be used in consolidated aquifers.

In New Jersey, the following heat carrier fluids were deemed acceptable for closed-loop systems: potable water, and solutions of calcium chloride, ethanol, potassium acetate, potassium carbonate, sodium chloride and propylene glycol. Direct circulation systems are not prohibited, but permits are issued on a case-by-case basis and the refrigerant fluid would need to be approved by a regulator.

In New York State, a comprehensive manual on GSHP systems has been prepared (Collins *et al.*, 2002). There are no specific regulations for either horizontal or vertical systems, although a strong 'Position Paper' has been published regarding system installation and testing. The use of closed-loop systems is preferred to open loop, due to a perceived lower environmental impact (Geoexchange, 2003), and guidance tends to be based on International Ground Source Heat Pump Association (IGSHPA) guidelines, especially as regards closed-loop borehole grouting. For indirect closed-loop systems, water, propylene glycol solution and potassium acetate solution are permitted as carrier fluids (with scope for alternatives to be approved). Ground loops must be tested at 100% greater than operational pressure, and must be equipped with an automated shut-off mechanism in the event that more than 5% of the carrier fluid is lost via leakage. For direct circulation systems, the refrigerant must be non-toxic and non-hazardous, while the loop must be pressure tested at 500 psi (34.5 bar) and equipped with cathodic protection in soils of pH < 5.

13.7 Promoting technology: subsidy

In the foregoing sections, we have considered what concerns might arise among legislators over the widespread use of ground-sourced heating and cooling (see also Younger, 2006). We would hope, of course, that the industry does not become overregulated, such that the 'baby is thrown out with the bathwater'. The current author would argue that the potential benefits in terms of savings in CO_2 emissions from space heating far outweigh concerns over possible groundwater pollution from a few litres of ethylene glycol solution.

The purpose of subsidies is to stimulate growth in a new market, where the product

- has a high initial capital cost and where payback times may be large.
- has not yet fully gained consumer confidence, either due to unfamiliarity with the technology or due to conceptual obstacles.

In the 1990s, an effective system of subsidies renewed the Swedish market for ground source heat following a slump in energy prices. In Norway, the uptake in ground source heat has been slower. Midttømme (2003) blames the fact that, while Sweden provided grants of up to 26 000 SEK (around 2000 GBP) to households installing GSHPs, the Norwegian state has focussed more on subsidising air-sourced heat pumps. The result was that ground- and sea-sourced heat pumps accounted for around 74% of the heat pump market share in Sweden in 2002, while air-sourced heat pumps were 90% of the Norwegian market.

In the United Kingdom, two systems of government subsidy are currently in force (and bear in mind that this is just a 'snapshot' of a growing market – things will probably have changed by the time you read this):

- The Low Carbon Buildings Programme may provide grants to individuals, communities and businesses intending to install renewable energy technology. At present, house owners can claim up to £1200 or 30% of eligible costs towards a GSHP scheme (LCBP, 2007).
- Within the tax system, businesses can claim 100% of the cost of renewable energy infrastructure (again, including GSHPs) against tax during the first year. This is the so-called Enhanced Capital Allowances scheme.

Often, however, in order to take advantage of such subsidies and tax breaks, it is necessary to utilise either an officially approved heat pump and/or an accredited installer. This should provide authorities with a means of ensuring that heat pumps meet certain standards of reliability and efficiency, and that installers operate within a certain code of professional competence. To apply for a grant under the Low Carbon Buildings Programme, it is also necessary to demonstrate that you have already taken measures to achieve a reasonable standard of building thermal performance (insulation, heating control and other energy efficiency measures).

Similar subsidy schemes exist in other nations, including Switzerland (Rybach, 2003) and Ireland (EGEC, 2006). The Irish 'Greener Homes' scheme offers (as of 2007) e4300 to horizontal closed-loop and well-based open-loop schemes and e6500 towards vertical closed-loop systems (SEI, 2007). The scheme maintains a register of installers and a list of approved products.

13.8 The final word

I would argue that subsidy plays an important role in the first few years of growth of a market for a socially and environmentally desirable technology. However, I have tried to argue, during the course of this book, that ground source heat not only has overwhelming environmental advantages, but can also compete economically with conventional fossil fuels in many cases. It is a technology that can be attractive to both environmentalists and petrol-heads – and everyone else too! I would thus hope that, at some point in the not-too-distant future, the technology will have gained sufficient consumer confidence to stand alone as an unsubsidised product. If we are to achieve this goal, I would argue that

- Drilling costs must come down, especially for closed-loop systems. Drilling for ground source heat needs to become tailored to the market. Rigs need to be more compact and installation procedures effectivised and standardised to the extent that variable geology permits.
- The price of heat pumps would be expected to decrease as demand increases, with a tendency to their becoming available as 'off the shelf' products rather than specialist hardware.
- New developers of housing stock need to explore ways of recovering the costs of installing ground source heat solutions. Can the costs of an individual system be passed on to the purchaser and included in a mortgage or finance plan? Should the site owner or a community association provide ground source heat to an entire housing complex as a form of 'mini-utility' company?
- The ground source heating and cooling option needs to be considered at a far earlier stage of most new developments.
- Designers and installers need to operate with integrity. Ground source heat should not be a form of 'snake oil' to solve all ills. Realistic expectations must be communicated to potential customers. Professional standards and a code of conduct should be encouraged.

In both the industrialised and developing world, one can get very excited about the possibilities that ground source heat can offer. We have already noted (Chapter 6) that GSHPs do not need to be powered by electricity: indeed, vapour-compression systems powered by gas or diesel motors can be purchased. It is certainly conceivable that we could develop efficient and reliable vapour compression heat pumps that are driven by biogas motors or waterpower, or even high-temperature gas absorption heat pumps running on biomass fuel. We can envisage these providing access to the 'thermogeological heat store' to rural communities that are not connected to reliable mains electricity.

In the industrial world, we must cease to regard heat fluxes, waste streams, water consumption and ventilation as separate processes. We must integrate them at a domestic and industrial level. Water and wastewaters are not simply fluxes of fluid; they also represent flows of heat that can be utilised. Could water companies recover heat from water pumped from a groundwater supply well, before it enters a distribution main (where it might then re-absorb heat from the soil before it arrives at the consumer's property)? Could householders recover heat from mains water supplies? Can we recover waste heat from sewage? Could a food-processing industry recover heat from an influx of groundwater in order to heat its offices and thereafter use the resulting cold water to chill a food store, before the water is utilised for washing or processing purposes?

At an even larger scale, we can conceive of regional management of ground source heat in the same way that we have achieved (via, e.g. the EU's Water Framework Directive) the regional management of groundwater resources. The work of Ferguson and Woodbury (2006) has indicated that a given block of aestifer has a finite capacity to support extraction or rejection of heat, in exactly the same way that a given aquifer can only support a finite amount of groundwater abstraction. From the 1970s to today, hydrogeologists have been increasingly interested in the concept of artificial aquifer recharge: re-injecting water to aquifers during periods of surplus and in locations where the aquifer can accept excess head. In other words, they are manipulating and optimising the hydraulic storage that the subsurface represents. Can we begin to apply the same ideas to the earth's enormous thermal storage: can we recharge heat to the aestifer/aquifer during times of summer excess, to re-abstract it in winter (yes, we can … via UTES schemes: see Chapter 10)? One could even envisage an aquifer system being managed in terms of its thermal fluxes as well as its groundwater fluxes, such that waste heat from major office, commercial and hotel complexes in a city centre could be injected to an aquifer, migrate down-gradient to residential areas where small houses could re-abstract the heat via GSHPs. If so, would the environmental impact be acceptable and would this type of thermal management be compatible with the aquifer's value as a water resource? Utopian, perhaps, but not wholly implausible in our more permeable aquifers.

The possibilities that the science of thermogeology and the technology of ground source heating/cooling offer are manifold, and there is no shortage of research tasks. The main challenge is to realise that these tasks can only be tackled meaningfully in a multidisciplinary manner, with seamless collaboration between the geologist, the architect, the engineer and the regulator. Those involved in marketing the GSHP should do so with integrity, realising that it represents a powerful tool in humankind's array of energy technologies, but acknowledging that it is not appropriate for all customers, that it poses significant challenges in terms of skills integration and that it requires a significant investment of capital 'up-front' in the expectation of long-term environmental and monetary benefits. As Heap (1979) wisely observed, we should protect against '*the twin dangers of excessive enthusiasm for apparent novelty and of undue scepticism for a concept for which worthwhile applications have been slow in developing*'.

References

Albu, M., Banks, D. and Nash, H., 1997, *Mineral and Thermal Groundwater Resources.* Chapman & Hall, London, 447 pp.

Allan, M. and Philippacopoulos, A., 2000, Performance characteristics and modelling of cementitious grouts for geothermal heat pumps. *Proceedings of the World Geothermal Congress 2000,* Kyushu – Tohoku, Japan, 28 May–10 June 2000, 3355–3360.

Allen, A., Milenic, D. and Sikora, P., 2003, Shallow gravel aquifers and the urban 'heat island' effect: A source of low enthalpy geothermal energy. *Geothermics* **32**: 569–578.

Allen, D.J., Brewerton, L.J., Coleby, L.M., Gibbs, B.R., Lewis, M.A., MacDonald, A.M., Wagstaff, S.J. and Williams, A.T., 1997, The physical properties of major aquifers in England and Wales. *British Geological Survey, Hydrogeology Group, Technical Report* **WD/97/34**; Environment Agency R&D Publication 8.

Alliant Energy, 2007, Alliant Energy Geothermal website. http://www.alliantenergygeothermal.com/stellent2/groups/public/documents/pub/geo_wrk_des_clo_001239.hcsp#P-4_0, last accessed January 2007.

Ampofo, F., Maidment, G. and Missenden, J., 2004, Underground railway environment in the UK. Part 3: Methods of delivering cooling. *Applied Thermal Engineering* **24**: 647–659.

Ampofo, F., Maidment, G.G. and Missenden, J.F., 2004, Review of groundwater cooling systems in London. *Applied Thermal Engineering* **26**: 2055–2062.

Andersson, O., 1998, Heat pump supported ATES applications in Sweden. *IEA Heat Pump Centre Newsletter* **16**(2): 20–21.

Angelino, G. and Invernizzi, C., 1996, Potential performance of real gas Stirling cycle heat pumps. *International Journal of Refrigeration* **19**(6): 390–399.

Arnold, D., 2000, The Equitable Building – the genesis of modern air conditioned buildings. *Chartered Institution of Buildings Services Engineers (CIBSE) Conference,* Dublin, 2000.

Arup, 2005, Ground storage of building heat energy: Overview report. *Arup Geotechnics Report* **O-02-ARUP3**, May 2005.

ASHRAE, 2002, Methods for determining soil and rock formation thermal properties from field tests. *ASHRAE Research Summary* **1118-TRP**. American Society of Heating, Refrigerating and Air-Conditioning Engineers.

ASME, 1980, *The Equitable Building Heat Pump System.* Information booklet from dedication ceremony 8/5/80. Oregon Section of the American Society of Mechanical Engineers (ASME).

Athey, R.E., 1999, Graham Corporation – evolution of a heat transfer company. *Heat Transfer Engineering* **20**(3): 6–11.

Bakema, G., 2001, Well and borehole failures in UTESS. State of the art 2000 (Draft 2). IF Technology bv report **2/9805/GW**, Arnhem, Netherlands, November 2001, 48 pp.

Bakema, G. and Snijders, A., 1998, ATES and ground-source heat pumps in the Netherlands. *IEA Heat Pump Centre Newsletter* **16**(2): 15–17.

Ball, D.A., Fischer, R.D., Talbert, S.G., Hodgett, D. and Auer, F., 1983, State-of-the-art survey of existing knowledge for the design of ground source heat pump systems. *Battelle Columbus Laboratories Report* **ORNL/Sub 80-7800/2**, Columbus, OH, USA, 75 pp.

Banks, D., 1992, Aquifer management – an introduction to hydrogeology. *Geology Today* **No. 6 (March/April 1992)**: Earth Reference i–iv.

Banks, D., 1997, The spas of England. In: Albu, M., Banks, D. and Nash, H. (eds.), *Mineral and Thermal Groundwater Resources*. Chapman & Hall, London, 235–280.

Banks, D., 1998, Predicting the probability distribution of yield from multiple boreholes in crystalline bedrock. *Ground Water* **36**(2): 269–274.

Banks, D., 2004, *Monitoring of Fresh and Brackish Water Resources.* Contribution E4.18.03, Encyclopaedia of Life Support Systems (EOLSS), UNESCO.

Banks, D., *in press.*, Potable water strategies in Southern Mudug, Somalia, with special reference to the local economics of motorised borehole systems for watering nomadic livestock. Accepted for publication in *Hydrogeology Journal.*

Banks, D. and Robins, N., 2002, *An Introduction to Groundwater in Crystalline Bedrock.* Norges geologiske undersøkelse, Trondheim, 63 pp.

Banks, D., Davies, C. and Davies, W., 1995, The Chalk as a karstic aquifer: the evidence from a tracer test at Stanford Dingley, Berkshire. *Quarterly Journal of Engineering Geology* **28**: S31–S38.

Banks, D., Gandy, C.J., Younger, P.L., Withers, J. and Underwood, C., *in press*, Anthropogenic thermogeological "anomaly" in Gateshead, Tyne and Wear, UK. In preparation for *Quarterly Journal of Engineering Geology and Hydrogeology.*

Banks, D., Morland, G. and Frengstad, B., 2005, Use of non-parametric statistics as a tool for the hydraulic and hydrogeochemical characterization of hard rock aquifers. *Scottish Journal of Geology* **41**(1): 69–79.

Banks, D., Parnachev, V.P., Frengstad, B., Holden, W., Karnachuk, O.V. and Vedernikov, A.A., 2001, The hydrogeochemistry of the Altaiskii, Askizskii, Beiskii, Bogradskii, Shirinskii, Tashtipskii and Ust' Abakanskii Regions, Republic of Khakassia, Southern Siberia, Russian Federation, data report. *Norges geologiske undersøkelse report* 2001.006, Trondheim, Norway, 45 pp. + appendices.

Banks, D., Parnachev, V.P., Frengstad, B., Holden, W., Karnachuk, O.V. and Vedernikov A.A., 2004b, The evolution of alkaline, saline ground- and surface waters in the southern Siberian steppes. *Applied Geochemistry* **19**: 1905–1926.

Banks, D., Skarphagen, H., Wiltshire, R. and Jessop, C., 2004a, Heat pumps as a tool for energy recovery from mining wastes. In: Gieré, R. and Stille, P. (eds.), *Energy, Waste and the Environment: a Geochemical Perspective, Geological Society Special Publication* **236**: 499–513.

Banks, D., Steven, J. and Cashmore, G., 2006, Hot rooms from cold stone. *GeoScientist* **16**(2): 4–7.

Barenbrug, A.W.T., 1974, *Psychrometry and Psychrometric Charts*, 3rd Edition. Cape and Transvaal Printers Ltd., Cape Town, SA.

Barker, J.A., 1986, Modelling of low enthalpy geothermal schemes. In: Downing, R.A. and Gray, D.A. (eds.), *Geothermal Energy – The Potential in the United Kingdom*. British Geological Survey/HMSO, London, 124–131, Chapter 7.

Barratt, T., 1914, Thermal conductivity Part II: Thermal conductivity of badly conducting solids. *Proceedings of the Physical Society of London* **27**: 81–93.

Bhatti, M.S., 1999, Evolution of automotive heating: Riding in comfort, Part 1. *ASHRAE Journal*, **August**: 51–57.

Beltrami, H. and Harris, R.N., 2001, Foreword: Inference of climate change from geothermal data. *Global and Planetary Change* **29**: 149–152.

Belyanin, N.M., Zvonarev, I.N., Pakh, E.M., Ponomarev, V.V., Selyatitskii, G.A., Senderzon, E.M. and Yavorskii, V.I. (eds.), 1969, Геология месторождений угля и горючих сланцев СССР. Том 7 *[Geology of Coal and Oil Shale Deposits of the USSR, Volume 7 – in Russian]*, Nedra, Moscow, 912 pp.

Biblioteca ETSEIB, 2003, Antecedentes históricos de la refrigeración [*Historical Antecedents of Refrigeration – in Spanish*]. Biblioteca ETSEIB (Escola Tècnica Superior d'Enginyeria Industrial de Barcelona), Fons Històric de Ciència i Tecnologia. Available at website http://bibliotecnica.upc.es/bib240/serveis/fhct/expo_et.asp, last accessed April 2007.

Bierschenk, W.H., 1963, Determining well efficiency by multiple step-drawdown tests. *International Association of Scientific Hydrology Publication* **64**: 493–507.

Blue Diamond, 2007, *HDPE specifications.* Available at website http://www.bdiky.com/specifications/specifications.htm, last accessed January 2007.

Boardman, B., Darby, S., Killip, G., Hinnells, M., Jardine, C.N., Palmer, J. and Sinden, G., 2005, *40% House.* Environmental Change Institute, University of Oxford, ISBN 1 874370 39 7, 126 pp.

Bodri, L., Cermak, V. and Kukkonen, I.T., 2001, Climate change of the last 2000 years inferred from borehole temperatures: data from Finland. *Global and Planetary Change* **29**: 189–200.

Bouma, J., 2002, Heat pumps – better by nature. *IEA Heat Pump Centre Newsletter* **20**(2): 10–27.

Boyle, G., 2004, *Renewable Energy*, 2nd Edition. Oxford University Press, Oxford, 452 pp.

BP, 2005, *BP Statistical Review of World Energy (June 2005), and Associated Web Pages.* http://www.bp.com, last accessed February 2006.

Brassington, F.C., 2007, A proposed conceptual model for the genesis of the Derbyshire thermal springs. *Quarterly Journal of Engineering Geology and Hydrogeology* **40**: 35–46.

Bredehoeft, J.D., Papadopulos, S.S. and Cooper, H.H., 1982, Groundwater: The water-budget myth. The Geophysics Study Committee, *Studies in*

Geophysics: Scientific Basis of Water Resource Management, National Academy Press, Washington DC, 51–57, Chapter 4.

Brekke, E., 2002, Termisk responstesting [*Thermal Response Testing – in Norwegian*]. Norwegian University of Technology and Science (NTNU), Trondheim, Institutt for Geologi og Bergteknikk, Prosjektoppgave, Autumn 2002.

Buxbaum, T., 2002, *Icehouses*. Shire Publications, Haverfordwest, 32 pp.

Calm, J.M. and Hourahan, G.C., 2001, Refrigerant data summary. *Engineered Systems* **18**(11): 74–88.

Carbon Trust, 2006a, *What are degree days?* Available at website http://www.carbontrust.co.uk/KnowledgeCentre/degree_days/what_are.htm, last accessed December 2006.

Carbon Trust, 2006b, Degree days for energy management – a practical introduction. *Carbon Trust Technology Guide* **CTG004**, HMSO, London, 22 pp.

Chapman, D.S., 2001, In recognition of Art Lachenbruch's contributions. *Global and Planetary Change* **29**: 189–200.

Claesson, J. and Eskilson, P., 1987a, Conductive heat extraction by a deep borehole: analytical studies. In: Eskilson, P. (ed.), *Thermal Analysis of Heat Extraction Boreholes.* Doctoral dissertation, Department of Mathematical Physics, University of Lund, Sweden.

Claesson, J. and Eskilson, P., 1987b, Conductive heat extraction by a deep borehole: thermal analysis and dimensioning rules. In: Eskilson, P. (ed.), *Thermal Analysis of Heat Extraction Boreholes.* Doctoral dissertation, Department of Mathematical Physics, University of Lund, Sweden. Later re-published in *Energy* **13**: 509–527 (1988).

Clark, A., 2006, *Lord Kelvin and the Age of the Earth.* Example from course ME201/MTH 281 'Applied Fourier Series and Boundary Value Problems'. Department of Mechanical Engineering, University of Rochester, New York State, USA.

Clark, L., 1988, *The Field Guide to Water Wells and Boreholes*. Geological Society of London Professional Handbook Series, Wiley, Chichester, UK.

Clauser, C. (ed.), 2003, *Numerical Simulation of Reactive Flow in Hot Aquifers: SHEMAT and Processing SHEMAT*, Springer, Berlin/Heidelberg/New York, 332 pp.

Clauser, C. and Huenges, E., 1995, Thermal conductivity of rocks and minerals. In: Ahrens, T. (ed.), *Rock Physics and Phase Relations*, American Geophysical Union, Washington, 105–126.

Cla-val, 2007, *Tube Seal Valve. Double Function: Injection + Pumping*. Information brochure, Cla-val Europe, Romanel/Lausanne, 2 pp.

Climate Master, 2004, *Commercial Case Study 6: Dunston Innovation Centre.* Climate Master, Oklahoma, USA, 2 pp.

Clyde, C.G. and Madabhushi, G.V., 1983, Spacing of wells for heat pumps. *Journal of Water Resources Planning and Management* **109**(3): 203–212.

Collins, P.A., Orio, C.D. and Smiriglio, S., 2002, *Geothermal Heat Pump Manual*. New York Department of Design and Construction, August 2002.

Connelly, W.M., 2005, Map of annual mean insolation at the Earth's surface, simulated using the HadCM3 model. Submitted to www.Wikipedia.com on 24 October 2005.

Cooper, H.H. and Jacob, C.E., 1946, A generalised graphical method for evaluating formation constants and summarizing well field history. *Transactions of the American Geophysical Union* **27**: 526–534.

Cullimore, D.R. and McCann, A.E., 1977, The identification, cultivation and control of iron bacteria in groundwater. In: Skinner, F.A. and Shewan, J.M. (eds.), *Aquatic Microbiology*, Academic Press, New York, 219–261.

Curtis, R.H., 2001, Earth Energy in the UK. *Geo-Heat Center (GHC) Bulletin* **22/4**, Klamath Falls, Oregon, USA, December 2001.

D'Accadia, M.D., 2001, Optimal operation of a complex thermal system: A case study. *International Journal of Refrigeration* **24**(4): 290–301.

DEFRA, 2005, *Sustainable Energy*. Available at website http://www.defra.gov.uk/environment/energy/, last accessed May 2007, UK Department for the Environment, Food and Rural Affairs.

De Marsily, G., 1986, *Quantitative Hydrogeology. Groundwater Hydrology for Engineers*. Academic Press, New York, 464 pp.

Den Braven, K., 1998, Survey of geothermal heat pump regulations in the United States. *Proceedings of 2nd Stockton International Geothermal Conference*, 16–17 March, 1998. Available at website http://intraweb.stockton.edu/eyos/energy_studies/content/docs/proceedings/DENBR.PDF.

Deng, Z., 2004, *Modeling of Standing Column Wells in Ground Source Heat Pump Systems*. PhD Thesis, Oklahoma State University, December 2004.

Deng, Z., Rees, S.J. and Spitler, J.D., 2005, A model for annual simulation of standing column well ground heat exchangers. *HVAC&R Research* **11**(4): 637–656.

Dickson, M.H. and Fanelli, M., 2004, *What Is Geothermal Energy?* Istituto di Geoscienze e Georisorse, CNR, Pisa, Italy. Available via the International Geothermal Association, http://iga.igg.cnr.it/geo/geoenergy.php, last accessed September 2007.

DoE, 2000, *Thermal Energy Storage for Space Cooling*. Federal Energy Management Program, *US Department of Energy report* **DOE/EE-0241**, December 2000, 34 pp.

Domenico, P.A. and Schwartz, F.W., 1990, *Physical and Chemical Hydrogeology*. Wiley, New York, 824 pp.

Drage, K., 2006, *Coefficient of Performance: What Is It? What can be Expected?* Presentation given at Meeting of the UK Ground Source Heat Pump Association, 'Ground Source Heat Pumps – Exploding the Myths', 15 November 2006, Flitwick, UK.

Driscoll, F.G., 1986, *Groundwater and Wells*. Johnson Filtration Systems, St Paul, USA, 1089 pp.

DTI, 2006, UK energy and CO_2 emissions projections. Updated projections to 2020. Department of Trade and Industry, February 2006, 69 pp.

Dudeney, B., Demin, O. and Tarasova, I., 2002, Control of ochreous deposits in mine water treatment. In: Nuttall, C.A. (ed.), *Proceedings of the Conference on Mine Water Treatment: A Decade of Progress*, Newcastle, 11–13 November 2002, 195–202.

EA, 2005, Groundwater levels in the chalk-basal sands aquifer of the Central London Basin: May 2005. Unpublished report, Environment Agency (Thames Region).

EA, 2006, Groundwater levels in the chalk-basal sands aquifer of the London Basin: June 2006. Unpublished report, Environment Agency (Thames Region).

EA, 2007, *Ground Source Heat Pump Systems.* Groundwater Protection: Policy and Practice GP3 Part 4, Legislation and policies. Draft for Discussion February 2007.

Earth Energy, 2005, *Dunston Innovation Centre.* Fact sheet produced by Earth Energy Ltd., Falmouth, Cornwall.

ECA, 2006a, ECA Energy Technology Criteria List – Technology: Heat Pumps. Ground Source: Brine to Water. Enhanced Capital Allowance scheme, HMSO 2006.

ECA, 2006b, ECA Energy Technology Criteria List – Technology: Heat Pumps. Ground Source: Brine to Air. Enhanced Capital Allowance scheme, HMSO 2006.

EGEC, 2006, *Key Issues for Renewable Heat in Europe (K4RES-H) – Key Issue 3: Regulations for Geothermal Energy.* European Geothermal Energy Council (EGEC)/European Renewable Energy Council (EREC). Available at website http://www. erec-renewables.org/documents/K4RES-H/D16-geothermal.pdf, last accessed December 2006.

Eggen, G. and Vangsnes, G., 2005, Heat pump for district cooling and heating at Oslo Airport Gardermoen. *Proceedings of 8th IEA Heat Pump Conference* 2005, 7 pp.

Energi-spar, 2007, *Freezium Calculator.* Available at website http://www.energi-spar.no/Freezium%20calculator.htm, last accessed January 2007.

Energie-Cités, 2001, *Geothermie – District Heating Scheme, Southampton, United Kingdom.* Energie-Cités in collaboration with the City of Southampton. Available at website http://www.energie-cites.org/db/southampton_140_en.pdf, last accessed March 2007.

Energy Saving Trust, 2004, Domestic ground source heat pumps – design and installation of closed loop systems. Energy Efficiency Best Practice in Housing Series, Energy Saving Trust, UK, 19 pp.

Engineering Tool Box, 2007, Available at website http://www.engineeringtoolbox.com, last accessed January 2007.

Environment Canada, 2002, Earth energy explored in the Yukon. *Your Yukon,* **272**, Available at website http://www.taiga.net/yourYukon/col272.html, last accessed September 2007.

EREC, 2006, *Key Issues for Renewable Heat in Europe (K4RES-H) – Geothermal Tables.* European Renewable Energy Council (EREC) website http://www.erec-renewables.org/documents/K4RES-H/D2-geothermal-tables.pdf, last accessed December 2006.

Eskilson, P., 1987, *Thermal Analysis of Heat Extraction Boreholes.* Doctoral dissertation, Department of Mathematical Physics, University of Lund, Sweden.

Eskilson, P., Hellström, G., Claesson, J., Blomberg, T. and Sanner, B., 2000, *Earth Energy Designer (EED) Version 2 (Software Parameter Database).* Blocon software, Sweden.

EST, 2005, The development of community heating in Southampton. *Energy Saving Trust Brochure* **CE158**: 3 pp.

EU, 2000, Directive 2000/60/EC of the European Parliament and of the Council of 23 October 2000 establishing a framework for Community action in the field of water policy.

EU, 2002, Directive 2002/91/EC of the European Parliament and of the Council of 16 December 2002 on the energy performance of buildings.

Everett, J.D., 1860, Observations of deep-sunk thermometers. The reduction of observations of deep-sunk thermometers, 1846–1859. *Royal Greenwich Observatory* RGO 18/169, cited by Rambaut (1900).

Everett, J.D., 1882, Summary of results contained in the first fifteen reports of the underground temperature committee. *Report of the 52nd Meeting of the British Association for the Advancement of Science, York 1882*, 72–90.

Everett, J.D., Thomson, W., Clerk Maxwell, J., *et al.*, 1986, Ninth Report of the Committee, appointed for the purpose of investigating the rate of increase of underground temperature downwards in various localities of dry land and under water. *Report of the 46th Meeting of the British Association for the Advancement of Science, Glasgow 1876*, 204–211.

Everett, J.D., Thomson, W., Clerk Maxwell, J., *et al.*, 1877, Tenth Report of the Committee, appointed for the purpose of investigating the rate of increase of underground temperature downwards in various localities of dry land and under water. *Report of the 47th Meeting of the British Association for the Advancement of Science, Plymouth 1877*, 194–199.

Everett, J.D., Thomson, W., Symons, G.J., *et al.*, 1880, Thirteenth Report of the Committee, appointed for the purpose of investigating the rate of increase of underground temperature downwards in various localities of dry land and under water. *Report of the 50th Meeting of the British Association for the Advancement of Science, Swansea 1880*, 26–29.

Everett, J.D., Thomson, W., Symons, G.J., *et al.*, 1882, Fifteenth Report of the Committee, appointed for the purpose of investigating the rate of increase of underground temperature downwards in various localities of dry land and under water. *Report of the 52nd Meeting of the British Association for the Advancement of Science, York 1882*, 72–74.

Ferguson, G. and Woodbury, A.D., 2004, Subsurface heat flow in the built environment. *Journal of Geophysical Research* **109:** B02402, 9 pp.

Ferguson, G. and Woodbury, A.D., 2005, Thermal sustainability of groundwater-source cooling in Winnipeg, Manitoba. *Canadian Geotechnical Journal* **42**: 1290–1301.

Ferguson, G. and Woodbury, A.D., 2006, Observed thermal pollution and post-development simulations of low temperature geothermal systems in Winnipeg, Canada. *Hydrogeology Journal* **14**: 1206–1215.

Fernández Alonso, P., 2005, *Hydrogeological Review of Borehole Heat Exchanger Design*. MSc Hydrogeology thesis, Department of Earth Sciences, University of Leeds, September 2005.

Fetter, C.W., 2001, *Applied Hydrogeology*, 4th Edition. Pearson Education, Prentice-Hall, NJ, USA.

Franzson, H., Tómasson, J., Sverrisdóttir, G. and Sigurðsson, F., 1997, Geothermal energy in Iceland. In: Albu, M., Banks, D. and Nash, H. (eds.), *Mineral and Thermal Groundwater Resources*, Chapman & Hall, London, 205–234.

Gabrielsson, A., 1997, Solvärmesystem med säsongslager: En simulerings- och kostnadsstudie [*Solar Heating Systems with Seasonal Storage: A Simulation and Cost Study – in Swedish*]. Statens Geotekniska Institut (SGI) Rapport **52**, Linköping, Sweden, 79 pp.

Gabrielsson, A., Bergdahl, U. and Moritz, L., 2000, Thermal energy storage in soils at temperatures reaching 90°C. *Journal of Solar Energy Engineering* **122**: 3–8.

Gehlin, S., 2002, Thermal response test: method development and evaluation. *Luleå University of Technology, Doctoral Thesis* 2002: 39. Available at website http://epubl.ltu.se/1402-1544/2002/39/index-en.html, last accessed October 2007.

Gehlin, S., Hellström, G. and Nordell, B., 2003, The influence of the thermosiphon effect on the thermal response test. *Renewable Energy* **28**: 2239–2254.

Geoexchange, 2003, *Geothermal Heat Pump Regulations Summaries.* Available at website http://www.geoexchange.com/publications/regs/, compiled 2003 by the Geothermal Heat Pump Consortium (GHPC) and accessed February 2007.

Goodman, R., Jones, G.L.I., Kelly, J., Slowey, E. and O'Neill, N., 2004, *Geothermal Resource Map of Ireland*. Sustainable Energy Ireland.

Greco, D., 2002, *Fourier and the Theory of Heat.* The Victorian Web website http://www.victorianweb.org/science/fourier.html, last accessed 2006.

Green, K., 2007, Peter Ritter von Rittinger: Inventor of the world's first known heat pump. *GeoOutlook* **4**(1): 20–21.

Gringarten, A.C., 1978, Reservoir lifetime and heat recovery factor in geothermal aquifers used for urban heating. *Pure and Applied Geophysics* **117**(1–2): 297–308.

Grove, D.B., 1971, U.S. Geological Survey tracer study, Amargosa Desert, Nye County, Nevada, Part II – an analysis of the flow field of a discharging-recharging pair of wells. *U.S. Geological Survey Report* **USGS-474-99**: 56 pp.

Guernsey, E.N., Betz, P.L. and Skan, N.H., 1949, Earth as a heat source and storage medium for the heat pump. *ASHRAE Transactions* **55**: 321–344.

Gustafson, G., 2002, Strategies for groundwater prospecting in hard rocks: Probabilistic approach. *Norges Geologiske Undersøkelse Bulletin* **439**: 21–25.

Güven, O., Falta, R.W., Molz, F.J. and Melville, J.G., 1986, A simplified analysis of two-well tracer tests in stratified aquifers. *Ground Water* **24**(1): 63–71.

GVP, 2006, *Larderello.* Available at website http://www.volcano.si.edu/world/volcano.cfm?vnum=0101-001, last accessed December 2006. Hosted by the Global Volcanism Program of the Smithsonian Institution in Washington, DC.

Haenel, R. (ed.), 1980, *Atlas of Subsurface Temperatures in the European Community.* Commission of the European Communities, Brussels-Luxembourg.

Haenel, R. and Staroste, E. (eds.), 1988, *Atlas of Geothermal Resources in the European Community, Austria and Switzerland.* Commission of the European Communities publication EUR 11026, Brussels-Luxembourg.

Haldane, T.G.N., 1930, The heat pump – an economical method of producing low-grade heat from electricity. *Journal of the Institution of Electrical Engineers* **68**: 666.

Halliday, D. and Resnick, R., 1978, *Physics*, 3rd Edition. Wiley, New York, 1131 pp.

Halozan, H. and Rieberer, R., 2003, Direct-evaporation ground-coupled heat pumps. Abstract ICR0461 submitted to *21st International Institute of Refrigeration (IIR) Congress of Refrigeration (Sector: Heat Pumps)*, 17–22 August 2003, Washington DC.

Hantush, M.S., 1964, Hydraulics of wells. In: Chow, V.T. (ed.) *Advances in Hydroscience, Volume 1*, Academic Press, New York/London, 281–432.

Heap, R.D., 1979, *Heat Pumps.* E. & F.N. Spon Ltd., London, 155 pp.

Hellström, G., 2006, Market and technology in Sweden. *Proceedings of the 1st Groundhit (Ground Coupled Heat Pumps of High Technology) Workshop*, Brussels, May 18, 2006. Available at website http://www.geothermie.de/groundhit/, last accessed October 2007.

Henderson-Sellers, B., 1984, *Engineering Limnology.* Pitman, London.

Henson, R., 2006, *The Rough Guide to Climate Change.* Rough Guides, London, 341 pp.

Herschel, A.S. and Lebour, G.A., 1877, Experiments to determine the thermal conductivities of certain rocks, showing especially the geological aspects of the investigation. *Report of the 47th Meeting of the British Association for the Advancement of Science*, Plymouth 1877, 90–97.

Herschel, A.S., Lebour, G.A. and Dunn, J.T., 1879, Experiments to determine the thermal conductivities of certain rocks, showing especially the geological aspects of the investigation. *Report of the 49th Meeting of the British Association for the Advancement of Science*, Sheffield 1879, 58–63.

Hill, D. and Parker, J., 2005/06. Presentations given by David Hill of the Metropolitan Housing Association and John Parker of Powergen at meetings of the UK Ground Source Heat Pump Association.

Himmelsbach, T., Hötzl, H. and Maloszewski, P., 1993, Forced gradient tracer tests in a highly permeable fracture zone. In: Banks, S.B. and Banks, D. (eds.), *Hydrogeology of Hard Rocks, Memoirs of the XXIVth Congress International Association of Hydrogeology*, Ås, Oslo, Norway, 237–247.

Hostetler, S.W., 1995, Hydrological and thermal response of lakes to climate: Description and modeling. In: Lerman, A., Imboden, D.M. and Gat, J.R. (eds.), *Physics and Chemistry of Lakes*, 2nd Edition. Springer-Verlag, Berlin, 63–82.

IEA, 2001a, City Hall Zurich. *IEA Heat Pump Center Newsletter* **19**(3): 6.

IEA, 2001b, Use of heat pump in Polar Circle. *IEA Heat Pump Center Newsletter* **19**(3): 5.

IGSHPA, 1988, *Closed-Loop/Ground-Source Heat Pump Systems – Installation Guide.* International Ground Source Heat Pump Association, Oklahoma, USA.

IGSHPA, 2007, *International Ground Source Heat Pump Association: History.* Available at website http://www.igshpa.okstate.edu/about/about_us.htm, last accessed April 2007.

Ingersoll, L.R., Zobel, O.J. and Ingersoll, A.C., 1948, *Heat Conduction with Engineering and Geological Application.* McGraw Hill, New York, 278 pp.

Ingersoll, L.R., Zobel, O.J. and Ingersoll, A.C., 1954, *Heat Conduction with Engineering, Geological and Other Applications (revised edition).* University of Wisconsin Press, Madison, 325 pp.

IPCC, 2007, *Climate Change 2007: The Physical Science Basis. Summary for Policymakers.* Contribution of Working Group I to the Fourth Assessment Report of the Intergovernment Panel on Climate Change. Intergovernment Panel on Climate Change, Geneva, 18 pp.

Janio de Castro Lima, J. and Braga Paraguassú, A.B., 2004, Linear thermal expansion of granitic rocks: Influence of apparent porosity, grain size and quartz content. *Bulletin of Engineering Geology and the Environment* **63**: 215–220.

Jones, H.K, Morris, B.L., Cheney, C.S., *et al.*, 2000, The physical properties of minor aquifers in England and Wales. *British Geological Survey, Hydrogeology Group, Technical Report* **WD/00/04**; Environment Agency R&D Publication 68.

Kapetsky, J.M. and Nath, S.S., 1997, *A Strategic Assessment of the Potential for Freshwater Fish Farming in Latin America.* COPESCAL Technical Paper No. 10. Rome, FAO, 1997, 128 pp.

Katsoyiannis, I.A., Hug, S.J., Ammann, A., Zikoudi, A. and Hatziliontos, C., 2007, Arsenic speciation and uranium concentrations in drinking water supply wells in Northern Greece: Correlations with redox indicative parameters and implications for groundwater treatment. *The Science of the Total Environment*, **383**: 128–140.

Kavanaugh, S.P. and Rafferty, K., 1997, *Design of Geothermal Systems for Commercial and Institutional Buildings.* American Society of Heating, Refrigeration and Air-Conditioning Engineers (ASHRAE), 167 pp.

Kazmann, R.G. and Whitehead, W.R., 1980, The spacing of heat pump supply and discharge wells. *Ground Water Heat Pump Journal* **1**(2): 28–31.

Kelley, I., 2006, Ground-source heat pumps deliver both high efficiency and reliability – good news for both contractors and their customers. *Wisconsin Perspective* September–October 2006: 14–16.

Killip, G., 2005, *Emission Factors and the Future of Fuel* (Background material M for the '40% House Report' of Boardman *et al.*, 2005). Environmental Change Institute, University of Oxford, March 2005.

Kipp, K.L., 1997, Guide to the revised heat and solute transport simulator: HST3D – Version 2. *United States Geological Survey Water-Resources Investigations Report* **97-4157**: 149 pp.

Kruseman, G.P., De Ridder, N.A. and Verweij, J.M., 1990, *Analysis and Evaluation of Pumping Test Data*, 2nd Edition. International Institute for Land Reclamation and Improvement, Wageningen, Netherlands.

Lachenbruch, A. and Marshall, B.V., 1986, Changing climate: Geothermal evidence from permafrost in the Alaskan Arctic. *Science* **234**: 689–696.

Laval, B., 2006, *Civil 542, Physical limnology, 2.0D budgets.* Department of Civil Engineering, University of British Columbia, Canada. Available at website http://supercritical.civil.ubc.ca/~blaval/Courses/CIVL542/Civil542_2_Budget.pdf, last edited 25 January 2006.

Law, R., Nicholson, D. and Mayo, K., 2007, Aquifer thermal energy storage in the fractured London Chalk. A thermal injection/withdrawal test and its interpretation. *Proceedings of the 32nd Workshop on Geothermal Reservoir Engineering*, Stanford University, California, 22–24 January 2007, Paper SGP-TR-183.

LCBP, 2007, *Low Carbon Buildings Programme.* Available at website http://www.lowcarbonbuildings.org.uk/home/, last accessed February 2007.

Less, C. and Andersen, N., 1994, Hydrofracture: State of the art in South Africa. *Applied Hydrogeology* (now *Hydrogeology Journal*) **2**(2): 59–63.

Lewis, C., 2000, *The Dating Game: One Man's Search for the Age of the Earth.* Cambridge University Press, Cambridge, UK, 258 pp.

Lian, Z., Park, S., Huang, W., Baik, Y. and Yao, Y., 2005, Conception of combination of gas-engine-driven heat pump and water-loop heat pump system. *International Journal of Refrigeration* **28**(6): 810–819.

Lienau, P. (ed.), 1998, *Geothermal Direct Use Engineering and Design Guidebook*, 3rd Edition. Oregon Institute of Technology, Geo-Heat Center, Klamath Falls, USA.

Lienhard, J.H. (IV) and Lienhard, J.H. (V), 2006, *A Heat Transfer Textbook*, 3rd Edition. Phlogiston, Cambridge, MA, USA, 749 pp.

Linacre, E., 1992, *Climate and Data Resources: A Reference and Guide.* Routledge, London, UK, 365 pp.

Linacre, E. and Geerts, B., 1997, *Climates and Weather Explained.* Routledge, London, UK, 432 pp.

Lippmann, M.J. and Tsang, C.F., 1980, Groundwater use for cooling: Associated aquifer temperature changes. *Ground Water* **18**(5): 452–458.

Logan, J., 1964, Estimating transmissibility from routine production tests of water wells. *Ground Water* **2**: 35–37.

Lund, J., 2005, Hitaveita Reykjavikur and the Nesjavellir geothermal co-generation power plant. *Geo-Heat Center (GHC) Bulletin (Klamath Falls, Oregon, USA)*, **June 2005**: 19–24.

Lund, J., Sanner, B., Rybach, L., Curtis, R. and Hellström, G., 2004, Geothermal ground-source heat pumps: A world overview. *Geo-Heat Center (GHC) Bulletin (Klamath Falls, Oregon, USA)*, **September 2004**: 1–10.

Majorowicz, J., Grasby, S.E., Ferguson, G., Šafanda, J. and Skinner, W., 2006, Paleoclimatic reconstructions in western Canada from borehole temperature logs: Surface air temperature forcing and groundwater flow. *Climate of the Past* **2**: 1–10.

Mands, E. and Sanner, B., 2001, In-situ determination of underground thermal parameters. *Proceedings of UNESCO/IGA International Summer School on Direct*

Application of Geothermal Energy/International Geothermal Days, Germany 2001 (Bad Urach), Supplement, 45–54.

Manning, D.A.C., Younger, P.L., Smith, F.W., Jones, J.M., Dufton, D.J. and Diskin, S., 2007, A deep geothermal exploration well at Eastgate, Weardale, UK: A novel exploration concept for low-enthalpy resources. *Journal of the Geological Society of London* **164**: 371–382.

Matte, M., 2002, Large heat pump faculty in rock cavern – efficient system for district heating and cooling. In *Norwegian Urban Tunnelling.* Norsk Forening for Fjellsprengningsteknikk (Norwegian Tunnelling Society), Grefsen, Oslo, Chapter 10, Accessed via Internet at http://www.tunnel.no.

McCray, K.B., 1998, Guidelines for the construction of vertical boreholes for closed loop heat pump systems. *Proceedings of 2nd Stockton International Geothermal Conference*, 16–17 March, 1998. Available at website http://intraweb.stockton.edu/eyos/energy_studies/content/docs/proceedings/MCCRA.PDF, last accessed October 2007.

McCray, K.B., 1999, *Guidelines for the Construction of Vertical Boreholes for Closed Loop Heat Pump System* (3rd printing). National Ground Water Association, Westerville, OH, 41 pp.

Midttømme, K., 2003, Sverige satser mer på vannbåren varme *[Sweden focuses more on water-sourced heat – in Norwegian].* Published on-line at the website of the Geological Survey of Norway, http://www.ngu.no, last accessed October 2007.

Mikler, V., 1993, *A Theoretical and Experimental Study of the 'Energy Well' Performance.* MSc thesis, Pennsylvania State University, College Park, USA.

Misstear, B., Banks, D. and Clark, L., 2006, *Water Wells and Boreholes.* Wiley, Chichester, UK, 498 pp.

Murphy, K.P. and Phillips, B.A., 1984, Development of a residential gas absorption heat pump. *International Journal of Refrigeration* **7**(1): 56–58.

Neal, W.E.J., 1978, Heat pumps for domestic heating and heat conservation. *Physics in Technology* **9**: 154–161.

Nordell, B. and Hellström, G., 2000, High temperature solar heated seasonal storage system for low temperature heating of buildings. *Solar Energy* **69**: 511–523.

Nowak, T., 2006, The market for geothermal heat pumps in Europe. *Proceedings of the 1st Groundhit (Ground Coupled Heat Pumps of High Technology) Workshop*, Brussels, May 18, 2006. Available at website http://www.geothermie.de/groundhit/, last accessed October 2007.

O'Connor, J.J. and Robertson, E.F., 1997, *Jean Baptiste Joseph Fourier.* University of St Andrews 'MacTutor History of Mathematics' website http://www-history.mcs.st-andrews.ac.uk/Biographies/Fourier.html, last accessed October 2007.

O'Connor, J.J. and Robertson, E.F., 2000, *Robert Boyle.* University of St Andrews 'MacTutor History of Mathematics' website http://www-history.mcs.st-andrews.ac.uk/Biographies/Boyle.html, last accessed October 2007.

O'Connor, J.J. and Robertson, E.F., 2003, *William Thomson (Lord Kelvin)*. University of St Andrews 'MacTutor History of Mathematics' website http://www-history.mcs.st-andrews.ac.uk/Biographies/Thomson.html, last accessed October 2007.

O'Neill, Z.D., Spitler, J.D. and Rees, S.J., 2006, Modeling of standing column wells in ground source heat pump systems. *Proceedings of Tenth International Conference on Thermal Energy Storage (ECOSTOCK 2006)*, The Richard Stockton College of New Jersey, 31 May–2 June 2006.

Odling, N., Banks, D. and Misund, A., 1994, A 3-dimensional, time-variant, numerical groundwater flow model of the Øvre Romerike aquifer, Southern Norway. *Norges Geologiske Undersøkelse Bulletin* **426**: 77–94.

Orio, C.D., Chlasson, A., Johnson, C.N., Deng, Z., Rees, S.J. and Spitler, J.D., 2005, A survey of standing column well installations in North America. *Transactions of the American Society of Heating, Refrigerating and Air-Conditioning Engineers (ASHRAE)* **111**(2): 109–121.

Owen, M. and Robinson, V.K., 1978, Characteristics and yield in fissured chalk. In: 'The Thames Groundwater Scheme', *Proceedings of Conference 'The Thames Groundwater Scheme', Reading University*, 12–13 April 1978, Institution of Civil Engineers, 33–49.

Park, C., Synn, J.H., Shin, H.S., Cheon, D.S., Lim, H.D. and Jeon, S.W., 2004, Experimental study on the thermal characteristics of rock at low temperatures. *International Journal of Rock Mechanics and Mining Science* **41**(3), Paper 1A 14: 1–6.

Parker, J.C.W., 2007, 600,000 hours and still going strong. Ground source heat pumps in social housing – a measured case history. *Unpublished information submitted to the Building Services Research and Information Association (BSRIA)/Carbon Trust.*

Parnachev, V.P., Banks, D., Berezovsky, A.Y. and Garbe-Schönberg, D., 1999, Hydrochemical evolution of Na-SO_4-Cl groundwaters in a cold, semi-arid region of southern Siberia. *Hydrogeology Journal* **7**: 546–560.

Perl'shtein, Z., Gulyi, S.A. and Buiskikh, A.A., 2000, Use of heat pumps to improve the bearing capacity of frozen soils. *Soil Mechanics and Foundation Engineering* **37**(3): 96–104.

Pollack, H.N., Hurter, S.J. and Johnson, J.R., 1993, Heat flow from the earth's interior. Analysis of the global data set. *Reviews of Geophysics* **31**: 267–280.

Prestwich, J., 1885, On underground temperatures, with observations on the conductivity of rocks, on the thermal effects of saturation and imbibition, and on a special source of heat in mountain ranges (abstract). *Proceedings of the Royal Society of London* **38** (1884–1885): 161–168.

Prestwich, J., 1886, On underground temperatures, with observations on the conductivity of rocks, on the thermal effects of saturation and imbibition, and on a special source of heat in mountain ranges. *Proceedings of the Royal Society of London* **41**: 1–116.

Price, M., 1996, *Introducing Groundwater*, 2nd Edition. Chapman & Hall, London, 278 pp.

Rabemanana, V., Durst, P., Bächler, D., Vuataz, F.D. and Kohl, T., 2003, Geochemical modelling of the Soultz-sous-Forêts Hot Fractured Rock system: Comparison of two reservoirs at 3.8 and 5 km depth. *Proceedings of the European Geothermal Conference*, 25–30 May 2003, Szeged, Hungary, Paper No. 0-1-08.

Rafferty, K., 2001, Design aspects of commercial open-loop heat pump systems. *GHC Bulletin (Geo-Heat Center, Klamath Falls, Oregon)* March 2001: 16–24.

Rafferty, K.D. and Culver, G., 1998, Heat exchangers. In: Lienau, P. (ed.), *Geothermal Direct Use Engineering and Design Guidebook*, 3rd Edition. Oregon Institute of Technology, Geo-Heat Center, Klamath Falls, USA, 261–277, Chapter 11.

Rakhesh, B., Venkatarathnam, G. and Srinivasa Murthy, 2003, Performance comparison of HFC227 and CFC114 in compression heat pumps. *Applied Thermal Engineering* **23**: 1559–1566.

Rambaut, A.A., 1900, Underground temperature at Oxford in the year 1899, as determined by five platinum-resistance thermometers. *Philosophical Transactions of the Royal Society of London, Series A* **195**: 235–258.

Rawlings, R.H.D. and Sykulski, J.R., 1999, Ground source heat pumps: A technology review. *Building Services Engineering Research and Technology* **20**: 119–129.

Richardson, R.N., 2003, The cooling potential of cryogens: Part 1 – The early development of refrigeration and cryogenic cooling technology. *EcoLibrium (Australian Institute of Refrigeration Air Conditioning and Heating)*, May 2003: 10–14.

Rollin, K.E., 1987, Catalogue of geothermal data for the land area of the United Kingdom. 3rd Revision: April 1987. Investigation of the geothermal potential of the UK. British Geological Survey, 16 pp + Tables.

Rouse, W.R., Oswald, C.J., Binyamin, J., *et al.*, 2005, The role of northern lakes in a regional energy balance. *Journal of Hydrometeorology* **6**: 291–305.

Rosén, B., Gabrielsson, A., Fallsvik, J., Hellström, G. and Nilsson, G., 2001, System för värme och kyla ur mark – En nulägesbeskrivning [*Systems for ground source heating and cooling – a status report – in Swedish*]. *Varia* **511**, Statens Geotekniska Institut Linköping, Sweden.

Russell, H.W., 1935, Principles of heat flow in porous insulators. *Journal of the American Ceramic Society* **18**: 1–5.

Rybach, L., 2003a. Geothermal energy: sustainability and the environment. *Proceedings of the European Geothermal Conference*, 25–30 May 2003, Szeged, Hungary, Paper No. 1-2-03.

Rybach, L., 2003b, Regulatory framework for geothermal in Europe, with special reference to Germany, France, Hungary, Romania, and Switzerland. *United Nations University, Geothermal Training Programme, International Geothermal Conference IGC2003 Short Course, September 2003, Reykjavík, Iceland*, 43–52.

Rybach, L. and Sanner, B., 2000, Ground-source heat pumps. The European perspective. *GHC Bulletin (Geo-Heat Center, Klamath Falls, Oregon)*, March 2000: 16–26.

Sæther, O.M., Misund, A., Ødegård, M., Andreassen, B.T. and Voss, A., 1992, Groundwater contamination at Trandum landfill, Southeastern Norway. *Norges Geologiske Undersøkelse Bulletin* **422**: 83–95.

Šafanda, J. and Rajver, D., 2001, Signature of the last ice age in the present subsurface temperatures in the Czech Republic and Slovenia. *Global and Planetary Change* **29**: 241–257.

Sanner, B., 2001, Some history of shallow geothermal energy use. *Proceedings of UNESCO/IGA International Summer School on Direct Application of Geothermal Energy/International Geothermal Days Germany 2001 (Bad Urach).*

Sanner, B., 2002, Geothermal heat pumps and geothermal district heating systems – the German experience. UNESCO/IGA International Summer School on Direct Application of Geothermal Energy, Greece 2002. Available at website http://www.geothermie.de/egec-geothernet/ci_prof/europe/germany/Sanner2.doc.PDF, last accessed October 2007.

Sanner, B., 2006, *Description of ground source types for the heat pump.* Available at website http://www.geothermie.de/oberflaechennahe/description_of_ground_source_typ.htm, last accessed December 2006.

Sanner, B. and Hellström, G., 1998, UTES with borehole heat exchangers in Central and Northern Europe. *IEA Heat Pump Center Newsletter* **16**(2): 24–26.

Sanner, B. and Knoblich, K., 1998, New IEA-Activity ECES Annex 12: 'High Temperature Underground Thermal Energy Storage'. *Proceedings of The Second Stockton International Geothermal Conference*, 16–17 March 1998, Richard Stockton College of New Jersey, USA.

Sanner, B. and Nordell, B., 1998, Underground thermal energy storage with heat pumps: An international overview. *IEA Heat Pump Center Newsletter* **16**(2): 10–14.

Sanner, B., Hellström, G., Spitler, J. and Gehlin, S., 2005, Thermal response test – current status and world-wide application. *Proceedings World Geothermal Congress 2005*, 24–29 April 2005, Antalya, Turkey.

Sanner, B., Reuss, M., Mands, E. and Mueller, J., 2000, Thermal Response Test – Experiences in Germany. *Proceedings of TERRASTOCK 2000 (Stuttgart)*, 177–182.

Sanyal, S.K., 2005, Cost of geothermal power and factors that affect it. *Proceedings World Geothermal Congress 2005*, 24–29 April 2005, Antalya, Turkey, 10 pp.

SAP, 2005, *Building Regulations Standard Assessment Procedure.* Available from UK Buildings Research Establishment website http://projects.bre.co.uk/sap2005/pdf/SAP2005.pdf, last accessed October 2007.

Schaefer, L.A., 2000, *Single Pressure Absorption Heat Pump Analysis.* PhD dissertation, Georgia Institute of Technology, USA, May 2000, 176 pp.

SEI, 2007, *Sustainable Energy Ireland.* Available at website http://www.sei.ie/, last accessed February 2007.

SEPA, 2004, *Decommissioning Redundant Boreholes and Wells.* Scottish Environment Protection Agency, 7 pp.

Shonder, J.A. and Beck, J.V., 2000, A new method to determine the thermal properties of soil formations from in-situ field tests. *Oak Ridge National Laboratory (US) Report* **ORNL/TM-2000/97**, Oak Ridge National Laboratory, Tennessee, USA.

Shook, M.J., 2001a, Predicting thermal breakthrough in heterogeneous media from tracer tests. *Geothermics* **3**: 573–589.

Shook, M.J., 2001b, *Predicting Thermal Velocities in Fractured Media from Tracer Tests.* Unpublished paper, Idaho National Laboratory, USA.

Signorelli, S., Bassetti, S., Pahud, D. and Kohl, T., 2007, Numerical evaluation of thermal response tests. *Geothermics* **36**: 141–166.

SINTEF, 2007, *Fjernvarme-/fjernkjøleanlegg - Oslo Lufthavn (Gardermoen).* IEA Heat Pump Programme – Annex 29. Ground-Source Heat Pumps Overcoming Market and Technical Barriers. Available at web site http://www.energy.sintef.no/prosjekt/Annex29/, last accessed March 2007. SINTEF, Trondheim, Norway.

Skarphagen, H., 2006, Grunnvarme – Varmeveksling med grunnen *[Ground source heat – heat exchange with the ground: in Norwegian]*. Distance learning module developed for the University of Oslo.

Smith, D.G. (ed.), 1981, *The Cambridge Encyclopedia of the Earth Sciences.* Cambridge University Press, Cambridge, UK, 496 pp.

Smith, M., 2000, Southampton energy scheme. *Proceedings World Geothermal Congress, Kyushu – Tohoku, Japan,* 28 May–10 June 2000.

Snow, D.T., 1969, Anisotropic permeability of fractured media. *Water Resources Research* **5**(6): 1273–1289.

Solik-Heliasz, E. and Małolepszy, Z., 2001, The possibilities of utilization of geothermal energy from mine waters in the Upper Silesian Coal Basin. *Proceedings of International Scientific Conference 'Geothermal Energy in Underground Mines',* 21–23 November 2001, Ustroń, Poland, 81–88.

Spencer, B.B., Wang, H. and Anderson, K.K., 2000, Thermal conductivity of IONSIV® IE-911™ crystalline silicotitanate and Savannah River waste simulant solutions. *Oak Ridge National Laboratory (Chemical Technology Division) Report* **ORNL/TM-2000/285**: 37 pp.

Spitler, J.D., 2000, GLHEPRO – A design tool for commercial building ground loop heat exchangers. *Proceedings of the Fourth International Heat Pumps in Cold Climates Conference,* 17–18 August 2000, Aylmer, Québec.

Spitler, J.D., 2005, Ground source heat pump system research – past, present and future. *International Journal of HVAC&R Research* **11**(2): 165–167.

Stefansson, V. and Axelsson, G., 2003, Sustainable utilization of geothermal energy resources. *United Nations University, Geothermal Training Programme IGC2003 – Short Course,* Orkustofnun, Grensásvegur 9, September 2003, IS-108 Reykjavík, Iceland.

Sterne, L., 1761, *The Life and Opinions of Tristram Shandy, Gentleman,* Vol. IV. Dodsley, London.

Stiles, L, 1998, Underground thermal energy storage in the U.S. *IEA Heat Pump Center Newsletter* **16**(2): 22–23.

Stine, W.B. and Geyer, M., 2001, *Power from the Sun*. Available via website http://www.powerfromthesun.net/book.htm, last accessed October 2007.

Sumner, J.A., 1948, The Norwich heat pump. *Proceedings of the Institution of Mechanical Engineers* **158/1**: 22–29, with discussion, 39–51.

Sumner, J.A., 1976a, *Domestic Heat Pumps*. Prism Press, Dorchester, 117 pp.

Sumner, J.A., 1976b, *An Introduction to Heat Pumps*. Prism Press, Dorchester, 55 pp.

Sundberg, J., 1991, Termiska egenskaper i jord och berg *[Thermal properties of soils and rock – in Swedish]*. Statens geotekniska institut (SGI) Report, Linköping, Sweden.

SVEP, 1998, Lathund – för energidimensionering av ett-hålssystem *[Energy-dimensioning of single borehole systems for lazy blighters! – in Swedish]*. Svenska Värmepumpföreningen (SVEP – *Swedish Heat Pump Association*), Stockholm.

Taniguchi, M. and Uemura, T., 2005, Effects of urbanization and groundwater flow on the subsurface temperature in Osaka, Japan. *Physics of the Earth and Planetary Interiors* **152**: 305–313.

Tetens, O., 1930, Uber einige meteorologische Begriffe. *Zeitschrift fur Geophysik* **6**: 297.

Theis, C.V., 1935, The relation between the lowering of the piezometric surface and the rate and duration of discharge of a well using ground-water storage. *Transactions of the American Geophysical Union (Reports and Papers – Hydrology)* **16**: 519–524.

Theis, C.V., 1940, The source of water derived from wells: Essential factors controlling the response of an aquifer to development. *Civil Engineering* **10**(5): 277–280.

Thiem, A., 1887, Verfahress fur Naturlicher Grundwassergeschwindegkiten. *Polytechnisches Notizblatt* **42**: 229.

Thomson, W., 1852, On the economy of heating or cooling of buildings by means of currents of air. *Proceedings of the Royal Philosophical Society Glasgow* **3**: 269–272.

Thomson, W., 1864, On the secular cooling of the earth. *Transactions of the Royal Society of Edinburgh* **23**: 167–169 (read 28 April 1862).

Thomson, W., 1868, On geological time. Address delivered before the Geological Society of Glasgow, February 27, 1868, *Popular Lectures and Addresses* **2**: 10.

Tovey, K., 2006, Towards carbon neutrality. *Proceedings of the HSBC Partnership in Environmental Innovation Seminar 'Decarbonising Culture'*, 23 November 2006, University of Newcastle-upon-Tyne, UK.

UNDP, 2004, *World energy assessment. Overview. 2004 Update.* United Nations Development Programme, Bureau for Development Policy, New York, USA. Available at website http://www.undp.org/energy/docs/WEAOU_full.pdf, last accessed October 2007.

UNEP, 2003, *2002 Report of the Refrigeration, Air Conditioning and Heat Pumps Technical Options Committee*. United Nations Environment Programme (UNEP) Ozone Secretariat Nairobi, January 2003, 197 pp.

University of Alabama, 1999, Pond loops, lake loops or surface water heat pumps. *Outside The Loop* **2**(1): 3–4. University of Alabama, USA.

USGS, 2004, National field manual for the collection of water-quality data. *United States Geological Survey, Office of Water Quality. Handbooks for Water-Resources Investigations Book* **9**. Available at website http://water.usgs.gov/owq/FieldManual/index.html, last modified April 15.

Vasil'ev, G.P., 2003, Geothermal heat pump systems of heat supply (GHPS): operating experience, technical and ecological aspects of rational integration into power balance of Russia. *Proceedings of the International Geothermal Workshop, Sochi, Russia*, October 2003, Paper number W00035. Available at website http://pangea.stanford.edu/ERE/pdf/IGAstandard/Russia/IGW2003/W00035.PDF, last accessed October 2007.

Vasil'ev, G.P., Khrustachev, L.V., Rozin, A.G., Abuev, I.M, Gornov, V.F., Orlov, V.O. and Vorob'ev, N.B., 2001, Руководство по применению тепловых насосов с использованием вторичных энергетических ресурсов и нетрадиционных возобновляемых источников энергии [*Management of heat pump applications utilising secondary energy resources and non-traditional renewable energy resources – in Russian*]. ГУП "НИАЦ" *(Scientific research analytical centre of Moskom-architects)*, Moscow.

VDI, 2000, Thermal use of the underground: Fundamentals, approvals, environmental aspects. *Verein Deutscher Ingenieure. Richtlinien* **VDI 4640**, Blatt 1/Part 1, December 2000, Düsseldorf, 32 pp.

VDI, 2001a, Thermal use of the underground: Ground source heat pump systems. *Verein Deutscher Ingenieure. Richtlinien* **VDI 4640**, Blatt 2/Part 2, September 2001, Düsseldorf, 43 pp.

VDI, 2001b, Utilization of the subsurface for thermal purposes: Underground thermal energy storage. *Verein Deutscher Ingenieure. Richtlinien* **VDI 4640**, Blatt 3/Part 3, June 2001, Düsseldorf, 42 pp.

VDI, 2004, Thermal use of the underground: Direct uses. *Verein Deutscher Ingenieure. Richtlinien* **VDI 4640**, Blatt 4/Part 4, September 2004, Düsseldorf, 40 pp.

Vos, L., 2007, Underground thermal energy storage in the Netherlands. *Proceedings of 27th Annual Conference of the Irish Group of the International Association of Hydrogeologists*, 24–25 April 2007, Tullamore, 4–17 to 4–24.

Walsh, J.B., 1981, Effect of pore pressure and confining pressure on fracture permeability. *International Journal of Rock Mechanics and Mining Sciences & Geomechanics Abstracts* **18**: 429–435.

Waples, D.W. and Waples, J.S., 2004a, A review and evaluation of specific heat capacities of rocks, minerals, and subsurface fluids. Part 1: Minerals and nonporous rocks. *Natural Resources Research* **13**(2): 97–122.

Waples, D.W. and Waples, J.S., 2004b, A review and evaluation of specific heat capacities of rocks, minerals, and subsurface fluids. Part 2: Fluids and porous rocks. *Natural Resources Research* **13**(2): 123–130.

Ward, R.W., 1992, The constants of alpha-quartz (1992 update). *Proceedings of 14th Piezoelectric Devices Conference and Exhibition*, Kansas City (15–17 September 1992) **2**: 61–70.

Waters, A. and Banks, D., 1997, The Chalk as a karstified aquifer: Closed circuit television images of macrobiota. *Quarterly Journal of Engineering Geology* **30**: 143–146.

Weightman, G., 2003, *The Frozen Water Trade*. Harper Collins, London, 200 pp.

Westaway, R., Maddy, D. and Bridgland, D., 2002, Flow in the lower continental crust as a mechanism for the Quaternary uplift of southeast England: Constraints from the Thames terrace record. *Quaternary Science Reviews* **21**: 559–603.

Wheildon, J. and Rollin, K.E., 1986, Heat flow. In: Downing, R.A. and Gray, D.A. (eds.), *Geothermal Energy – the Potential in the United Kingdom*. British Geological Survey/HMSO, London, 8–20, Chapter 2.

White, R.R. and Clebsch, A., 1994, C.V. Theis: The man and his contributions to hydrogeology. In 'Selected contributions to ground-water hydrology by C.V. Theis, and a review of his life and work', *US Geological Survey Water-Supply Paper* **2415**. Available at website http://www.olemiss.edu/sciencenet/saltnet/theisbio.html, last accessed October 2007.

Whitlock, C.H., Brown, D.E., Chandler, W.S., DiPasquale, R.C., Meloche, N., Leng, G.J., Gupta, S.K., Wilber, A.C., Ritchey, N.A., Carlson, A.B., Kratz, D.P. and Stackhouse, P.W., 2000, *Release 3 NASA Surface Meteorology and Solar Energy Data Set for Renewable Energy Industry Use.* Rise & Shine 2000, the 26th Annual Conference of the Solar Energy Society of Canada Inc. and Solar, 21–24 October 2000, Halifax, Nova Scotia, Canada. Cited at website http://www.apricus.com/html/insolation_levels_europe.htm, last accessed October 2007.

Wickersham, G., 1977, Review of C.E. Jacob's doublet well. *Ground Water* **15**(5): 344–347.

Worsøe Schmidt, P., 1982, Modern trends in heat pump development. *International Journal of Refrigeration* **5**(2): 70–73.

Younger, P.L., 2000, Nature and practical implications of heterogeneities in the geochemistry of zinc-rich, alkaline mine waters in an underground F–Pb mine in the UK. *Applied Geochemistry* **15**: 1383–1397.

Younger, P.L., 2004, 'Making water': The hydrogeological adventures of Britain's early mining engineers. In: Mather, J.D. (ed.), *200 Years of British Hydrogeology, Geological Society Special Publication* **225**: 121–157.

Younger, P.L., 2006, Potential pitfalls for water resources of unregulated development of ground-source heat resources. *Proceedings of the Annual Conference of the Chartered Institution of Water & Environmental Management*, 13–15 September 2006, Newcastle-upon-Tyne, UK.

Glossary

Absolute Zero: The temperature at which a material has no utilisable thermal energy. Zero on the Kelvin scale or −273.15°C on the Celsius scale.

Aestifer: A body or stratum of rock or sediment that has adequate thermal properties (thermal conductivity, heat capacity) to permit the economic abstraction of ground source heat.

agl: above ground level.

Air-conditioning: The modification and control of the temperature, humidity, quality and circulation of air, usually in an interior space. Normally, the term refers to the cooling of circulating air, achieved by one or a combination of the following mechanisms: direct circulation of, or heat exchange with, a naturally cool fluid or medium; artificial chilling via a refrigeration (heat pump) cycle; evaporative cooling. Chilling of circulating air is often accompanied by dehumidification and removal of condensed water (as the air cools, the amount of water vapour it is able to contain decreases).

Air cycle heat pump: In an open, air-cycle heat pump (of the type first proposed by William Thomson in 1852), mechanical compression and expansion are performed on the fluid that actually delivers the heat effect – air. In Thompson's concept, air is allowed to expand into a reservoir, resulting in cooling. The cooled air absorbs heat from the environment through the reservoir walls and then is mechanically compressed into a second cylinder. From here, the warm air is released in a controlled manner to the space to be heated.

Albedo: A measure of the reflectivity of the earth's surface – the ratio of reflected sunlight (electromagnetic radiation) to incident light. The albedo of fresh snow can be as high as 80%, whereas the albedo of water is usually less than 10%.

Aperture: The width of a fracture opening in a rock, or the diameter of a pore space in a sediment.

Aquifer: A body or stratum of rock or sediment that has adequate hydraulic properties (transmissivity, hydraulic conductivity, storage) to permit the economic abstraction of groundwater.

ASHRAE: American Society of Heating, Refrigerating and Air-Conditioning Engineers.

asl: above sea level.

ASME: American Society of Mechanical Engineers.

ATES: Aquifer thermal energy storage. A form of UTES where heat is stored in the form of warm groundwater (carrying a cargo of waste heat from the summer cooling season) that has been injected to an aquifer. The warm groundwater is re-abstracted in winter to provide heating.

Asthenosphere: The plastic, flowing part of the earth's subsurface, below the more rigid *lithosphere*. The asthenosphere lies in the upper mantle.

Balanced scheme: A balanced ground source heat pump scheme will have approximately similar annual loads of heat extracted and rejected to the ground. An unbalanced scheme will be strongly biased, over the course of a year, towards net heating or net cooling.

bgl: below ground level.

Bivalent scheme: A ground source heating/cooling scheme that satisfies only a portion of a building's heating or cooling demand, and is supplemented by alternative heating or cooling arrangements.

BTES: Borehole thermal energy storage. The storage of surplus heat in the subsurface for re-extraction at times of high demand, specifically by means of closed-loop arrangements installed in boreholes. BTES is a type of UTES.

bwt: below well top.

Cementitious Thermally Enhanced Grout. The State of New Jersey, USA (GeoExchange, 2003), defines this as a mixture of 94 lbs (43 kg) of cement with 200 lbs (91 kg) dry fine silica sand, 1.04 lb (470 g) sodium bentonite and 21 oz (600 g) sulphonated naphthalene superplasticiser and 6.19 gal (23.4 L) water to give a target density of 2.18 kg L^{-1}.

Closed-loop ground source heating system: A system whereby heat is extracted from the ground by a chilled 'carrier' fluid, circulating within a buried heat exchanger – typically a subsurface loop of pipe installed in a trench or borehole.

Coefficient of performance (COP): The ratio of heating (or cooling) effect delivered by a heat pump to the electricity (or other primary energy) required to power the compressor. It is an instantaneous measure and will depend on the temperature of the heat source and heat sink at that moment in time (see Seasonal performance factor).

Coefficient of thermal expansion: The *linear coefficient of thermal expansion* (α_L) is the fractional increase in the length (L) of a bar of material (or in the dimension of an object) for every degree temperature (θ) rise. It is cited in K^{-1} and is given by the formula $\frac{1}{L}\frac{\partial L}{\partial \theta}$. For rocks, values of 3×10^{-6} to 9×10^{-6} K^{-1} are typical, depending on porosity and mineral composition. The volumetric coefficient of thermal expansion (α_V) is the fractional increase in volume per degree temperature rise. In general, the value of α_V is around three times α_L.

Confined aquifer: A transmissive stratum of sediment or rock that is overlain by a low permeability *aquitard* and where the groundwater head is higher than the top of the aquifer.

Darcy velocity: The flux rate of groundwater through an aquifer or porous medium per unit cross-sectional area (i.e. $m^3\ s^{-1}$ per m^2, or $m\ s^{-1}$). As groundwater can only flow through pore apertures or fractures, the Darcy velocity (v_D) is related to the actual average linear velocity (v) by the expression $v_D = vn_e$, where n_e is the effective porosity.

DX: Direct expansion or direct circulation heat pump system – a type of closed-loop system where refrigerant circulates in the ground and where the ground loop functions as the evaporator (or condenser) of the heat pump.

EA: Environment Agency (of England and Wales).

Emissivity: The ratio of energy radiated by a given real material to the energy radiated by an ideal black body at the same temperature.

EU: European Union.

Enthalpy: A measure of the heat content of a substance per unit mass, depending not only on temperature, but also on pressure and volume (Boyle, 2004).

Fahrenheit: An outdated system of measuring temperature, still used in the USA. On the Fahrenheit scale, water freezes at 32°F and boils at 212°F. To convert a temperature in Fahrenheit (θ_F) to a temperature in Centigrade (θ_C), use the formula:

$$\theta_C = \frac{5(\theta_F - 32)}{9}$$

Geothermal energy: In this book, we have chosen to restrict the term to relatively high temperature, high enthalpy heat occurring in the geosphere. As such, geothermal energy is typically accessed either via very deep boreholes or in locations with a naturally high geothermal gradient (e.g. Iceland), or both. This definition will annoy purists, who will argue that the science of geothermics includes thermogeology!

Geothermics: The study of geothermal energy.

Ground source heat: Very low enthalpy heat that exists in the subsurface at 'normal' temperatures. The type of heat that we are considering in the science of thermogeology.

GSHP: Ground source heat pump.

Head: A measure of the potential energy of water due to a combination of elevation (z) and pressure (P). Groundwater always flows from high head to low head. Head (h) is defined by the formula:

$$h = \frac{P}{\rho g} + z,$$

where g = acceleration due to gravity and ρ = fluid density.

HDPE: High-density polyethene.

HVAC: Heating, ventilation and air-conditioning.

Hydrogeology: The study of the occurrence, movement and exploitation of water in the geosphere. The study of groundwater.

ID: Internal diameter (of a pipe).

IGSHPA: International Ground Source Heat Pump Association (currently based in Oklahoma, USA).

Kelvin: The SI unit of temperature. A degree Kelvin is equal to a degree Centigrade, but the zero points are different: 0 K is absolute zero (−273.15°C), while 0°C is the freezing point of water. To convert a temperature in Centigrade (θ_C) to an absolute temperature (θ^o) in Kelvin, use the formula:

$$\theta^o = \theta_C + 273.15$$

Latent heat: The heat absorbed or released by a material purely by virtue of a change in phase – for example, from ice to water or from water to steam. The absorption or release of latent heat is not accompanied by any transfer of 'sensible heat' – that is, any change in temperature of the material.

Lithosphere: The upper, relatively rigid part of the earth's subsurface, floating on top of the plastic, ductile *asthenosphere*. The lithosphere comprises the earth's crust and the uppermost part of the mantle.

mOD: metres above Ordnance datum. A bizarre British way of saying 'metres above sea level'!

MDPE: Medium-density polyethene.

Monovalent scheme: A ground source scheme that fully satisfies a building's heating or cooling demand (see 'bivalent scheme').

Net radiation (R_n): The net radiative flux incident on one square metre of the earth's surface. It comprises the sum of short-wave insolation and long-wave atmospheric radiation, less reflection and less long-wave back radiation from the earth's surface. It varies throughout the day (it may be positive during the morning and negative during the night), but its annual average value is typically positive and several tens of W m^{-2} in temperate regions, exceeding 100 W m^{-2} in the tropics. See Box 3.5.

NGWA: National Ground Water Association (of the USA).

OD: Outer diameter (of a pipe).

Open hole: The practice of drilling into a well-lithified aquifer without any form of well screen or casing. The sides of the well or borehole are strong and stable enough to support themselves.

Open-loop ground-coupled heating system: A system whereby water is physically abstracted from a well or spring in an aquifer (or another natural water body: a flooded mine, a lake, a river or the sea). Heat is extracted directly from this flux of water with its advected 'cargo' of heat.

PE: Polyethene (alternatively named polythene or polyethylene).

PP: Polypropene (polypropylene)

Prophylactic heat exchanger: A heat exchanger (typically a plate heat exchanger) that transfers heat between a natural fluid (whose composition is not controlled and which may lead to problems of corrosion, biofouling or chemical incrustation) and a carrier fluid of controlled composition (e.g. anti-freeze solution). This ensures that the natural water does not enter a building loop or heat pump, thus protecting the evaporator from corrosion or fouling. However, the prophylactic heat exchanger may be subject to these problems and may need to be regularly cleaned, maintained or even (occasionally) replaced.

PVC: Polyvinyl chloride.

Screen temperature: The air temperature as measured in a Stevenson screen: a louvred, shaded, ventilated white box, mounted around 1.5 m above ground level.

SCW: Standing column well (see Chapter 11).

Seasonal performance factor (SPF): The ratio of heating (or cooling) effect, integrated over an entire heating or cooling season, to the amount of electricity (or other primary energy) consumed by the compressor over the same period. A *system* performance factor is the ratio of heat output to total electrical energy consumed by the compressor and any auxiliary devices (circulation pumps, etc.).

Sensible heat: The transfer of heat to or from a material that manifests itself as a change in temperature of the material (contrast with 'latent heat').

SEPA: Scottish Environment Protection Agency.

Sialic: A rock that is rich in *si*lica and *al*uminium minerals. This chemical composition is typical of the upper continental crust.

Specific heat capacity (S_C): This describes how good a material is at storing heat. It is defined as the amount of heat released from a unit mass of material, corresponding to a 1 K temperature change. It is measured in J kg^{-1} K^{-1}.

Specific installed thermal output: For a closed-loop scheme, this is the peak installed heating or cooling output of a ground source heat scheme, divided by the number of metres of borehole (vertical systems) or trench (horizontal systems) required to support the system. It is thus typically cited in W m^{-1}.

Specific thermal absorption: For a closed-loop scheme, this is the peak rate of heat transfer between ground and ground loop (i.e. the amount of heat extracted from or dumped to the ground), divided by the number of metres of borehole (vertical systems) or trench (horizontal systems) required to support the system. It is thus typically cited in W m^{-1}. In heating mode it is *lower* than the *specific installed thermal output* (because the electrical energy input to the heat pump also contributes to the heating system's total output). In active cooling mode it is *higher* than the *specific installed thermal output.*

Temperature: A measure of the thermal potential energy of a fluid or body. At a molecular level, temperature is related to the kinetic or vibrational energy of motion of the molecules. Temperature can be measured in degrees Fahrenheit (°F), degrees Centigrade (°C) or Kelvin (K).

Thermal conductivity: This describes how good a material is at conducting heat. It is measured in W m^{-1} K^{-1}. It is defined by Fourier's Law: a material of thermal conductivity 1 W m^{-1} K^{-1} under a temperature gradient of 1 K m^{-1} will conduct 1 W of heat through every m^2 of its cross section.

Thermal Grout 85: A specific grout mix of quartz sand, bentonite and water. The State of New Jersey (GeoExchange, 2003) defines it as mix of 54 lbs (24 kg) bentonite and 200 lbs (91 kg) fine dry silica sand, mixed with 17.5 gal (66 L) water to achieve a density of 13.1 lb kg^{-1} (1.57 kg L^{-1}).

Thermogeology: The study of the occurrence, movement and exploitation of low enthalpy heat in the relatively shallow geosphere. By 'relatively shallow', we are typically talking of depths of down to 200 m or so. By 'low enthalpy', we are usually considering temperatures of less than 30°C.

USEPA: United States Environmental Protection Agency.

UTES: Underground Thermal Energy Storage. Using the subsurface as a heat reservoir, to store excess or waste heat from cooling activities during the summer. This 'waste' heat can then be re-extracted during winter. The term encompasses both ATES and BTES schemes.

VDI: Verein Deutscher Ingenieure (German Association of Engineers).

Volumetric heat capacity (S_{VC}): This describes how good a material is at storing heat. It is defined as the amount of heat released from a unit volume of material, corresponding to a 1 K temperature change. It is measured in J m^{-3} K^{-1}.

Symbols

A note on the dimensionality of energy and power

The Joule is the SI unit of energy. It can be regarded as dimensionally equivalent to work. *Work* can be defined as *Force* × *Distance over which force is applied*.

The dimension of energy and work is thus $[M][L][T]^{-2}[L]$ or $[M][L]^2[T]^{-2}$.

1 Joule = 1 kg m^2 s^{-2}

The Watt is the SI unit of power, or rate of delivery of energy. The dimension of power is thus $[M][L]^2[T]^{-3}$.

1 Watt = 1 kg m^2 s^{-3}

Note, however, that when considering electricity, *Power* = *Current* × *Voltage*.

1 Watt = 1 V C s^{-1}

General symbols

A = cross-sectional area perpendicular to direction of heat flow or groundwater flow; $[L]^2$; typically in m^2.

e = 2.7182 (the base to which natural logarithms are calculated). Note that ln(10) = $\log_e(10)$ = 2.303.

g = acceleration due to gravity = c. 9.81 m s^{-2}; $[L][T]^{-2}$.

γ = Euler's constant = 0.5772.

L = distance coordinate; [L]; typically in m.

m = mass; [M]; typically in kg.

P = pressure; $[M][L]^{-1}[T]^{-2}$; typically in Pascals (Pa). 1 Pa = 1 Newton per m^2 (N m^{-2}).

r = radial coordinate; [L]; typically in m.

t = time; [T]; typically in s.

W = work performed; $[M][L]^2[T]^{-2}$; typically in Joules (J).

W = rate of work performed; $[M][L]^2[T]^{-3}$, typically in J s^{-1} or W.

x = distance coordinate; [L]; typically in m.

z = distance coordinate in the vertical direction; [L]; typically in m. Depending on the context this may be either depth below the earth's surface *or* elevation above some arbitrary datum (i.e. take care!).

Heat flow symbols

Å = internal (e.g. radiogenic) heat production per unit volume of earth's crust; $[M]^{-2}[L]^2[T]^{-3}$; typically in W m^{-3}.

α_L = linear coefficient of thermal expansion; typically in K^{-1} (see Glossary).
α_V = volumetric coefficient of thermal expansion; typically in K^{-1} (see Glossary).
α_{sw}, α_{lw} = short-wave and long-wave albedo of a lake or ground surface; typically expressed as a fraction or percentage.
C = cooling load delivered to building (i.e. rate of heat extraction from building); $[M][L]^2[T]^{-3}$; typically in J s^{-1} or W.
D = effective thickness of aestifer *or* depth of borehole over which heat extraction takes place; [L]; typically in m.
E = efficiency (usually expressed as a ratio or percentage).
E_b = 'black body' energy radiated; typically in W m^{-2}.
ε = emissivity: the ratio of energy radiated by a given real material to the energy radiated by an ideal black body at the same temperature.
F = flow rate of carrier fluid in closed-loop system; $[L]^3[T]^{-1}$.
F_{turb} = flow rate of carrier fluid in closed-loop system necessary to ensure turbulent flow; $[L]^3[T]^{-1}$.
G = rate of heat extraction from, or dumping to, a ground-coupled receptor; $[M][L]^2[T]^{-3}$; typically in J s^{-1} or W.
H = heating load delivered to building; $[M][L]^2[T]^{-3}$; typically in J s^{-1} or W.
H_{in} = heat input to a heat engine; $[M][L]^2[T]^{-3}$; typically in J s^{-1} or W.
$\bar{h}$ = local coefficient of heat transfer; typically W m^{-2} K^{-1}.
L_V = latent heat of vaporization; typically MJ kg^{-1} (or MJ L^{-1} for liquids).
λ = thermal conductivity; typically in W m^{-1} K^{-1}.
λ_g = thermal conductivity of grout in a closed-loop borehole; typically in W m^{-1} K^{-1}.
λ_{bh}, λ_{bv} = bulk horizontal and bulk vertical thermal conductivities of anisotropic rocks (see Box 10.1), typically in W m^{-1} K^{-1}.
Λ = thermal conductance, often in W m^{-2} K^{-1} (i.e. per m^2 of the material), or in W K^{-1} for an entire structure.
Q = Rate of heat flow; $[M][L]^2[T]^{-3}$; typically in J s^{-1} or W.
q = Specific rate of heat flow or transfer; rate of heat transfer per metre length of borehole; $[M][L][T]^{-3}$; typically in J s^{-1} m^{-1} or W m^{-1}.
q^* = Specific rate of heat flow; rate of heat flow per metre of cross-sectional area; $[M][T]^{-3}$; typically in J s^{-1} m^{-2} or W m^{-2}.
r_{array} = effective radius of a cylindrical or hexagonal array of closed-loop boreholes; [L]; typically in m.
r_b = radius of closed-loop borehole; [L]; typically in m or mm.
r_U = the radius of the U-tube in a closed-loop borehole [L]; typically in m or mm.
R = universal gas constant = 8.314 J K^{-1} mol^{-1}.
R = thermal resistance, typically in m^2 K W^{-1}.
R_b = thermal resistance of a borehole; typically in m^2 K W^{-1} per m of borehole, or K m W^{-1}.
R_n = net incident radiation on the earth's surface; $[M][T]^{-3}$; typically in J s^{-1} m^{-2} or W m^{-2}.
S_C = specific heat capacity; typically in J kg^{-1} K^{-1}.

S_{VC} = volumetric heat capacity; typically in J m^{-3} K^{-1}.
S_{VCwat} = volumetric heat capacity of water = c. 4.18 MJ m^{-3} K^{-1} (but is temperature dependent).
S_{VCaq} = volumetric heat capacity of a saturated aquifer; typically in J m^{-3} K^{-1}.
S_{VCcar} = volumetric heat capacity of a carrier fluid; typically in J m^{-3} K^{-1}.
SDR = the ratio of a pipe's outer diameter to its wall thickness.
σ = Stefan–Boltzmann constant = 5.67×10^{-8} W m^{-2} K^{-4}.
t_{earth} = hypothetical age of the earth; [T].
t_s = time taken for a closed-loop system to approach steady state (the time at which a steady-state approximation begins to be a better description of the loop's behaviour than a radial flow model); [T].
T_{the} = thermal transmissivity. The product of thermal conductivity and thickness of an aestifer; typically in W K^{-1}.
θ = temperature; typically in °C.
θ^o = absolute temperature in K.
θ_o = ambient temperature of groundwater *or* initial, far-field temperature of an aestifer.
θ_b = the average temperature of the carrier fluid in a closed-loop borehole (i.e. the average of the upflow and downflow fluid temperatures).
θ_i = initial temperature of the earth's molten interior (according to Kelvin's 'cooling earth' model).
θ_s = temperature of the earth's surface.
$\theta_{s,b}$ = the average steady-state temperature of carrier fluid in borehole.
θ_{sur} = temperature of the surface of a lake.
$\Delta\theta$ = a temperature change or differential; typically in K.
U-value = a buildings engineering designation for thermal conductance of a building component per square metre of area, often in W m^{-2} K^{-1}.
U = total thermal conductance of a building in W K^{-1}.
V_{aest} = volume of aestifer under consideration; $[L]^3$; typically in m^3.
ψ = current geothermal gradient; in K m^{-1}.

Groundwater flow symbols

b_a = aperture of a fracture; [L].
B = bleed rate, as a fraction of total fluid flow in a standing column well ($0 < B < 1$).
B = linear head losses in a pumping well; $[T][L]^{-2}$.
B' = linear well losses in a non-ideal pumping well; $[T][L]^{-2}$.
C = non-linear head losses in a non-ideal pumping well; $[T]^n[L]^{(1-3n)}$.
D = borehole depth or effective thickness of aquifer; [L]; typically in m.
D^* = dispersion coefficient; $[L]^2[T]^{-1}$.
h = groundwater head; [L]; typically in m relative to some arbitrary datum.
H = elevation of water table; [L]; relative to the base of an unconfined aquifer.
K = hydraulic conductivity; $[L][T]^{-1}$; typically in m s^{-1} or m d^{-1}.

L = separation; [L]; between abstraction and injection wells in a doublet scheme.
λ^* = effective thermal conductivity of the saturated aquifer material, modified to take into account effects of hydrodynamic dispersion.
μ, μ_W = the dynamic viscosity of a fluid or of water; $[M][T]^{-1}[L]^{-1}$; typically in kg s^{-1} m^{-1}.
n = power law coefficient in the Hantush–Bierschenk equation – Equation 7.18.
n = porosity; [dimensionless]; typically expressed as a fraction or percentage.
n_e = effective porosity; [dimensionless]; typically expressed as a fraction or percentage.
r_W = radius of a pumped well; [L].
Re = Reynolds number; [dimensionless].
ρ_W = density of water = c. 1000 kg m^{-3} = 1 kg L^{-1}; $[M][L]^{-3}$.
s = drawdown; [L]; typically in m.
s_a = available drawdown in a pumping well; [L].
s_W = drawdown in a pumping well; [L].
S = groundwater storage coefficient or storativity; [dimensionless].
S_S = specific storage; $[L]^{-1}$.
S_Y = specific yield of an unconfined aquifer; [dimensionless].
T = transmissivity (T_f = transmissivity of a fracture); $[L]^2[T]^{-1}$; typically in m^2 s^{-1} or m^2 d^{-1}.
u = a dimensionless parameter = $r^2S/4tT$, used in the Theis well function.
v = actual linear velocity of groundwater flow = v_D/n_e; $[L][T]^{-1}$; typically in m d^{-1} or m s^{-1}.
v_D = Darcy velocity (= rate of groundwater flux per unit cross-sectional area of aquifer); $[L][T]^{-1}$; typically m day^{-1} or m s^{-1}.
w = width of aquifer under consideration; [L].
$W(u)$ = the Theis well function; [dimensionless].
Z = rate of groundwater flow; $[L]^3[T]^{-1}$; typically in L s^{-1}, m^3 s^{-1} or m^3 d^{-1}.

Electricity flow symbols

F = Faraday's constant, the electric charge on 1 mole of electrons = 96 485 C mol^{-1}.
I = current, typically in Amps (A). 1 A = 1 Coulomb per second (C s^{-1}).
V = voltage or potential difference; typically in volts (V).

Units

Mass

1 kg = 2.2046 lbs
1 tonne (metric) = 1000 kg = 2205 lbs
1 tonne = 1.102 tons
1 lb (pound) = 16 oz (ounces)

Length

1 inch = 25.4 mm
1 foot = 0.3048 m
1 yard = 0.914 m

Area

1 m^2 = 10.76 ft^2
1 acre = 4046.9 m^2 = 0.40469 Ha
1 acre = 4840 square yards

Volume

1 acre-foot = 43 560 ft^3 = 1233 m^3
1 Imperial gallon = 4.5461 L
1 US gallon = 3.7854 L
1 m^3 = 35.31 ft^3
1 barrel (oil) = 42 US gallons
1 barrel (oil) = 158.99 L

Volumetric flow rate

1 Imperial gallon per minute = 0.07577 L s^{-1}
1 L s^{-1} = 13.2 Imperial gallons per minute
1 US gallon per minute = 0.06309 L s^{-1}
1 L s^{-1} = 15.85 US gallons per minute

$100\ m^3\ day^{-1} = 1.157\ L\ s^{-1}$
$1\ L\ s^{-1} = 86.4\ m^3\ day^{-1}$

Density

$1000\ kg\ m^{-3} = 1\ kg\ L^{-1} = 1\ g\ cm^{-3}$
$1\ lb\ ft^{-3} = 16.02\ kg\ m^{-3}$

Force

$1\ N = 1\ kg\ m\ s^{-2}$

Viscosity

1 centiPoise (cP) = 1 mPa s = 0.001 N s m^{-2}
1 centiPoise (cP) = 0.001 kg m^{-1} s^{-1}

Pressure

1 Pascal (Pa) = 1 N m^{-2}
1 atmosphere = 101 325 Pa
1 atmosphere = 406.8 inches of H_2O = 33.90 ft of H_2O (at 4°C)
1 atmosphere = 760 mmHg = 29.92 inches of mercury (at 32°F)
1 atmosphere = 10.33 m of water (at 4°C)
1 atmosphere = 14.696 pounds per square inch (psi)
1 kg m^{-2} = 9.81 Pa
1 bar = 0.9869 atmospheres
1 bar = 100 000 Pa
1 bar = 14.50 psi
1 torr = 1 mmHg = 133.32 Pa

Energy and heat

1 $calorie_{15}$ = specific heat capacity of 1 g water at 15°C = 4.1855 J
1 calorie (thermochemical) = 4.1840 J
1 International Table calorie = 4.1868 J
1 Joule = 9.48×10^{-4} British thermal units (Btu)
1 Joule = 10 000 000 ergs

1 kWh = 3 600 000 J = 3600 kJ = 3.6 MJ
1 kWh = 3412 Btu
1 Btu (international) = 1055.056 J
1 Btu = 0.000293 KWh
1 therm = 100 000 Btu = 105.5 MJ
1 Thermie = 1 000 000 calories = 4.1855 MJ
1 tonne oil equivalent = 44.76 GJ = 12433 kWh
1 tonne oil equivalent (alternative definition) = 41.87 GJ and 11 630 kWh

Power

1 Btu h^{-1} (international) = 0.2931 W
1 Btu s^{-1} (international) = 1055.1 W
1 ton (refrigeration) = 12 000 Btu h^{-1}
1 ton (refrigeration) = 3517 W

Specific heat capacity

1 Btu lb^{-1} $°F^{-1}$ = 4186.8 J kg^{-1} K^{-1} = 1 kcal kg^{-1} $°C^{-1}$
1 kJ kg^{-1} K^{-1} = 0.239 Btu lb^{-1} $°F^{-1}$
1 Btu ft^{-3} $°F^{-1}$ = 67.07 kJ m^{-3} K^{-1}
1 MJ m^{-3} K^{-1} = 14.91 Btu ft^{-3} $°F^{-1}$

Specific heat flux

1 Btu h^{-1} ft^{-2} = 3.154 W m^{-2}

Thermal conductivity

1 Btu h^{-1} ft^{-1} $°F^{-1}$ = 1.731 W m^{-1} K^{-1}
1 W m^{-1} K^{-1} = 0.5778 Btu h^{-1} ft^{-1} $°F^{-1}$

Thermal resistance of a borehole

1 h ft °F Btu^{-1} = 0.5778 K m W^{-1}
1 K m W^{-1} = 1.731 h ft °F Btu^{-1}

Index